THE A380 FLIGHT TEST CAMPAIGN

* * *

A pilot's view

ISBN 13 : 978-2-37301-005-3
EAN 13 : 9782373010053

www.editions-jpo.com
Computer graphics : Bastien Otelli

Claude Lelaie

THE A380 FLIGHT TEST CAMPAIGN

* * *

A pilot's view

Translation: Gerald TRAYNOR & Tony SMITH

Éditions JPO

To Denis Legrand, my dearly missed friend, with whom I would so loved to have shared this great adventure.

FOREWORD

The world of flight tests is not only a great industrial and human adventure, it can also be intensely exciting. There are so many stories to tell, some dramatic, some fascinating from a technical and scientific perspective, while others have a more humorous content, are more anecdotal. However, most flight test crews shy away from revealing much of what goes on for fear of the media blowing it all out of proportion. Even the most insignificant discrepancy noted on the A380 during its development period could easily have been distorted and become the lead item on a TV news broadcast, harming the aircraft's prestige for no valid reason.

Just after the A380's certification, a good friend of mine, and accomplished author and journalist, suggested we work together to publish an account of the aircraft's flight tests. I replied that for the time being that would be impossible. My view was that the harsh reality of flight tests hardly lent itself to prose. The only options open to me at the time were either to accept the proposal, but then refrain from being totally candid to avoid in any way harming the reputation of the aircraft (still awaiting its first delivery); or refuse, as in our profession we are duty bound to state the truth and the whole truth. I asked for advice from Jacques Rosay, Chief Pilot, and Fernando Alonso, Head of Development Flight Tests who were my two flight test companions at that time. Both endorsed my decision.

That being said, I had no intention of seeing all that we had done to bring about the emergence of this extraordinary aircraft fall into oblivion. I had already lived through many 'adventures' during the flight test phase on previous programmes, some amusing, others much less so. But over time you end up forgetting so much of it all, which is a great pity as there is so much there to fire the imagination of aviation enthusiasts. So this time, I decided to take notes on a daily basis. All the major events relating to the flight tests on the A380 were noted down in chronological order starting right from the very moment our test team took control of this extraordinary aircraft.

This book is being published six years after A380 certification. There are now things that can be said without harming the aircraft's excellent reputation. Admittedly, development problems did arise. It is also true that some crews had to perform some pretty delicate manoeuvres, but that's nothing new in our profession. Thanks to all the teams involved (not just the flight tests group, but also and above all the Design Office), we all worked together to bring the project through to completion. We took on a challenge, and world travel has been improved because of an outstanding aircraft that will serve as a reference for the next fifty years or so.

A number of people have already recounted the history of the aircraft's design better than I could ever do. That's why I have stuck to the story of the flight tests and final development. The focus here is mainly on the technical side of things rather than events that marked the Airbus and EADS organisation, although one chapter later in the book is devoted to programme planning so as to place certain issues in their context.

The tests are not recounted in chronological order but are grouped together by topic so as to be more readily understood. Many facts and details of only minor interest have been left out on purpose, otherwise the reader would have to wade through the equivalent of thousands of flying hours!

My daily logbook provided the basis for most of this book, but I have also referred to a host of other documents. A number of specialists were kind enough to help me retrieve further information and proof read the relevant chapters to make sure there were no technical errors. Everything asserted here can readily be checked out from other documentary sources and testimonies. Some people, I fear, may feel hurt by some of the comments made. However, I had no wish when taking my notes to twist the facts: what is expressed here is precisely how we felt at the time, whether we were right or wrong in the matter. Indeed, the flight tests crew, of which I was a member, is not exempt from criticism! We learn from our mistakes.

This book is also an opportunity to inform a broader public about the many testing techniques used to develop and certify commercial airplanes; much of the finer detail that adds little to our understanding of the main issues in testing the A380 have been left out. I hope out-and-out enthusiasts will forgive me for that.

Some of what follows is bound to appear rather complicated for those not familiar with the world of aeronautics. But my feeling was that those who work in or come close to the world of aviation in their daily lives, though lacking detailed knowledge of all the operations involved, may be keen to get to know the basic purposes of some systems, why we perform such and such a test and what difficulties we encountered.

I have sometimes dwelt rather more on the adventures I was personally drawn into, giving the book a subjective slant. I hope this account of my first-hand experience will give the reader a hands-on approach. The story's aim is to give an example of the life of a front-line test pilot, working alongside a group of friends, on a programme of this scale.

I invite you to accompany me on this journey through the history of the A380 flight tests. So first let's go back a few years...

MAIDEN FLIGHT

— What date's the first flight?

— 10:30 am.

It had become a standard joke in flight tests to misconstrue the question. Test pilots are not soothsayers and we had no way of predicting the date of the big day. The final choice remained very much up in the air until the last minute. Even if we had known the exact date, there would have been no question of making an announcement since any postponement would have meant a field day for the media. And there was no point creating additional problems.

Having mentioned that particular time so many times, we were determined not to miss that slot. So, come D-day on 27 April 2005, we just had to get on with it and take off at 10:30. That meant that we were all ready and strapped in some twenty minutes ahead of time, wearing parachute, helmet and life jacket and with a life raft available. We then just had to while away the minutes. Then suddenly a gong sounded announcing a failure. One of the four pumps on one of the plane's two hydraulic circuits was reported defective, a valve was shown to be in an abnormal position and we had no hydraulic fluid level reading. All that for a single failure. Gérard Desbois, our test flight engineer, the third man in the cockpit during flight tests, soon realised that it was a false alarm due to an electrical anomaly. In just two minutes everything was back to normal.

We taxied out slowly then lined up on Toulouse Airport's runway 32L (the longest, facing north) and waited for the crucial moment.

About three minutes before take-off, we were to give the go-ahead to 'release brakes' to our accompanying aircraft, a small Corvette twin-jet. This plane was to turn in behind us and then give us a starting order to accelerate so we would be in patrol formation on take-off. From that position, those aboard could observe the A380, detect possible anomalies and, of course, film the entire show.

In the A380, everyone remained calm and collected, though one of us suggested we take off immediately so as not to run the risk of a new failure emerging while waiting. We were all so keen to set off. Finally, we 'released' the Corvette just one minute ahead of schedule. The official take-off time was therefore 10:29 a.m., exactly when the brakes were released and the wheels quit the ground at exactly 10 hours 29 minutes and 34 seconds.

* * *

Prior to all this, there had been much speculation as to who was to be the pilot-in-command for the first flight of this magnificent aeroplane. Clearly, the title was to be a tremendous honour for the designated pilot. In our small Airbus flight tests world the betting was fierce and the odds were on Jacques Rosay and myself (although we never got to know who came in as favourite among the punters!). Each of us had all the necessary credentials through our experience in developing the Airbus family.

Jacques was Chief Pilot and 'A380 Project Pilot', meaning he was the leader of the test pilots associated with aircraft design. Since 1996, he had taken part in many meetings with the Design Office engineers and had also been on the front line in development work using the flight simulator.

Meanwhile, as Flight Tests Director, I supervised all the test teams and had an overview of all programmes concerning safety and overall cohesion. In that position, responsibility for any incident or accident rests firmly on your shoulders and there is no shirking responsibility by blaming others. The tradition at Airbus has it that the Flight Tests Director be the pilot-in-command for first flights on major programmes.

So who was it to be?

Having discussed the matter with Jacques, it became clear that there was only one solution, which was to share the honour. We therefore agreed that he would be the first one to fly the aircraft and ensure take-off, as he was so keen to do so. I would take over the controls on landing. There would therefore officially be two captains for this maiden flight of the A380: one for the first part of the flight and another for the second. We decided that I would occupy the left seat and he the right. Take-off on the first flight would be conducted from the right seat, which is completely unprecedented!

To the world of transport pilots, this decision appeared odd, absurd even. But that stems from lack of familiarity with the flight test profession. In flight testing you often find two experienced pilots on board both capable of performing the required tests and with each flying in turn. When a subjective decision is required, each expresses his own opinion. At any given time, the pilot flying may well be seated on the right. If the test proves difficult and he fails to complete his manoeuvre, this could lead to the aircraft suffering damage or even loss of control. In such cases, the real pilot-in-command will finally turn out to be the co-pilot, as everything will then depend on his reaction. This situation, where a co-pilot has such major responsibilities, remains unknown to the everyday air transport world. We were therefore hardly surprised to read a number of readers' letters to specialised reviews expressing their astonishment at this unheard of situation!

I derived some personal satisfaction from this joint approach in that we were able to show that we knew how to work together as a real team in flight tests. No private ego trips could interfere with our work.

I still wonder if any of those taking out bets on the choice of Captain foresaw such an outcome?

• See photo n°1 – page 229

* * *

Nobody nursed any doubts as to the final composition of the crew (apart from the pilots) for this first flight.

The test flight engineer had to be Gérard Desbois, the boss of the test flight engineers who had followed 'his' aircraft right through from the start of assembly.

There was a consensus among the flight engineer community in favour of choosing Fernando Alonso to be in charge of flight controls. He had headed the entire development team working on new aircraft and was recognised as a great specialist in that field. The same can be said for Jacky Joye for the engines, as he had the greatest experience in that department. We would not have wished to dispense with his skills when it came to monitoring the Rolls-Royce engines and the fuel system. Finally, we entrusted Manfred Birnfeld, a seasoned German flight engineer, to keep a constant eye on all the systems that, for the first time, were going to operate in a different environment.

Even though these choices were natural for the flight test managers and while the appointments were within my powers, we felt it was desirable, considering the issues at stake, to have them approved by Top Management, especially with respect to the media. So one evening, a few months before that first flight, I had a short meeting with Noël Forgeard, President of Airbus, and Charles Champion, the Programme Manager, to ask for their approval.

* * *

What do you actually do on a maiden flight?

Each manufacturer has their own traditions. Some proceed with a short flight with the landing gear down. Others adopt a longer flight with landing gear retracted and start to explore high airspeeds, low airspeeds and high altitudes. Airbus comes into this second category.

On a new programme, the maiden flight marks a major stage in an airplane's development. It remains highly symbolic and, although simple in principle, also generates a host of technical data, whether positive or negative. For the first time, assumptions about the aerodynamics and systems operations will be validated, or found to be unjustified. As from the first flight, you get to see a whole number of reasons to be satisfied or, conversely, be able to identify those issues on which you need to carry on working. Thus, for example, when exploring low airspeeds, the crew will immediately see whether the aircraft remains laterally stable, meaning that its axis naturally remains in the flight direction. On the A340-600, we detected problems during this manoeuvre as from the very first flight, without taking the plane to its limits. This inevitably led to substantial further development work. On the A380 however, there was no anomaly, confirmed by exploration through to minimum airspeeds as from the third flight.

The first real key issue for an ambitious first flight as envisioned at Airbus is retracting the landing gear, a procedure that sometimes leads to surprises. Indeed, this manoeuvre had been performed many times on the ground, but in flight minor deformations of the aircraft can, for example, lead to a locking anomaly. This explains why some manufacturers prefer to leave the landing gear down on the maiden flight. Other tests usually planned by Airbus for a first flight include operations using the slats (mobile control surfaces on the wing leading edge) and flaps (mobile control surfaces on the wing trailing edge); assessing behaviour at high airspeed and low airspeed; and, weather conditions permitting, evaluation of fuel consumption when cruising at high altitude. Of course landing is also a key moment, as aerodynamic behaviour close to the ground can never be perfectly reproduced on the simulator.

During first testing of fly-by-wire aircraft, the crew spends a great deal of time checking the flight characteristics and validating the handling systems. Basically, there are two types of laws incorporated in the software for the aircraft controls – normal law and direct law (other types of laws will be encountered in the following chapters). Under normal law, the aircraft is fully stabilised and if the pilot sets it to a certain attitude, it will remain there. In direct law, the control surfaces are deflected in proportion to the movement of the stick, and the aircraft will behave like a plane with conventional flight controls. Fly-by-wire flight controls incorporate a host of feedback systems that can only be validated in flight, despite the wealth of simulation facilities available nowadays. This explains why the first take-off and landing are performed under direct law. Once in flight, normal law is only engaged after a large number of prior checks.

* * *

So how did we proceed?

The units used in what follows are the nautical mile (NM) and knots (kt), Mach and feet. The nautical mile (1.15 standard miles or 1.852 kilometres) is defined as a minute of arc along a meridian of the Earth and is used in shipping and aviation especially when working with charts. A knot is the speed of one nautical mile per hour. The Mach is speed as a percentage of the speed of sound. Meanwhile the familiar 'foot' is used by aviators to measure altitude rather than a metric equivalent.

The initial objective was ambitious: retracting the landing gear, a few checks at medium altitude up to a maximum airspeed of 340 kt, climbing to high altitude, then flight up to Mach 0.89 before finally descending to start exploring low airspeeds and flying back to Toulouse.

Obviously, the first stage was the take-off. But for the crew, apart from the symbolism involved, this didn't really make for a technical event. With all the knowledge already acquired on the aircraft, nothing unexpected was likely to happen. The first comment from Jacques, after take-off, was just to say that all the settings were exactly as predicted on the simulator. The next remark during that first climb was to mention the absence of vibration. That was interesting and a promising sign because vibrations, whether mild or more severe, are often felt when the flaps are extended for take-off. That was the first real discovery on the flight, as no simulator or wind tunnel can give valid information on this phenomenon.

Once the slats and flaps had been retracted, the next 'serious' step was to bring up the landing gear and see what happened. As I explained earlier, the many operations conducted on the ground could provide no guarantee as to real performance. In flight, the aircraft deforms slightly and you can never be sure that the settings validated on the ground remain satisfactory. The doors that close all the landing gear openings have to fit together perfectly and lock into place. If the landing gear and these doors do not close as planned, a warning is issued during the retraction sequence.

And that is precisely what happened. On the screen monitoring the landing gear status, one of the doors of the right hand fuselage landing gear was indicated as being open. In fact, the indicator light clearly showed that closure and locking could not be confirmed and therefore a doubt subsisted. The crew aboard the accompanying aircraft following us had observed the entire manoeuvre and had seen nothing untoward. On closing in, they confirmed that the door appeared to be fully closed. There followed a lengthy concerted discussion between the A380 and accompanying aircraft crews and the telemetry room. From that discussion, it transpired that although the door appeared to be in place, there was a risk that the lock had engaged incorrectly. In that case, as the valve sending hydraulic pressure to hold the door in position closes at high speed (280 kt or a little more than 320 mph), the door would not be held in position and could open if we were to fly faster. Caution prevailed and we decided not to explore high airspeeds. Another consequence was that we were prevented from flying at high altitude as, in order to climb, relatively high airspeeds had to be attained. We found out what had actually happened back on the ground and realised we had made the right decision, as the lock had not engaged correctly. On arrival at our parking position and facing the cameras, the real specialists saw what had occurred. The right landing gear door had opened by itself, but very slowly. I turned back to take another look. But the journalists present hadn't cottoned on to a thing!

Deprived of high speed and high altitude, we therefore resorted to plan B with one of the substitute programs we had prepared. This was done with direct and then normal flight laws.

We also explored low speeds in landing configuration and these tests showed the aircraft's excellent behaviour.

Finally, we were ready for landing. The runway was all ours as all commercial air traffic had been interrupted, although we did have to wait for a few minutes for a helicopter medical evacuation that naturally had priority.

The approach, under direct law, gave us our first impressions - that of a very stable aircraft that, oddly enough, despite its mass, reacted to turbulence much like a light aircraft. We had only let a select few into the secret but, unless we encountered any particular obstacle, we had planned to initiate a go-around at an altitude of 300 feet and perform a fly-past in front of the official stand followed by a round tour of the city before the final approach. Landing was conducted extremely cautiously starting with the flare, or round-out, to increase the angle of attack and adopt a greater tangent to the ground at about 100 feet, before very gradually sailing down to the ground. The goal here was not to demonstrate the short landing distance but to bring the plane down fairly gently and get a 'feel' for the aircraft.

After landing, we made a short lap of honour along the taxiways to the north and south for all to admire our fine aircraft, before heading over to the official stand for our media duties.

* * *

For the first flight, as indeed for all flights involving any special risks, safety and rescue equipment remain available to the crew except during the take-off and landing exercises. We all settled into special seats that accommodate a back parachute, seated atop an inflatable life raft with a survival kit. We also wore a helmet and life jacket. We sometimes spent as much as six hours at a time in these seats. So for the A380, we had requested improvements be made to the equipment which had previously been used during the A340-600 programme and had turned out to be particularly uncomfortable.

If the worst came to the worst, emergency evacuation of the aircraft was planned to take place through a 'tunnel' that crossed the forward cargo hold. This led from the floor of the main deck though to a location where a pyrotechnical cutting system can make an opening in the main cargo door. There was also a ramp so that all crew members could get to this tunnel from their work station. The idea was to just drop down into the hole. However, steps were also installed in case of inverted flight.

* * *

Handling the media wasn't really our specialty but we had to get on with it all the same.

Thanks to Barbara Kracht who managed all relations with the press and television, we were afforded considerable protection. Her thorough knowledge of the aviation world, her availability and the excellent relations we maintained with her entire team greatly helped us to manage any difficulties in that area.

More than a year in advance of the flight, some journalists were keen to get their scoop with the names of the crew members! By common agreement, we refused to let on. The identities were only finally disclosed one month before the big Reveal ceremony, where the aircraft was shown to the Heads of State of the countries that had taken part in the Airbus adventure and to the media.

During the entire period before the maiden flight, interviews mainly focused on the two pilots. We tried to get across the message that we were working as a team with the engineers, but we were wasting our breath! They were only interested in the pilots.

Everything started to change after the first flight when people began to discover a real team. Above all, Michel Drucker's prime time national TV chat show broadcast on a Saturday evening the day before the Paris Airshow opened at le Bourget, contributed to showing what the flight engineers' work involved. This included a short film made with cameras installed inside the aircraft during the maiden flight. Spectators finally realised that it wasn't just the pilots running the tests.

The film, which was widely broadcast after the programme itself, caricatured our profession and our behaviour to some extent. It contains much light-hearted banter and some amusing sequences, but little detail on the delicate and highly technical tests that would have been hard for non-professionals to grasp. But I was delighted after the programme was over to see that our working methods as a team had been clearly highlighted.

Obviously, before this A380 programme, we had all undergone interviews, but the real novelty now was the sheer number of them and their being practically non-stop. Earlier, we had spent some time to prepare for interviews, but that was now no longer possible given their number. Finally, for the spectators, it wasn't that bad as there was a certain freshness and spontaneity in what we had to say.

On the big day, we were completely isolated from journalists and cameramen right until the end of the flight. The only other people we had around us were the teams that had been following us for a number of weeks to prepare a big film on the history of the A380. We felt completely at ease with them. It had been agreed in advance that they would first exercise their own discretion in filming and editing and that finally everything would be reviewed by Airbus before being released.

* * *

How many times did we have to face questions about our feeling afraid before that first flight? Even at Airbus the question cropped up frequently.

I obviously can't answer on behalf of the others, but I sincerely believe they were on the same wavelength as me.

Technically speaking, a maiden flight is extremely simple. Obviously, the unexpected can always occur, but personally, I never felt the slightest apprehension before or during that flight. We remained in a limited flight envelope that had been reasonably identified by all the wind tunnel and simulator tests. My main concern was to know whether the landing gear was going to lock in the retracted position as the entire programme for that first flight depended on it.

Stress does arise during flight tests, but usually on flights when we go to the limits of the aircraft's performance envelope or perform especially delicate manoeuvres. Thus, we sometimes get into situations that could never have been predicted by the wind tunnel or the simulator as we quit the boundaries that characterise these forecasting tools. For instance, the third flight was stressful because we ran the first stalling operations, reducing speed down to the limit before complete loss of control. We couldn't know what was going to happen and all we had were predictions subject to considerable caution. For example, the aircraft could pitch up, actually go into a roll or even fail to respond. With the A380, everything went well for the aerodynamics, but abnormal stresses were felt on the tail unit that, though still remote from those presaging breakage, were nevertheless worrying. That's what a stressful flight is all about. But this is run of the mill in our profession.

You need to bear in mind that with our accumulated experience in test flights (at that time, more than 7,000 hours in my case, for example), even if you are always on the lookout, the stress during test flights is not as great as people from outside our small world tend to imagine. Engine shut-down is a pretty ordinary occurrence for us while, on an airliner and as seen by the media, it is virtually considered tantamount to an accident!

* * *

A few organisers worried that our arrival at the airport for the flight may have been delayed by big traffic jams. We were ready to come in early to avoid them, but within reason, as we had to be fresh and well rested for the flight. Some fairly far-fetched solutions had even been envisaged such as a motorcycle escort to open up the way, or a helicopter trip from the small aerodrome of Lasbordes on the other side of the city!

In the end we just kept things simple. I personally chose to stay at the Sofitel Hotel right next to the aerodrome hoping to have time enough to get some rest. This proved to be a monumental error as in the middle of the night my neighbour had his television on full blast and I had to go round to knock on his door for him to turn it down!

The others stayed at home. Jacques had simply stowed his folding bicycle away in his car just in case! Much to the delight of the cameramen, he cycled round the A380 after his arrival.

* * *

But let's move back a bit in time. What did we do with the aircraft before the maiden flight?

During the final days of its assembly, and above all as from the first electrical power-up, our flight engineer, Gérard Desbois, closely followed everything that happened on the aircraft and its equipment.

As soon as the plane was reckoned to be in flight-worthy condition, it was handed over to the flight test team. This transfer of responsibility is a highly formalised event called 'the hand-over' where the aircraft's exact status has to be checked through very thoroughly. Obviously, there are thousands of differences compared to the configuration that would finally be certified but everything has to be very accurately reviewed, from major issues to even the slightest detail. For example, not having all the evacuation slides installed is no hindrance to testing and the fact merely has to be recorded without further discussion. Conversely, discrepancies may be encountered that call for remedial action before the maiden flight. The review process for all these issues starts well upstream of works on the aircraft being completed as it all takes a number of days and nobody wants to lose time. As soon as the aircraft is ready, it has to be handed over to flight tests.

Hand-over took place on 6th April 2005. We were running about three months late on the planned schedule that had initially set the flight for mid-January. But when you look at what happens on similar programmes, not just at Airbus, such a delay on a project on that scale is finally quite minor.

As from that hand-over, we had allocated a full week to get the aircraft up in the air, but in reality, due to a number of unexpected difficulties, that maiden flight only took place twenty-one days later.

A number of mishaps cropped up during this preparation work for the flight. On 7 April 2005 for example, trouble emerged with the Auxiliary Power Unit (APU), a small turbine that generates energy for the electrical and pneumatic systems. It cut out by itself and stopped generating. The first engine start-ups took place on the next day and ground runs were continued through to Saturday 9 April with a number of anomalies emerging, including hydraulic and air system leakages, vibrations on some of the engines and excess oil consumption on all of them. We taxied back slowly from the run-up area to the apron with the engines running. The brakes worked fine.

Considering all the difficulties encountered, the aircraft stayed under works up to Wednesday 13 April. We then conducted another ground run to come to the conclusion that, while some issues had been resolved, others remained outstanding, even to the extent of being compounded by new problems. Despite all this, we were able to perform two taxiing tests up to 60 kt.

On 15 April, taxiing tests were resumed, again at 60 kt, with the normal braking system, then in degraded mode (emergency braking). The aircraft was then again brought back under works for a few days to correct all the anomalies encountered.

Then on 20 April we resumed acceleration stop tests up to a speed of 100 kt. One of the tests simulated the loss of both the aircraft's hydraulic systems with no real problem.

Friday 22 saw the final rehearsal under the exact configuration for the maiden flight with parachute, helmet and so on. Considering the number of anomalies observed, it took us one hour ten minutes to get the aircraft ready whereas it usually only takes a few minutes. The acceleration stop was performed at 142 kt airspeed and 130 kt ground speed, the difference being due to the prevailing wind. This speed was almost as high as that planned for take-off on the maiden flight.

Apart from a few minor details, we were apparently ready. However, Dominique Chatrenet, Head of the flight controls division in the Design Office, noticed that there was a minor discrepancy in the flight control software. An 'electronic filter' on the stick controls appeared to be missing and so reprogramming was deemed necessary. This led to a further day being lost: but the outcome was that weather-wise, the slot turned out to be much more favourable, so there were no complaints from us!

As with previous programmes, the mechanics team's final preparation of the aircraft took place over the last two days. Meanwhile, we had two briefing sessions. The first, for the crews of the A380 and the accompanying aircraft, the telemetry room and Design Office engineers, was purely of a technical nature: re-examining the aircraft's limitations, final validation of the test schedule and of alternative programmes in case things didn't happen as expected.

On the same day, the major briefing session took place with all those concerned, including the security teams, Airbus communication, air traffic control, Civil Aviation Authority people and so on. This was the final meeting to make absolutely sure that nothing had been overlooked across the entire organisation developed over several months.

• See photo n°2 - page 229

* * *

That first flight was rounded off the next day by French President Jacques Chirac's visit and walkabout.

MY FELLOW TEAM WORKERS

The flight tests team proved to be extremely solid and demonstrated keen solidarity throughout the A380 programme: information circulated freely among us. Which is just as it should be given that we were facing a huge task that required an optimal share-out of responsibilities and great mutual trust. The difficulty here was not essentially technical but rather related to the way the programme was organised.

Before looking at the global organisation of Airbus during the A380 development period, I would first like to introduce my immediate associates.

* * *

Fernando Alonso was the Head of Development Flight Tests, which meant he supervised all the flight engineers in charge of development and certification under his command. We shared the same opinions on pretty much all the important issues so things were easier for all of us. If either one of us spoke out he could be assumed to have expressed our common point of view. We were also interchangeable when it came to a certain number of meetings. Needless to say, we still had our own specialities.

Fernando had worked in aeronautics, first in the United States, at McDonnell Douglas (now Boeing). He then joined Airbus on flight testing to develop flight controls under orders from Gilles Robert, one of the great flight engineers who had retired by the time of the A380.

I had a great time working with Fernando. He was tremendously supportive and helped me defend the flight tests side against a few 'big managers' at the time who were unwilling to address a number of issues of concern to us. We didn't always win out but we never backed off (with the exception of one episode as recounted later).

Fernando is a master of organisation and he succeeded me in flight tests management though with broader powers stemming from a re-organisation.

• See photo n°3 – page 230

* * *

Jacques Rosay was Chief Pilot. He is no doubt more diplomatic than I am, but he knows how to stand his ground when needed. There is no way you can occupy such a position without showing the necessary determination in certain circumstances.

Jacques had accomplished enormous work with the Design Office since 1996 when the A3XX, later to become the A380, was launched technically. As designated test pilot, he spent an incredible number of hours in definition meetings and on the simulator to get the project moving. The Design Office always appreciated working with him, even though at times, he had to turn down some of their proposals. Their work together finally led to the emergence of really easy to use pilot-airplane interfaces. For example, you can enter a flight plan into the aircraft computer intuitively without the need to pore over an instruction booklet. In some respects, the aircraft is comparatively easier to use than a cell phone! Such design improvements require a tremendous amount of work to synthesise the issues involved and there is no room for improvisation.

As Chief Pilot, Jacques managed to appoint the crews in an effective and tactful way. You need the right pilot for the right flight, but you also need to share out the cake and make the most of every opportunity to help the younger elements make further progress.

Of course, I chose him to be Chief Pilot at the end of 2000 also because he was such an excellent pilot. It would have been unthinkable to have someone occupy such a position without knowing how best to execute all the development and certification tests on our aircraft.

• See photo n°4 – page 230

* * *

When still professionally active, Jacky Joye was probably the most competent engine test engineer in the world. Over thirty years of testing on both civil and military programmes, he had accumulated tremendous experience. In the engine field, his views were always considered to be the reference, not just for flight tests, but also by all those intervening in our programmes. Although he went into retirement after completion of the A380 programme, he remains an authoritative advisor.

Jacky also has a great sense of humour and a marvellous turn of phrase, hitting the nail right on the head whether in French or (having spent a long period in the United States) perfect English.

Taking part in a test flight with him is a delight. Firstly because he's a great professional who prepares his flights to perfection, but above all because he has an extraordinary capacity for adaptation when faced with the unexpected. Such instances are a common occurrence in flight tests and the correct response is then to elucidate the matter and refrain from over-reacting in a way likely to compound matters and lead to a specific danger. That's where the skills of 'great engineers' like Jacky come in, quickly figuring out what's needed to sort things out and move forward.

• See photo n°5 – page 231

* * *

Gérard Desbois, a big figure in the flight tests world, was the chief of the test flight engineers. He's well known for his bluntness and volubility! He also has a true pioneering spirit and his first objective is always to see his aircraft fly. He's absolutely crazy about aircraft and everything to do with them so it's no surprise to learn that he's the proud owner of a Piper J3.

Obviously, he's a truly great professional, but he also has great writing skills, having published two books recounting some of his flying adventures in a humorous vein. He has also regularly contributed to an 'aero-medical' review, firstly under the pseudonym AAA (though his inimitable style soon gave the game away!). Neither did he hesitate to pen some blunt e-mails to send 'upstairs' as he put it when he considered development of the A380 wasn't moving forward fast enough for his liking.

His excellent relations with the first line team (implementation and maintenance) helped ensure we kept to a good schedule of flights during development.

• See photo n°6 – page 231

* * *

I first met Harry Nelson when I was looking for a test pilot to conduct flights on planes coming off the production line in Hamburg. Harry had held a number of management positions, the most recent being at the ATR Pilot Training Centre in Toulouse and then at the Toulouse centre for the British aeronautics industry. I immediately realised that he was a 'real manager'.

My deputy Manfred Weidensee had been extremely supportive when I took on flight tests management in 1994 as he knew everybody, having lived through the Airbus adventure right from the start. When he went into retirement in 2002, I was pressed by those around me to take on a non-aircrew German successor like Manfred. Maybe I didn't try hard enough, but in any case I failed to find one and decided instead to choose Harry, an English pilot. This wasn't just my obstinacy but because right from the start I knew he was the man for the job.

I greatly appreciated working with Harry. He was my confidant on all matters and hardly a decision was made without prior discussion with him. Through his very British imperturbability, he also managed to calm things down when we were getting close to potentially explosive situations, which happened from time to time.

While Harry supported me in many areas, he stayed in the background on matters concerning the A380 so as not to interfere with the actions we were pursuing with Fernando.

In the autumn of 2005, I appointed him as manager of the Acceptance Department responsible for the completion of aircraft coming off the assembly line, from their maiden flight through to delivery to the customer.

* * *

I can't name all the many engineers who worked to develop this aircraft, but I shall give a few names of the aircrews involved. I hope all those whom I've missed out, and the ground engineers who accomplished such great work, will forgive me for not mentioning them by name.

Didier Ronceray, the big specialist in flight controls, was always on the front line, resolving the mass of problems that can arise on such an aircraft under development. The Design Office relied enormously on his pragmatic analyses to make progress.

Robert Lignée was in charge of performances and especially all brake-related issues. He became a reference in this field and also actively took part in a working group with the US Federal Aviation Administration (FAA) to review the rules that calculate landing distances for all transport aircraft.

Bernard Masdoumier worked with the Design Office to ensure development of the autopilot and automatic landing, as he had done on other programmes. It's worth noting that, while he still retained his function as an engineer, following on from the A380 programme we had him take his test pilot qualification, considering his experience as a pilot.

Sylvie Loisel and Jean-Christophe Lair (the latter a test pilot) took part in developing the navigation systems.

Jean Piatek headed the entire team in charge of cabin development. At the time the A340 and A330 were developed, the cabin aspect had been neglected on the flight tests side. There was no way we could fail in this area on the A380, as our customers would never have forgiven us. That's why we brought all the specialties concerned in cabin development under a single entity to ensure full coordination. Jean Piatek, who has exercised almost all the professions in flight tests (general aviation at Socata and development of ATR regional transport aircraft), pulled off this task with his entire team, both on the ground and in flight.

All the in-flight certification aspects were covered by test pilot Hugues Van der Stichel, aka VDS. It may appear strange to entrust this task to a pilot, but there are two good reasons for this. Firstly, a number of certification tests call on issues relating to ease of handling, so we need one of our pilots to ensure a permanent interface between Airbus and the certification authority's pilot. Secondly, certification gives you an overall view of all the problems and so it's an ideal position for someone who appears destined for promotion. Moreover, as predicted, when I quit flight tests fin 2007, Hugues replaced Fernando to head the Development Department.

• See photo n°7 – page 232

* * *

It would also be impossible here for me to mention all the test pilots who took part in this adventure. You will find the names of some of them as the story of testing unfolds in the following chapters. Thousands of hours were spent in the various development simulators and in flight, as well as in design meetings. I'd just like to refer simply to Peter Chandler who responded rapidly to come in as Jacques Rosay's deputy in developing the operational functions of the A380, and who was later appointed Chief Pilot for civil aircraft under Jacques' orders.

* * *

To wind off this chapter, just a word on how we worked. My office door always remained open. I'd made no changes to the office lay-out when I took over my functions and the desk of my assistant, Christine Hubert, and mine were positioned so we were constantly facing each other. The door was only ever closed when we had to address confidential matters, an extremely rare occurrence. Christine lived through both the highs and the lows on the programme and I must say that I found it pretty agreeable to be able to pass comment out loud on some issues, either in all seriousness or jokingly, with no regard for political correctness. It helps the general atmosphere at work to be able to let off steam from time to time!

• See photo n°8 – page 232

CREWS

Flight tests crews are a mix of different specialties. Alongside the two pilots and the test flight engineer in the cockpit are one or more engineers and possibly cabin specialists too. Let's take a look at what their work involves during development and certification - they have slightly different tasks when it comes to production flights.

* * *

Experienced test pilots have highly specific functions: taking part in the development of a new programme or new item of equipment; activities related to certification for the flight side; analysis of accidents or incidents with the Product Safety Department; relations with airlines, and so on.

Where Airbus flights are concerned, there is hardly any real specialisation for the pilots apart from some very specific areas or certain tests involving special risks. The pilots fly on all types of aircraft and perform both development flights and production line flights (from the maiden flight through to the acceptance flight with the customer).

At Airbus, near complete experience is only acquired after five to eight years of practice. This is the average time it takes to cover the full spectrum of these delicate tests, not just those involving particular risks, but also those requiring finer judgement.

* * *

The flight test engineer is a specialist in a particular field - flight controls, engine, braking or autopilot for example. He coordinates all the work with the Design Office for development and certification. Depending on the volume of work, several engineers may be needed to cover all the corresponding tests.

The engineers working in these different fields define together the instrumentation they are to share on board any new aircraft. This instrumentation is extremely powerful, providing a host of charts and calculations in real time so that, in general, each test point can be validated. Development is essentially governed by the time factor, so everything has to be endorsed as quickly as possible. The teams of systems designers following the flight in the telemetry room can come in to help the flight test engineer on specific or unexpected issues, and also validate certain tests.

For many of the A380 tests, the flight test engineers used an extremely powerful tool known as ADIS, an acronym of AFDX and Digital Injection System (the AFDX or Avionics Full DupleX is the 'on-board Internet'). It can be used to change the characteristics of laws (software programmes) on different systems, like flight controls, braking, autopilot, etc., that can then be modified during the flight. The logic behind this set-up is that the aerodynamic models available are never perfect. This is especially true as we get closer to the ground where you encounter what is called the ground effect, like a kind of 'air cushion'. The situation is much the same when you reach the limits of manoeuvrability for the aircraft and you then need to be capable of refining the settings of software in flight by analysing the tests in real time. Having reviewed the initial results, it may be decided to repeat a particular manoeuvre after adjusting one or more parameters. For obvious safety reasons, all such settings that can be used in flight have previously been fully tested on the simulator.

During major programmes, the leading flight test engineers in each domain head the IFTT (Integrated Flight Tests Team). The IFTT is responsible for final development in a particular specialty. Here, Flight Tests, the Design Office, the data analysis team and Programme Management all take part.

• See photo n°9 – page 233

* * *

For the last thirty years there has not been a third person in the cockpit of a commercial airliner, the flight engineer post being obsolete. However, at Airbus testing the test flight engineer was retained for other reasons.

Every test flight engineer is assigned to a particular aircraft that they stick to for many years. That way they see the aircraft emerge right from the very earliest days and are familiar with all the different special features dating from its construction. In addition, they coordinate the installation of new computers on request from flight engineers and design offices, as well as programming modifications. They therefore know everything about what's going on on the aircraft and keep a record of special features and instructions that is reviewed at each pre-flight briefing.

The test flight engineer plays a really essential role, as a host of minor warnings can crop up (after equipment replacements, for example) and the crew can hardly get to know them all on such extremely complex aircraft with their different design features and sometimes differing computer versions. So the test flight engineer is in a position to confirm whether such and such an anomaly is already known or not and whether it has been communicated to system managers, thus saving valuable time. Because test flight engineers represent a tiny 'caste' (with just one per development aircraft), there is a ready exchange of information between them.

In flight, the flight engineer is in the jump seat, between and slightly to the rear of the two pilots. Gérard Desbois, the head engineer, says he found himself seated between the ox and the donkey, like the baby Jesus (though he never told us which was which!)

The other good reason for having a test flight engineer on board is the help he can provide in managing certain systems or during especially delicate tests. A good and recurring example of this is when we use the engines in dissymmetrical fashion: the engineer has to ensure fuel transfers to maintain good lateral stability for the aircraft, leaving the pilots to focus on their tests. He sometimes also takes over responsibility for the engines. This can occur during VMCG tests (minimum control speed on ground), for example, where one of the external engines is cut off on the ground by the co-pilot while it is on maximum thrust. Here, the test flight engineer is entrusted with relight just after take-off, according to the specific instructions given him by the engine manufacturer.

* * *

In the flight tests field, the latest skill to have emerged is that of cabin specialist. This function became necessary when the first A340s came off the production line in 1992. Prior to this, the A300, A310 and A320 aircraft made their maiden flights from Toulouse and were then ferried to Hamburg for the cabin to be fitted out. During the return flight from Germany, the flight engineer had time to check out the entire installation and, if necessary, technicians could make last checks on additional flights. The new system involved the cabin being fitted out before the maiden flight.

In addition to this, cabin equipment became much more complex with the A340 generation and it seemed rational to define a new job description. This also meant upgrading the technicians to flight engineer status as, in principle, it was out of the question for ground crew to take part in a maiden flight, even on a production aircraft. Much later, a ministerial decree (signed the day he left government by Prime Minister Jean-Pierre Raffarin!) gave official recognition to this profession.

As we will see in the CABIN COMPLETION chapter, most of the work was done on the ground, but some issues can only be addressed in flight. On an aircraft just off the production line, for example, you have to check the noise near the doors (sometimes requiring new settings), the toilets, the galleys, etc. Cabin specialists have a long list of inspections to be made during the first flight. Given the amount of work to be done in just a few hours, large aircraft like the A380 and A340-600 sometimes have two or three cabin specialists on board.

During development of a new aircraft, much of their work is routine, except that the problems encountered have to be analysed by the Design Office, and a ready solution is not always immediately forthcoming. So they take part in the first flights with the cabin installed, then those with passengers. Their remit is to scrutinise everything and ensure that no undetected problem emerges once commercial operations begin. In final test flights with passengers on board, there are two cabin crews, one taking care of on-board services and the other comprising our technicians entrusted with checks on the aircraft. The service side is assigned to airline personnel (Lufthansa for the A380) who help by giving us their first impressions and a pretty clear indication of how the installations will satisfy our customers. These crews are delighted to play a part in the adventure and this first contact with the aircraft means they can better prepare entry into service with their airline.

* * *

Now we've seen the various specialties involved, let's take a look at how a typical test flight is organised, apart from development of the cabin. In general, there are five crew members: two pilots, one test flight engineer and two flight test engineers. Depending on the nature of the tests, extra members can be brought on board for various reasons, where more complex issues have to be faced or when a member of the certification authority is present, for example.

The flight test engineer prepares a test sequence giving details of the manoeuvres to be performed. These are planned to satisfy either a Design Office request or the long list of EASA-approved check points for certification. The specific flight conditions are forwarded to the test flight engineer for the aircraft to be prepared, such as the computers to be installed, fuel replenishment, or adjust the weight and centre of gravity position.

Before the flight, the flight test engineer holds a briefing session with the entire crew and the specialists concerned from the Design Office and Telemetry, to explain the manoeuvres planned in his test sequence. This sequence is sometimes amended in light of the remarks made by the attendees. Specific issues relating to the aircraft - kept up to date by the test flight engineer - are then reviewed, especially when new computers have been installed. Both the engineer responsible and the captain sign the original of the test sequence document jointly. After the briefing session, if necessary, one of the pilots discusses those manoeuvres that may raise air traffic issues with air traffic controllers.

Once in flight, the pilots share out the tests and, usually, each of them performs a series of manoeuvres. To judge the handling qualities of the aircraft, both pilots can perform the same manoeuvres. This is sometimes the case when tuning the flight controls, especially if doubts subsist as to the result. After each test, pilots and engineers exchange their views, each having their own vision, the former with their sensations and the latter with their charts and curves. If the outcome is satisfactory you move on to the next issue. Otherwise, discussion ensues to see whether a repeat can be made using a slightly different technique or a new setting on the computers.

The test flight engineer monitors all systems and assists the pilots as required according to the level of difficulty of the test.

On return from the flight, a debriefing session is held with the Design Office, obviously to report back on everything that worked well, but above all to discuss what proved unsatisfactory. This is a crucial moment where pilots, flight engineers and the Design Office people exchange ideas for the future.

* * *

How are the crews chosen for each flight?

At the time of the A380, Chief Pilot Jacques Rosay appointed the pilots. Whatever the programme, this task is a delicate one. Obviously it involves naming pilots perfectly qualified for each flight. However, this does not necessarily mean having the best two, since there is a real need for experience to be handed down from the more experienced pilots to the younger ones. This relates not just to conducting a particular type of test, but also, on the first flights of a programme, knowledge of the new aircraft. It should be added that, right from the start, all test pilots await their turn impatiently: to avoid exciting too much rivalry, Jacques had to show considerable sagacity and equanimity in his choices.

The problem is pretty much the same for the flight test engineers but easier to resolve as, for any given speciality, there are relatively few engineers but a large number of flights to be conducted. The younger ones can therefore be rapidly confronted with the reality of testing.

* * *

Back in the mid-1990s, Airbus set up a training school for its test pilots and flight engineers.

Prior to that, all crew performing flight tests in France had to follow a one-year training period at the Ecole du Personnel Navigant d'Essais et de Réception (EPNER) at Istres or in an equivalent training college in Great Britain or the United States. The trainees learn to prepare and conduct a substantial share of the tests for final development and certification on civil and military aircraft. Due to a boost in production, Airbus needed to train new crew fairly quickly to cater for aircraft coming off the A320, A330 and A340 assembly lines. All the flight test professionals agreed that it was pointless for an engineer to follow such a long training period to ensure the first flights of production aircraft, together with a certain number of straightforward tests. In 1994 and 1995, the authorities, the unions and the manufacturers worked together to reconsider French legislation and two levels of tests, A and B, were created, with A corresponding to the most delicate ones. For level B, training within the company was authorised. That is how we came to set up the Airbus Flight Test School and, come the A380's maiden flight, about 50 engineers and 20 pilots had already been trained in this school.

Right from the start, it was decided that the engineers hired would come only from the pool of in-house candidates, the idea being to encourage young people and get them on the career ladder. This is the principle adhered to since.

To recruit the pilots, we looked wider afield as there were not enough potential candidates at Airbus. Apart from hiring already qualified pilots, those chosen to be trained for the tests were selected either from among instructors from the Training Centre that qualifies airline pilots for Airbus customers, or from the airlines. A few engineers with considerable experience as pilots were also qualified.

In-house training includes several weeks of theoretical courses mainly covering aerodynamics, flight mechanics, engines and, above all, test techniques. With a few exceptions, all their flights on Airbus, whether on production or development airplanes, are conducted during real tests, with the trainees being accompanied by experienced crew. For some techniques that are difficult to demonstrate effectively on Airbus, a light aircraft equipped with a sophisticated measurement installation is also used. The training is rounded off by exams including an in-flight test with instructors from EPNER. The entire course takes a full year but the trainees are relatively quickly authorised to get on with their future job as if they were already qualified.

As far as recruitment and training for A tests is concerned, Airbus finds pilots already trained for flight tests ‘on the market’ from the main traditional training schools, including the French EPNER. As a general rule, it is best that they have also had experience as airline pilots so they have a clearer understanding of our customers' concerns. Airbus also sometimes hires flight engineers trained by EPNER and working at the Flight Tests Centre (named DGA Flight Tests since 2010) or for the administration.

As this level B internal training is well established and recognised as being of high quality, it has also opened up the possibility of preparing B-level pilots and engineers to perform A-level tests. So an additional training module has been set up for some of them. This comes in two parts: first, taking part in a large number of typical A type test flights to cover everything that can be handled with an aircraft manufacturer; then additional training at EPNER to cover tests that cannot be run at Airbus. Here they discover other types of aircraft, from military twinjet trainers to gliders, and they can become familiar with the effects of 'defects' on flight controls using a specially equipped flying test bed.

Test flight engineers are also qualified internally after their level B flight test engineer training.

Cabin specialists first take part in the same training as all airline personnel, after which they are given additional training for their activities with the manufacturer.

Thus, in ten years a highly effective training system for crew members has been set up that caters for all Airbus flight test needs.

THE PEOPLE INVOLVED IN DEVELOPMENT

This book will essentially focus on what happened in flight, with explanations on the difficulties encountered and the changes that proved necessary on the aircraft. Of course, the flight crew performs the most visible part of development and certification tests during flights, with their cohort of pleasant or not so pleasant surprises. But they have to be seconded by several other teams to make good progress. So let's take a look at how we were organised. In so doing, we shall stick to the technical side of things. Clearly, a number of other entities (procurement management, finance, etc.) played a fundamental role in the entire programme but we shall only deal here with flight tests and those departments directly involved.

* * *

Engineers from the Design Office, working on each of the specialties concerned by a flight, take part in the preliminary briefing to finalise the test sequence. Above all, they must be present for the debriefing, after landing, to be informed of the outcome as quickly as possible. When all goes well, we all take note and move on quickly to the next issue. If a difficulty arises, discussions ensue as to whether the engineers had predicted it, and whether they had a ready-made solution or had to go back to the drawing board. In all events, they have to pull out all the stops to find a solution!

During the test period, our contact with the Design Office was certainly not just limited to flights. Regular meetings took place between the specialists to address current issues. Has a solution been found? Are we ready to try it out on the simulator? When can we hope to see it installed in a computer on board the aircraft? Such meetings also enable the team to define modifications to be implemented, because while some may appear fine on paper, in practice they may create operational difficulties. Close collaboration between the Design Office and crew is essential to steady progress.

In reality, exchanges with Design Offices do not start with the flights themselves. Aircraft definition meetings begin very early, right at the start of the programme, to ensure that the first concepts are satisfactory operation-wise. The flight tests team stands in more or less as the customer's representative. Thus, for the A380, the first discussions took place back in 1996.

The Design Office depends on Technical Management but also has very close links with Programme Management and particularly with the Chief Engineer and his specialised engineers. This is important for overall technical coherence of the programme, developing the best solutions in the various fields and maintaining effective coordination of the schedule.

• See photo n°10 – page 233

* * *

Tight-knit co-operation between Programme Management and Flight Tests was ensured by an entity known as MAP or 'Mise Au Point' that can be translated as engineering support to flight tests. For once the French term stuck as it seemed perfectly suitable so no English equivalent was sought. This group was no new organisation invented for the A380 but rather the continuation of a team that had proven its worth since development of the A340 and A330.

MAP's mission was to help the flight tests team prepare the aircraft, for both short and long term operations. For the A380, a 200-strong team managed by Mark Cousin included a large number of specialists working in permanent contact with their Design Office counterparts. Each of them had to be kept abreast of all the latest developments planned on the systems in their fields of skills and thus be in a position to optimise preparation of the aircraft for each flight and brief the working parties.

In addition, each flight test engineer expressed their requests for the flights to come and it was the MAP that launched the preparation of the aircraft in line with these demands. When this was not possible,

MAP reported the difficulties and tried to get the best trade-off to make rapid progress.

The usual point of contact for flight tests with MAP to prepare the aircraft in the short term was the test flight engineer.

* * *

The many simulators involved in the A380's development are very different from those installed in pilot training centres. The cockpits for the latter have to be strictly identical to that of the aircraft itself and movement is reasonably well reproduced in standard flight (though not extreme) conditions. Conversely, the cockpits for development simulators can be different in many respects from the ones on the aircraft, as long as this has no incidence on the tests themselves. They are generally static. The software for the simulator used for final validation has to be representative of the aircraft's behaviour from the aerodynamic perspective up to the limits of the normal flight envelope and even sometimes beyond, as for example at very high speed.

The simulation team has its own pilots who work full time on the various types of Airbus. They are entrusted with initial engineering support, firstly with simulators as such, and then systems, for example to do the preliminary work for the flight controls laws or the autopilot. The test pilots only intervene when an operational judgement is needed to pursue development.

When it comes to validating the principles that do not depend on the type of aircraft, we use a research simulator known as Epopée that represents an Airbus but without all its operational functions. All research projects initially start up on this simulator and then, once the decision is made to install a new device on a particular model, a development campaign is launched on a simulator representative of the chosen aircraft.

The first simulator really devoted to a type of aircraft is known as Aircraft Minus 1. Right at the start, the aerodynamic characteristics are not always available. That's why the simulator prepared for the A380 incorporated a simple aerodynamic model of the A340-600, our biggest aircraft of the time. This meant we could establish a rough definition, for example for how data was to be presented to the pilots, and also validate the new logics for the flight controls. Later, it was equipped with aerodynamic models of the A380.

The simulators that really allow progress to be made in development are known as Aircraft Zero and the A380 had three of them. They are equipped with practically all the aircraft's computers, enabling correct global integration of the systems. Some of them, however, can be simulated when their characteristics are not compatible with the tests on the simulator. This is the case with the cabin pressurisation system, for example. The aerodynamics must always be as close to the knowledge of real conditions as possible and it is important for them to be re-calibrated as from the start of tests, either thanks to specific flights (the first of this type took place late in May 2005), or using certain test phases. These simulators continue to be used well after the aircraft are brought into service as all the computers are under permanent development and new versions come out regularly.

One of these latter simulators can be connected to a full-scale test bench that includes the hydraulic, mechanical and electrical systems. The flight controls surfaces are represented by metal masses, but with all the real servo-controls. The landing gear is also replaced by such masses, with all the aircraft actuators installed, whether for the landing gear itself or the trap doors. The motors commanding the leading edge slats and flaps are installed. One of the main goals of all this is to produce a realistic representation of the hydraulic flows on all systems. This huge mechanical installation explains this simulator's name of Iron Bird. A preliminary validation of the flight characteristics is conducted before the maiden flight on the basis of calculations and wind tunnel models. Next, after recalibration using the flight test results, this simulator is used to refine the settings, resolve any problems that may arise, and also provide a demonstration of failures on the various systems, as with flight controls or the autopilot; this all has to be performed for certification purposes when the test is not conducted in flight.

• See photo n°11 – page 234

For the A380 programme, the Iron Bird was only available just a few months before the maiden flight, whence a few difficulties with the schedule both to conduct all the checks to ensure that the tests could be launched safely, and to conduct a first exploration of the aircraft's characteristics to prepare certain flight controls settings.

For the first time in the development of our programmes, we were given the option of installing the simulator cockpit on a mobile system, just like in training centres. I shall return later, in the chapter on flight controls development, to the reasons for this choice and on our (interesting!) observations on how this motion was used.

Pilots responsible for development generally spend an impressive number of hours on simulators. Jacques Rosay, the leading pilot for the A380, admitted to me one day that he reckoned he had spent at least 1,000 hours in them over a period of about ten years.

* * *

All pre-certification test flights are monitored from the telemetry rooms located in Toulouse, providing the aircraft is flying close enough to one of the message reception centres. Apart from the one located in Toulouse, there are three centres located at Martigues, close to Marseille and Istres Aerodrome (the base of the French Flight Test Centre) to cover tests over the Mediterranean and at Istres, and Bordeaux, and Saint-Nazaire for the Atlantic and western France. These centres each have between one and four antennae (like Toulouse) as there may well be two or three development aircraft in flight at any time in the same region. In addition, for some especially delicate tests requiring a very large number of parameters to be kept under surveillance, two different messages are sent down to the ground by the aircraft. These messages are then relayed on by satellite from the non-Toulouse reception centres to the main control room.

The Toulouse telemetry centre has three rooms, access to which is reserved only for the specialists needed for flight monitoring. The conversations on board the aircraft are retransmitted via the telemetry message, which is encrypted and no outside organisation can hack its content. However, the answering messages from the telemetry room are radioed back to the aircraft and can therefore be readily understood by anyone interested and who knows the few frequencies we use. This fact worried me well before the maiden flight and I asked for at least one of the radio frequencies used by the telemetry team to be encrypted. I was afraid that some super sleuth journalist might guess at any potential difficulties we encountered just by listening in to the response from the telemetry room and use the information to blow things out of proportion (as is sometimes their wont) or, worse, that we could be victims of industrial or state espionage! Unfortunately, this coding system never worked and we had to give up the idea just before the first flight.

On board the aircraft, we tried to hold our tongue since there were plenty of people listening in. We would certainly have been more open in what we had to say without this constant monitoring, but sometimes we still let off steam!

The flight is run from the aircraft by the flight test engineers and, usually, conversations with the telemetry room are kept to a minimum. However, this didn't stop us asking for information or an opinion from the specialists down on the ground. They too were there to intervene if they detected a potential problem before we did.

There is one type of flight where the intervention of the telemetry room specialists is truly essential: this involves what are known as 'flutter' flights (a special chapter is devoted to these tests later on). On board, we don't have the computers that would be needed to analyse the stability of the structure subjected to vibrations, so it is the specialists on the ground who check this out and give us instructions as to whether to pursue tests or abandon them.

In order to maintain some measure of discipline in the room, an engineer responsible for telemetry systems is appointed as what might be translated from the French as 'Chief Listener' (a kind of Master of Ceremonies). He holds the microphone and ensures the interface between the aircraft and the specialists on the ground (who generally have not acquired radio speaking voices). However, in some special circumstances, when more complex explanations are required, he may hand the mike over to another engineer.

* * *

There would be no useful purpose in conducting test flights if the results were not then carefully analysed, so there is a team that thoroughly reviews all the data. To simplify matters, the team uses the measurements to back up the views given to the Design Office by the flight test engineers who, in turn, rely on the impressions of the pilots and the results obtained on board. The engineers in charge of extracting data perform an in-depth analysis of all the tests conducted and prepare a summary report for the Design Office. So, according to the precise results, you can check whether the certification or contractual conditions are met or not and, in the event of difficulties, the specialists will dispose of quantified data to make decisions. When the time comes, once everything is in order on a given topic, the report is handed over to the certification authorities.

Considering the large number of types of Airbus that have been certified over the last twenty years, a good many computer tools have been developed to extract the useful data as rapidly as possible - because Design Office engineers are obviously extremely impatient to get their hands on all the results. A good deal of common sense is vital to correctly interpret and rapidly detect anomalies of any nature. That's why true specialists are needed in each discipline.

Alongside this whole team in charge of data extraction, another group has to prepare the installation of measurement systems on board the aircraft, including telemetry, either for each flight, or for campaigns requiring specific instrumentation. Like the aircraft itself, this equipment requires maintenance and preparation work before each departure as everything must be in perfect working order. You never leave for a test flight without having checked that the recorders operate correctly and that telemetry monitoring is ensured. Such installation work can lead to delays, in the same way as failures on an aircraft lead to flights being postponed.

* * *

For a programme like this to be completed within the allotted timescale, it is essential for flights to follow on from each other in a sustained manner. As from 2002, based on experience acquired during the A340-600 flight test campaign, we argued a need to be able to fly six days out of seven, outside periods of work on the aircraft, and make two flights a day or one long flight (in the case of cabin tests where stabilisations of the parameters can often be very time-consuming). The seventh day was to be devoted to maintenance or minor works and, for organisational and labour legislation reasons, this usually turned out to be a Sunday. But we wanted to retain flexibility and be able to fly on that day too as required by the current campaigns and according to weather forecasts. Considering the usual rate of cancellations of test flights for various reasons, including minor modifications, we reckoned we could keep to a rate of about 90 flying hours a month per aircraft. Of course, this was not possible every month due to major works being undertaken to prepare new tests. For each of our development aircraft, this rate required five first line teams, three during the week and two over the week-end, in order to ensure servicing and implementation.

We had asked for complete integration of the various national entities so as to facilitate all operations. Indeed, on previous programmes, each national (French, German, British or Spanish) company had to agree before any intervention on an aircraft. Too much time had been lost with that system and it was out of the question to proceed in the same way with the A380. Finally, this request was accepted and everything worked fine under orders from Gabriel ('Gaby') Quadri. This first line arrangement worked perfectly and the flight schedule was respected.

* * *

The organisational core on a daily basis for flight tests is what we call Operations or OPS. A number of technicians ensure overall co-ordination, which is never an easy matter when you consider the special profile of some flights, weather conditions, aircraft failures or the installation of measuring equipment that can delay take-offs.

The control room is always the scene of frenetic activity. The pilots dispose of a storyboard that gives their availability for flights, considering their various meetings, simulator sessions, etc. Each morning, the engineers write up on another board the flight(s) they intend to have made the next day according to the schedule planned for each programme. In the afternoon, the Chief Pilot, or his representative, appoints the pilots for each of the flights.

Operations draw up flight plans for the various air traffic organisations and above all they have to resolve any problems that may arise in real time. For example unavailability of a control centre, negotiating a new slot for the flight when delayed, etc. The Operations managers don't just take care of aircraft undergoing tests but also those coming off the production line.

This team also has to take responsibility operationally for the preparation of all missions abroad. Difficulties here were exacerbated by the fact that for the A380 programme the personnel on stopover aerodromes were always worried about seeing such a huge aircraft land on their pitch. The parking stand, the possible landing runways and various taxiing routes all had to be determined in advance and in truth there was often much exaggeration involved. We had to explain that the likelihood of damaging a runway or stand was much less than with a B777-300 as the latter has just two main landing gears while the A380 has four for just 50% more mass. Technically, this reasoning was irrefutable but, unfortunately, some people construed it as being just another sales pitch! The Operations Manager also had plenty of other problems before actual take-off, like going through immigration checks, customs, and booking the bus to transport the crews and mechanics. There was never room for improvisation. The people on the organisational side really had their work cut out and were truly part of the great adventure!

* * *

You can't work to ensure final development of an aircraft in flight without effective support from air traffic control. In France we are lucky to have a system that is unique, with controllers who are specialised in tests and who belong to what is known as CER (standing for ATC for Tests and Acceptance). During flight, apart from take-off and landing in Toulouse, we are in contact with one of their control centres: Toulouse, Bordeaux, Mont-de-Marsan, Brest, Aix-en-Provence, Istres and more rarely Paris or Reims. The aircraft being tested is assigned its own controller who ensures we have complete freedom to manoeuvre, relieving us of concern for other aircraft in the sky, even to the extent of co-ordinating with his or her colleagues to re-route a commercial aircraft to give us a clear sky. When we follow a conventional flight profile, for example for the maiden flight of an aircraft that has come off the production line, as performed hundreds of times a year, no prior discussion with the controller is needed as, like us, he knows the sequence of events by heart. However, if special manoeuvres are required, we usually have a brief telephone conversation with him before the flight in addition to the explanations given in the test description sheet provided.

Decidedly, the CER helped us optimise our flying hours and save time. Two obstacles remained however. Firstly we had to pay quite a high price for this service, although, compared with a day lost for an A380, this can be considered minor. In this context, it is worth noting the ongoing dispute between Airbus and its US rival on the matter of direct and indirect subsidies to the aviation industry and it is perhaps regrettable that the CER service is not provided free of charge. The CER also suffered from understaffing, this having reached chronic proportions in the years running up to the A380. The controllers had plenty of good will and always sought to help us but it was impossible to ignore the rules on working hours. Special efforts were made for the A380 so no delay resulted from this side of things.

The controllers of the Toulouse-Blagnac aerodrome, just like the entire Civil Aviation management team, also gave us valuable assistance throughout this period. The climate of confidence between controllers and pilots made it possible to perform unusual manoeuvres on the runway to make the most of the tests. For example, we sometimes took off from a runway and landed back on it in the opposite direction in order to save time and avoid the brakes being heated up by a long period taxiing. On landing, this meant fitting the Airbus in between commercial aircraft as they came in facing us without holding them up too long, or providing us with exclusive use of one of the two runways. Mutual trust was needed for us to be able to work together successfully; a quick take-off is possible with a large aircraft but we sometimes had to remain stationary for fifteen or so seconds to start a calculation or for much longer for a special test. It sufficed for us to announce what our intentions were.

DISCOVERING THE AIRCRAFT

The first flights of a new aircraft are intended to start opening out the flight envelope, meaning you explore the minimum and maximum speeds, quickly check essential performances and make sure vital systems work correctly.

For each new programme, all the designers and managers are impatient to obtain key results so as to see right away whether the aircraft is going to be a success, or whether a great deal of extra work is going to be required to bring it up to scratch. If all goes well, this initial exploration is conducted over ten or so flights. In the case of the A380, in just seven flights, less than one month after the first flight take-off, the main results were acquired and we were all convinced that the aircraft was going to be a success.

* * *

During flight # 2, we thought we could explore the flight envelope as planned on the maiden flight. Remember, we had been unable to accelerate to maximum speed due to a landing gear door that failed to lock in the up position. The first thing to do at the start of this second flight was thus to check that this anomaly had indeed been corrected. As from take-off, on retraction of the landing gear, we felt reassured: theoretically we could continue with the flight as planned.

But then, as we gradually accelerated towards the maximum low altitude speed of 340 kt, at about 305 kt we had an oscillation of the pressure regulation valves. Then, suddenly, these valves opened leading to rapid depressurisation of the aircraft. That meant a huge shock for our ears as the cabin altitude sped up by about 40 metres per second. There were indeed four pressurisation computers, but they were affected by the same ailment: the outside pressures measured by various sources to be fed to these computers were compared and, exceeding 305 kt, the disparity between some of them (whose value had been set too low) triggered a safety mechanism. The systems then declared the pressurisation to have failed and completely opened out the valves that were intended to regulate cabin pressure. This, clearly, was a design error.

The pressurisation system had to be thought out again as even if, at the end of flight # 2, it was clear that specific modifications were required, other tests showed later that it was affected by other weaknesses.

Meanwhile, we had a major problem: it was impossible to accelerate to maximum speed and climb as we had doubts about the behaviour of pressurisation at higher altitudes. There was much discussion between us all, with the designers of this system pushing us to continue tests without any changes being made but keeping the airspeed down. Faced with this uncertainty, we refused and demanded that contactors be installed in the cockpit so that the valves could be closed manually in the event of an anomaly. Initially, this demand was rebuffed by some people at the Design Office. However, we stuck to our guns and refused to climb to higher altitudes without these controls and a solution had to be found.

* * *

We had to fall back on the alternative programme for this flight # 2. This included checks on the flight controls characteristics and an evaluation of the aircraft's handling with the flaps deflected so as to start development of the computers.

Then, on our return, we had a nasty surprise. The two landing gears under the wings came out perfectly but those under the fuselage remained blocked in the retracted position. We discussed what to do. Should we 'recycle' the landing gear, meaning retract and then lower it again with the risk of seeing it remain blocked in the up position? Or should we perform an emergency extension that would lead to keeping all the landing gear doors open? Finally, two minutes later, as we pondered the advantages and drawbacks of these solutions, both landing gears came down with no further intervention. It appears that one of the small doors on the wing landing gear failed to recognise its locked position, which interrupted the entire extension sequence, and then vibrations finally enabled a sensor to detect the real position of that door.

By then, we had clocked up some 10 flying hours.

* * *

A procedure was considered to avoid the depressurisation anomaly and was tried out, unsuccessfully, on flight # 3. As we could neither climb to high altitude nor explore high speeds due to this problem, we again had to fall back on 'plan B' and the flight was devoted to exploring low speeds. The goal was to determine the minimum flight speed with different flaps deflections. To do so, we had to go through stall procedures where the aircraft is taken to the limit of loss of control on deceleration down to the minimum speed.

Light aircraft pilots are familiar with these exercises as they have to show they know how to handle the situation to obtain their licence. On transport aircraft, for testing, stall exercises have to be performed to check the aircraft's behaviour when you get down to very low speeds. More than a thousand are performed before certification.

What keeps an aircraft up in the air is the lift exerted on the wings, this being proportional to the angle made between the air flow (the speed) and the wing (the axis of the aircraft). When this angle (angle of attack) increases, lift too increases before suddenly collapsing as the flow over the wings starts to go into a disorganised movement. That's when stalling occurs.

On transport aircraft, when the leading edge slats and the flaps are not deflected, stalling gives rise to stronger and stronger vibrations (buffeting). This becomes extremely off-putting when the angle of attack starts to reach 12 to 14 degrees and that is when the crew put an end to the test for fear of overstepping the limit and causing damage.

With the slats and flaps deflected, the angle of attack reached will be higher, between 17 and 21 degrees. Now, on stalling, there may be a small pitch-down movement of the nose in something like a salute as seen on smaller aircraft. Sometimes the stall is not simultaneous on both wings. If one dips before the other, the aircraft will move into a roll. Amateur pilots are well familiar with this characteristic. Other phenomena can also be encountered, as with sliding sideways and flat, with no bank – the aircraft then goes into sideslip. Finally, in some instances, there may be a pitch-up event, meaning an increase in angle of attack not commanded by the pilot.

Thus, stall, which amounts to a partial loss of control as the aircraft no longer fully responds to the pilot's orders, can take different forms. However, the pilot can always react to recover normal flight conditions.

Measurement of the stall speed is absolutely essential to calculate the aircraft's performances as, for obvious safety reasons, there must be a margin between stalling and minimal flying speeds, whether on take-off or on landing.

A large number of tests are performed in the wind tunnel and, generally, the stall speed estimations are correct, but what the wind tunnel will never tell you is what it will feel like. It's only on a scale 1 model that you will see what it's like! It is therefore up to the flight test crews to really find out what happens with the different deflections of the slats and flaps.

Since the A320, Airbus aircraft have been equipped with protections on the electrical flight controls preventing them from going as far as a stall situation. Indeed, stalling has always been a major cause of accidents in commercial aviation and this explains why Airbus installed these devices, thus hugely improving safety. But for stall tests, all protections have to be cut out. That is precisely what the flight engineer does using his ADIS system – he replaces the normal law with direct law. The latter also lets the pilot really feel how the aircraft reacts, especially its stability.

There are a huge number of precautions taken to ensure safety during the first stall procedures. Thus, all crew members wear an aviation helmet and a parachute. The emergency evacuation door is also prepared as for the maiden flight. Later, having gone through many other stall procedures and having been reassured as to the aircraft's behaviour, the entire set of safety systems is set aside.

The stall routine is always the same. The aircraft is first perfectly balanced to fly straight with the stick released. Then the pilot gently decelerates to about one knot per second. The second pilot or the flight engineer announces the angles of attack. Finally, either the pilot sees an anomaly (for example movement in pitch or roll) and pushes gently on the stick to resume speed, or, as is more often the case, one of the engineers to the rear announces "break" when he has seen a reduction in lift on his monitoring curves and pursuing deceleration would no longer be appropriate. The pilot's action is then the same as previously described to get back to normal flight conditions.

During the third flight performed with Jacques Rosay, Fernando Alonso and Didier Ronceray, we decided to commence stall operations by configurations with the slats deflected as there would be a lower level of vibration on the structure and we didn't want to start off straight away with an extremely severe configuration. We therefore pursued a number of such exercises with several flap deflections. After a few tests, just as we were in the process of recovering the aircraft, the sensors measuring the stresses on the structure on the horizontal tail unit warned us that these were too high. There was therefore a doubt as to whether this could be a problem of structural resistance. Given this uncertainty, we called things to a halt after the twelfth stall.

But we were bringing back positive results. Firstly, the aircraft's behaviour was perfect, with no significant roll or lateral movement. Also, measurement by the on-board test computers of the coefficient of lift showed a gain, meaning much lower approach speeds than planned could be envisaged together with a reduction in take-off and landing distances. Of course, all of this had to be backed up later by fine tuning after analysis in the Design Office. The results were then finally confirmed.

* * *

What then was to be done on flight # 4? High and low speeds were forbidden to us and we could not climb higher as the decision to install a closing system for the pressurisation valves had not yet been made.

The flight was therefore devoted to anemometer calibration, meaning setting the sensors that measure outside pressure and allow the altitude and airspeed to be calculated. On previous programmes, we had used a Russian supplied laser system, but unfortunately it was impossible to find spare parts and there was no other such device available on the market. We therefore had to resort to a very old empirical fly-past technique. The method is simple: the aircraft flies in stable conditions at different speeds at very low altitude over the runway and photos are taken to determine the height of the sensors above the ground. The correlation between the ground pressure and the pressures recorded on board at the same instant and at that height allow the calibration curve to be found in relation to speed. These fly overs were made at the Istres aerodrome.

* * *

Two days later on 12 May 2005, two pressurisation valve cut-off switches were installed. These were not manual controls but their activation was intended to lead to closure of the valves. In addition, a temporary modification to the computers was made to avoid opening in the conditions of flight # 2. With a little luck, we could set off on our first flight at high altitude. This was flight # 5.

The first thing to do was obviously to check the effectiveness of these two new controls in different flight and cabin pressure conditions. Everything worked fine. These controls were to be used again a number of times despite the correction made to the computers, but for other reasons.

We climbed by stages up to 36,000 ft. At different altitudes, we firstly accelerated up to the maximum speed with the direct law of the flight controls, and then repeated the same test with the normal law. We thus had a good idea of the aircraft's handling characteristics. The results were positive and final development would only concern minor adjustments.

The characteristics at 36,000 ft were especially interesting as the aircraft was fairly close to the optimal cruising altitude for its present weight (in fact slightly beneath). In just a few minutes we were able to note a number of extremely positive results.

Having performed our two accelerations up to a maximum of Mach 0.89, we stabilised the aircraft for about six minutes at the optimal cruise speed of Mach 0.85 to check fuel consumption. The values turned out to match the forecasts that had been made before the flight. So there was every reason to be optimistic: the aircraft, in its final version, should indeed have the expected cruise performances.

For this flight, as it involved a new flight envelope, of course we had all the safety equipment with us: parachute, life raft, life jacket, etc. Having run through these few tests, I struggled out of my harness. I was keen to get an idea of the noise level and vibrations to the rear of the aircraft at Mach 0.85, but also at Mach 0.89. The result was a very pleasant surprise, with no vibrations up to Mach 0.89 and a very much reduced noise level as compared with other aircraft with the cabin not fitted out. At that moment, I was convinced the aircraft would be a success because passengers were sure to appreciate that extra comfort.

The last positive point was the ease with which you accelerated up to maximum Mach as compared with other Airbus. We had the intuition that the drag increased little with the Mach and that it would thus be possible to cruise at a higher Mach without this being too penalising in terms of fuel consumption.

Towards the end of the flight, we had decided to climb to the maximum altitude of 43,000 ft. This took a long time as the aircraft was still heavy and over the recommended weight to cruise at such an altitude.

To finish off, we performed a first test of the anti-stall protection system at low altitude. The stall angle of attack was not reached so there was no risk of damaging the structure due to major loads on the horizontal tail unit. We stuck to a test with slow deceleration, without dynamic manoeuvring.

* * *

Flight # 6 was devoted to measuring fuel consumption in cruise at different Mach and altitudes.

Usually, these tests are not applied to the first development aircraft as the latter is intended to finalise the flight controls and various systems. In addition, that aircraft does not generally have the engine instrumentation needed to provide a fine analysis of performances. On the A380 programme, the second development aircraft off the production line was delayed by several months rather than coming out just two months after the first one as initially scheduled. Programme Management wanted to be reassured as to cruise flight performances. This was understandable as commercial contracts included clauses with penalties where the aircraft failed to live up to its promises. It was therefore urgent to know precisely how we stood.

During this flight, a large number of stabilisations at different speeds and altitudes were performed and we saw that the aircraft complied with the guarantees given to the airlines, even if there was only a slim margin.

However, there was one unpleasant surprise: the temperature regulation failed to work correctly. At the end of the flight, the blow air was frozen and the temperature in the cabin dropped down to 0°C. The crew was not kitted out with fleece jackets so the flight had to be cut short although it brought in a good number of results. But something clearly had to be done for the next flights at altitude!

* * *

Flight # 7 took place on 17 May 2005. This was intended to check operation of the flight controls laws with a centre of gravity position well to the rear. We conducted various manoeuvres with direct and normal laws in the entire flight envelope at low altitude. Special attention was devoted to stall protection, but, once again, to avoid exceeding maximum loads on the horizontal tail unit, we kept to low speed for an evaluation with little dynamics.

Overall, the results were satisfactory, even if it was clear that flight controls law tunings were needed.

Towards the end of the flight, when retracting the landing gear, a red light remained on telling us that the right landing gear was not locked in the up position. After several retraction and extension tests, we had to face the fact that we couldn't lock it in the retracted position. We therefore had to set the landing gear in the extended position to complete the tests scheduled in that configuration and fly back down to the aerodrome.

* * *

After seven flights, we had completed what can be called an initial exploration of the aircraft's flight characteristics. In the above summary, much detail of what was tested out during these flights is missed out, but many systems were verified. That applies for example to extension of the landing gear in emergency, and the back-up hydraulic and electrical systems.

Overall, the results surpassed our expectations. Firstly the aircraft was well born and secondly we had not discovered any problem likely to delay development and certification.

OPTIMISING THE AERODYNAMIC CONFIGURATION

Right from the start of design of an aircraft, the engineers do their very best to optimise performances globally, both to reduce take-off and landing distances and to ensure minimal fuel consumption at the planned cruise speed. For this an empty weight that is as low as possible and good aerodynamic characteristics are required. To achieve this optimisation, design offices now dispose of considerable resources. Firstly, powerful computers allow the stresses in each element of the structure to be determined and thus optimise the weight. They can also gain a fairly precise idea of the aerodynamic flow in different flight conditions, either by computations or using wind tunnels. But this is not always sufficient as it remains difficult to understand some elements this way as, for example, the ground effect already mentioned. Other aerodynamic phenomena are sometimes only detected in flight as they relate to a scale effect or the flexibility of the aircraft.

At flight testing, we were subject to very strong pressure from Airbus Management to reduce lead-times. So, with Fernando Alonso, when the overall schedule was being drawn up, we put the choice to them bluntly: either we do the mere essentials to validate the original aerodynamic configuration, postponing optimisation to a later date, or we be given a bit more time to choose the various settings enabling us to obtain a definition we would not have to go back on. We reckoned that the first solution required three months and the second, four. Obviously, we were pushing for the latter in the long term interests of the programme and, luckily, our advice was followed.

What does aerodynamic optimisation involve? It mainly means having the configuration that most intelligently optimises performances on take-off, landing and in cruise. In practice, it was above all important to concentrate on what are known as 'low speed' performances, meaning take-off and landing, bearing in mind that for cruise flight, modifications could be made later, on the condition however that the contract clauses relating to fuel consumption be respected right from the start.

* * *

As from the first flights, a number of cruise flight fuel consumption measurements were made. The engines were not yet up to delivery standard and although not availed of certain improvements that were already in process, they were clearly identified. It became quite clear on this issue that there were no grounds for concern as the contract clauses for delivery were going to be respected. All cruise performance measurements were planned, as was customary, for the second test aircraft. It didn't seem necessary to pursue the matter further at this stage in development.

* * *

So we focused on optimisation of performances on take-off and landing.

For landing, the solution is extremely simple: it is enough to have as low an approach speed as is possible. As this is determined by a margin percentage on the stall speed, reducing the latter means you can land more slowly and thus reduce the landing distance.

For take-off, the study is much more complex. Firstly, the aircraft must be capable of taking off in complete safety, even with a failure on one of the engines. It must also be capable of passing over obstacles at varying distances from the end of the runway after take-off (still with a failed engine). Optimisation has to be enacted for all aircraft weights, and all altitudes and temperatures likely to be encountered in operations. Of course, trade-offs are required as it remains impossible to benefit from optimal performances in all conditions. On take-off, crews have the choice between three flaps and slats deflections. The highest deflection, known as configuration 3, gives the greatest lift and allows for take-off at lower speeds. Thus the necessary runway length is minimised. Conversely, minimum deflection of the flaps, called 1+F, induces less drag and a better rate of climb with an engine cut off, but paying the price of increased take-off speed and therefore a longer runway distance. It is well suited to situations where obstacles far away from the runway have to be over-flown. Configuration 2, with a deflection of the flaps between 1+F and 3, is used to cater for intermediate cases.

In times past, pilots disposed of tables for each runway on each aerodrome they used, enabling them to choose the optimal flaps deflection setting according to the aircraft weight, the temperature and the wind. These tables also gave the speeds to adopt on take-off. Nowadays, practically all aircraft are equipped with a PC to make these calculations.

As has just been seen, take-off performances depend, among other things, on the rate of climb with a failed engine. But all speeds the pilot is required to respect must also have a margin in relation to those that could endanger the aircraft. Thus, before each take-off, the crew has to determine three speeds.

V1 is the decision speed. If there is a failure on an engine before V1, the take-off has to be interrupted. If the engine failure occurs after V1, take-off must be pursued as the remaining runway may not be sufficient to stop the aircraft. To simplify matters, bear in mind that computation of that speed relies especially on acceleration and braking criteria.

V2 is the minimum take-off safety speed with one engine inoperative in the prevailing conditions. It must ensure both an adequate rate of climb and respect the safety margins in relation to the various speeds that will be considered below.

Between V1 and V2, there is a third speed known as VR or the rotation speed. It is when going into VR that the pilot pulls back the stick to take off. Calculating VR is extremely complex as you have to make sure that if using different rotation techniques in the phase the plane actually leaves the ground, it gets to the end of the runway at a height greater than 35 feet and at a speed at least equal to V2 in the event of a single engine failure.

There are several speeds to be taken into account to establish safety margins, starting with stall speed. You then need a margin in relation to the minimum take-off speed, called VMU (Minimum Unstick). As will be seen in what follows, this speed is obtained taking off with the rear part of the fuselage actually touching the runway and therefore the take-off takes place with the aircraft resting on the main landing gears and a skid fitted under the rear fuselage. For special cases of flight with a low aircraft weight, it may be necessary to provide a margin in relation to the control speed with a failed engine, whether on the ground or in flight.

* * *

So how did we optimise the aerodynamic configuration?

Firstly, we always assume that we are not going to encounter a problem of such major proportions as to undermine design of the aircraft and therefore stick to adaptations that were planned for when construction started. In the case of the A380, we had planned to consider local modifications in three domains: deflection of the flaps, deflection of the slats and finally the size of the strakes positioned on the engine nacelle on the fuselage side.

Various deflections of the flaps were envisaged for configuration 3. These included 22 degrees (this being the value as the aircraft came off the production line) then 26 and 29 degrees which were later considered more appropriate by the aerodynamics specialists. Tests were therefore to focus mainly on comparing 26 and 29 degree deflections.

As far as the slats are concerned, a reduced deflection, initially by 20 degrees, was considered for the zone between the fuselage and the inner engines.

Each strake creates a vortex around it. As the vortex circulates over the wing's upper skin, it allows the flow to again adhere locally giving an extra surge of energy to boost the strength of the lift. Various strake sizes had been considered including broader or longer ones. Even doing without them altogether was contemplated.

* * *

As already stated, one of the major issues in optimisation is the search for the configuration that allows for the best rate of climb with one engine inoperative. There is no single solution as there are a large number of parameters involved including the weight of the aircraft, the airport altitude, the temperature, the height of obstacles and their distance in relation to the runway threshold. At this stage, the goal is not to calculate the aircraft's final performances (that was to be done later) but to compare those obtained in the various configurations considered to work towards the best choice.

Tests with an engine shut down and the others at maximum thrust were conducted with short climbs at around 8000 ft with about five to six measurement points at different speeds, for each deflection of the flaps and for a given aircraft weight. Such flights require tremendous concentration as, for each stabilisation, the speed must be held with considerable accuracy, with a maximum deviation of 1 kt for three minutes. The variation in heading must be reduced, down to less than five degrees. In addition, the weather conditions have to be near perfect, with only a gentle wind, no turbulence, a temperature that decreases with the altitude and so on.

To maintain the required level of concentration, the two pilots take it in turns to run the series of tests, each of which lasts about thirty minutes considering the heading and altitude repositioning between two measurements. Such flights can be quite long, from four to five hours for example, especially when weather conditions are good.

For the A380, these tests started with the ninth flight, on 20 May 2005, less than one month after the first take-off. The fourth and last flight of this type, # 12, used to choose the configuration, took place on 25 May.

We encountered several difficulties. First of all, flight # 10 had to be cut short due to bad weather. Flights 9 and 11 were also shorter than planned since oil consumption on the inboard engines and especially on engine # 2 was excessive. This was a familiar problem as we knew that the succession of idle and full throttle phases created minor leaks. This problem was later resolved by Rolls-Royce.

A number of flights were made with a heavy aircraft weight of up to 522 tonnes. The reason for these tests is that extrapolation from flights with lower aircraft weights needs to be treated with caution. For example, when the wings contain a large quantity of fuel, they will bend differently. But what is not so readily seen is that the shape can also change in torsion. Finally, these various deformations may affect the rate of climb and that is why validation is required. Such flights can sometimes involve considerable excess landing weights. As from flight # 10, we were 85 tonnes over the maximum landing weight.

Flight # 12 provided the opportunity for François Fabre, a pilot from the certification authorities to fly.

It was during flight # 9 that we discovered what was to become our main problem in development of the A380: the structural resistance of the fin. During certain turns with three engines at maximum thrust and the other one inoperative, we reached about 90% of the limit loads. Under the regulations, it is an accepted rule that, beyond limit loads, the structure may become definitively deformed; while as from 150% of the limit loads, structural failure is a possibility.

The margin on loads seemed to us to be too low. It was less than what we had seen on other programmes. That's why, immediately after that flight, we asked for the fin to be reinforced. The story of this modification is covered in a separate chapter.

* * *

The principles of stalling were explained in the chapter on DISCOVERING THE AIRCRAFT. On flight # 3 we had remained at a permissible overload on the horizontal tail unit. The experts spent a long time discussing the method to calculate loads. There were many things that seemed strange or irrational. Were we above or below the limit loads? We were pretty perplexed. Some of us thought that the assumptions on which the calculations were based were incorrect as it had been estimated that the margins to be adopted were to be similar to those for an A340-600, although that aircraft has a longer fuselage and as a result should have relatively less vibrations on the tail unit. Moreover, it appeared that apart from 'normal' loads, an additional margin had been added for buffeting. It was also argued that the safety margins had been reduced in relation to a possible failure. In all, the whole situation remained pretty confused over the next few days.

Three days after this flight, on Tuesday 10 May, the decision was made to perform load measurements on the ground (also known as 'static tests') on the rear section of the fuselage and the empennage to try to understand what was happening and, at the same time, calibrate the load measurement gauges.

On 24 May, before these ground tests, but after many computations, the specialists gave us permission to resume stall tests with a new limitation on the loads. However, considering the test programme under way, these tests were postponed to 2 June.

For this second stall flight, I was again with Jacques. This was flight # 18. On the first stall in 1+F configuration, the new limit was just reached. On the second stall with the flaps in position 2, we were almost there. Apparently, the empennage was getting too close to the wake of the stalled wing on recovery. We therefore called a halt after these two tests. To continue, we wanted an increase by at least 25% of the authorised loads, even if this meant having to reinforce the empennage.

In parallel with this, we were contemplating recovery techniques where, instead of pushing firmly on the stick during recovery, we simply keep it in neutral or very slightly forward so as to avoid bringing the empennage into the wing wake. On 7 June, on flight # 19, a series of tests was made with this technique: the stick was released just as stalling was reached. Handling was extremely delicate, but it meant that valid measurements could be made. Thus, 29 stalls were performed during that same flight.

This activity was interrupted so we could take part in the Paris Airshow at le Bourget. But as soon as we got back on Monday, the gauges used to measure loads on the empennage were calibrated on the ground through applying real loads to them. This led to a new turn of events as they were found to have been wired incorrectly! That should have been seen in the tests conducted during assembly of the aircraft, but these checks had been dispensed with to gain a day on the schedule. Cutting corners thus led to a further waste of time at the start of the test flights! But good humour prevailed as the real loads now proved to be about 40% less than the values initially calculated.

During this calibration, 'tufts' were installed. These are very light, small plastic cones hooked on to a short length of string that is in turn secured to the wing by aluminium tape – it sounds like DIY, but it actually works! These tufts are placed on the entire wing upper surface and on certain parts of the engine pylons to visualise the progress of airstream separation as seen in the air flow on different areas and more especially on the wing upper surface during stalling. All tests are filmed so the aerodynamics specialists disposed of a scale 1 wind tunnel!

• See photo n°12 – page 234

• See photo n°13 – page 235

Stalling flights were then conducted one after the other every day and various aerodynamic configurations were studied.

On 24 June, during flight # 32, we checked what happens with a reduction in deflection by 20 to 17 degrees of the slats between the fuselage and the inner engines. We quickly realised that there was greater lateral instability with lower values. With the flaps in a 26 degrees position, we even reached 13 degrees of sideslip on recovery. This was unacceptable and that configuration's fate was sealed – the slats would remain deflected at 20 degrees.

We then concentrated on the strakes, with various sizes selected by the aerodynamic specialists. Apart from the original shape, we also considered two different sizes and even doing without them altogether. Some measurements were made at different aircraft weights representative of take-off and others at intermediate weights.

On 27 June, during flight # 33, we decided against the no strake configuration as, without them, stalling was seen to take place earlier: there was more buffeting and a tendency to pitch up, meaning the pilot had to hold back the aircraft before getting into a stall. The decision to keep the strakes was obvious and not that unexpected in so far as the strakes were designed in the first place to keep the airstream adhering to the wing upper surface!

Some stalling flights were then made with two types of enlarged strakes, but no significant improvement was seen.

On 6 July, after 256 stalls, the decision was made to retain the original strakes and slats deflection. As far as the flaps were concerned, the choice for configuration 3 between 26 and 29 degrees could only be made after VMU tests.

This was still not the end of the stalling flights as the goal of this first series was confined to validating and optimising all the aerodynamic configurations, so there were plenty more to be gone through for certification. Several hundred stalling operations were performed to justify in-flight performances and behaviour. I made 69 stalls in a single flight but make no claim to the Airbus record!

* * *

As explained previously, the Minimum Unstick speed or VMU is obtained when the main landing gears and the skid installed under the rear part of the aircraft and in contact with the ground take to the air at the same time. The manoeuvre is spectacular and traditionally immortalised by photos and films in which you can see a shower of sparks shooting out from the skid as it scrapes against the runway before leaving the ground.

VMUs strike the imagination with their small firework display under the rear of the fuselage. But, considering all the development and certification tests, they are among the most harassing for the crew and there is a very real risk of damaging the aircraft. Few pilots can boast of having conducted such tests on different types of aircraft without having encountered more or less serious trouble. I admit that out of seven types of aircraft put through their paces in this situation, I had known two incidents including one during the first test on the A380. Indeed, there is a tradition at Airbus whereby the Flight Test Director or the Chief Pilot, or sometimes the two of them together, perform these tests. If the aircraft is damaged, they alone are to blame!

For these VMU tests, as for those to finalise the flight controls during the rotation phase and others intended to measure the take-off distance, a skid is installed to the rear of the fuselage. The associated structural reinforcement is huge, with the A380 skid having to sustain loads of up to 160 tonnes. During the VMU tests, we sometimes reached 100 tonnes as we forced the skid down against the ground. Considering all the modifications needed, just one test aircraft is normally equipped with this skid.

There are a number of difficulties involved in performing these VMU tests. Firstly, the skid obviously has to be brought into contact with the runway structure as gently as possible as the aircraft, even though locally reinforced, is not scaled to withstand too violent an impact. The manoeuvre has to be piloted with great finesse. The level of risk is exacerbated during tests at maximum thrust as, on some aircraft, you can start to raise the nose at a speed close to the take-off speed and there is sometimes only one second left to bring the skid down gently against the ground before take-off. This test is always done last as the pilot has had the time to become more familiar with the aircraft.

When we conduct tests with very low thrusts, the aircraft sometimes actually leaves the ground but then refuses to climb, remaining captive to the ground effect. To picture things graphically, it's as if the airplane is carried by a cushion of air close to the ground but cannot climb higher. As an example, at the Istres test base in Provence, where we perform these tests, there were times when we got to a height of 100 feet after 3 miles of runway. If you realise that the aircraft will decidedly not go up, the test is interrupted and one of the pilots will gradually boost engine thrust.

A last difficulty that can lead to the aircraft being damaged relates to the fact that you have to maintain the aircraft's attitude precisely at the moment of take-off as the regulation prohibits its reduction below that for take-off. As a result, the crew can be led to slightly increase it to be sure that the test is valid. If, as soon as the wheels leave the ground, it is increased by more than one or two degrees, there is a risk of touching behind the skid and thus damaging the rear part. This risk was especially pronounced on the A380 as the skid had been secured to a metallic part, well to the fore of the entire rear part of the fuselage made of carbon fibre. Knowing the aircraft was likely to be damaged with this type of assembly, a double metallic skin had been installed to protect the zone made of carbon fibre, which is always difficult to repair.

This protection was a welcome addition on two occasions. On the first VMU flight, I made contact once on the first test and again on the second flight, while Jacques Rosay barely touched on his first test. This shows that even with extremely good preparation on a simulator, the slightly differing reactions of the aircraft can still hold surprises for the crew.

During all the VMU tests, we use a special device to turn off all the sound warnings, otherwise we would have our ears blasted with "stall", "thrust not set" and other audio messages that would make it difficult to concentrate on the work in hand. Only the visual warnings remain and the test flight engineer is given the task of checking them and announcing whether there is an anomaly that would prevent the test from being pursued safely.

For those who are interested in flying techniques, the share-out of tasks between the pilots during a VMU flight as conducted at Airbus will seem pretty strange.

One of the pilots, traditionally the one seated to the left, sets his seat to the very low position. He only looks at his artificial horizon and the stick is used just to control the longitudinal axis - that is to set the pitch of the aircraft. The sensitivity with which he handles this will determine the quality of the skid's contact and then how attitude is held just after take-off.

The other pilot sets his seat to a very high position, so as always to see the runway, even when the aircraft is resting on the skid. This is especially impressive on a short fuselage aircraft that will be at an extreme angle of attack with the nose well up on take-off. He uses the pedals to hold the aircraft to the runway then, after take-off corrects the bank angle to keep it at zero, if possible using the rudder. Finally, and above all, he is responsible for safety, meaning that he will take over complete control of the aircraft to boost power and ensure handling if he considers there is a safety issue.

A test flight engineer assists us for these tests. He is especially entrusted with setting the engines, which has to be done extremely accurately, especially when the thrust required is low.

So, who is the real captain? The one conducting the test and has in his hands the risk of damaging the aircraft? Or the one there to ensure safety? We never really decided on the matter. What's certain though is that, once again, it would appear that the success of the test is a team matter rather than a question of hierarchy.

* * *

Tests were scheduled at Istres for mid-July 2005, but on the afternoon of Friday the 8th, we received a fax from the Istres base telling us that everything had to be postponed as it was planned to install Xenon lights at the runway threshold to allow for the US space shuttle to land there in the event of diversion after take-off. This prohibition even remained in place right up to the shuttle's take-off and we knew from experience that this could take several days, or even weeks. We were outraged that priority should be given to the hypothetical diversion of an American shuttle and thus delay the A380. But maybe our American friends had upped the stakes financially!

We made strong protestations to the Colonel commanding the air base and the Flight Tests Centre Director. Finally, a compromise was found. Just by shifting the take-off point of departure by 1,000 ft would avoid the jet blast hitting the lights. This issue was settled quickly but is indicative of the sort of false problem that cropped up constantly.

The tests started on 13 July 2005. This was now flight # 41. The airplane weight for the first test was 526 tonnes. It was a success, but I pulled the nose a little too much just after take-off and touched the runway with the metallic skin intended to protect the carbon fibre behind the skid. The engineers nevertheless agreed that there was no need to interrupt proceedings.

That day certainly had its frustrations. Firstly, after the second test, we had to wait eighteen minutes to allow an Air France B747 cargo to land. We would have had had plenty of time to take off and land in that slot, but the matter was out of our hands.

After the third test, there was a further delay as there were 15 Mirage 2000 fighter jets ready to leave on a mission. Unfortunately, one of them burst a tyre on take-off (though this could have been our fault as we tended to project small stones onto the runway). We had to wait for a full hour before it came back to land, then a further thirty minutes for the wheel to be changed on the runway. A sum total of two hours lost time for us.

On the fourth test, it was already getting late and slight turbulence interfered with the measurements so we decided to call it a day.

A first conclusion that emerged from these tests on 13 July was that there was a need to finance a widening of the runway shoulders at Istres. Despite the presence of four sweepers that came into operation as soon as the runway was free, we were unable to clean it completely. Bear in mind that the A380 outboard engines are some 169 ft apart and the precautions taken on powering up were not enough to prevent projections of stones from the rocky plain of la Crau onto a runway almost 200 ft wide including the shoulders. The cost of widening had already been estimated to come to one million Euros. Programme Manager

Charles Champion was unhappy about this and like any good financier first argued that he didn't have the money for unexpected expenditures. But we warned him of the risks involved if nothing was done: the Mirage 2000 incident had made its mark and the Istres airbase could quite simply be closed to us. Finally, as we shall see in the chapter devoted to the runways, these works were completed quickly over the following weeks.

The next VMU flight took place the following day, 14 July, in splendid weather conditions. Our departure from Toulouse was slightly delayed by a data recording system failure. But on arrival at Istres, we learnt that we could do nothing between 7:45 and 8:30 a.m. as that slot had been allocated for the take-off to two KC135 refuelling aircraft that were to take part in the Bastille Day fly past over the Champs-Elysées. We protested vigorously: all of 45 minutes just to get two aircraft off the ground! But nothing could be done and we just had to put up with the military air controllers' decision.

There was no other traffic and the tests quickly followed on from each other. With Jacques, we reversed functions. Two factors helped go through things quicker. Firstly, it proved possible to replace the removable part of the skid (that wears fast and has to be changed regularly) without shutting down the engines. What's more, Jacques had gradually developed a skid touchdown technique that reduced wear with a shorter and less sustained contact. Replacements were thus less frequent than initially foreseen.

Finally, nine VMUs were performed during this second and last flight, that is a sum total of thirteen.

Examination of the structure showed that the first touch on the dual skin had slightly damaged the carbon fibre. The aircraft was grounded for the next three days for repair. In flight testing, we consider the price to pay for such tests reasonable in so far as a positive outcome is obtained.

Another series of eleven VMUs was conducted on 25 and 26 March 2006 in order to obtain all the results needed for certification. During this last campaign, there were a few tests with a minimal distance between the APU and the ground of 30 to 50 cm, which corresponds to a margin in attitude of half a degree, further proof of how difficult it was to run these tests. Also note that during a test with just 48% of maximum take-off thrust, at an aircraft weight of 440 tonnes, we were at 200 ft over a house situated just under 5 miles from the runway threshold. We were able to complete the test at 400 ft without go around!

- See photo n°14 – page 235
- See photo n°15 – page 236

* * *

What were the results of this optimisation?

First of all, as we saw previously, it was possible to make some decisions immediately. Thus, reduction in deflection of the slats was rejected just after the corresponding stalling flight, considering the poor results.

As far as the strakes were concerned, this modification was intended to reduce the stalling speeds for take-off configurations, but above all for approach so as to reduce landing distances. The final conclusion for all these stalling tests was that the initial size and shape were good. The aerodynamics specialists had got it right.

The choice of flaps deflection posed no particular problem for take-off configurations 1+F and 2 as the performances required were satisfactory. In approach, we had a speed well below what was predicted, and so no modification was needed. A choice still had to be made for configuration 3 between deflection at 26 or 29 degrees. A trade-off was found by examining the rates of climb with an engine shut down, the stalling speeds and VMUs. The decision to opt for 26 degrees was made on 27 July 2005, but in fact, there was very little real difference between the two deflections.

Not all the speeds necessary for optimisation of take-off performances were determined during these tests. The minimum control speeds with one engine shut down, both on the ground and in flight, still needed to be set. But in principle, these had little operational impact since, being pretty low, they only changed the take-off speeds for a low weight airplane taking off from a short runway. Considering the usual operating conditions for A380s, this was hardly penalising.

We had run ahead of our forecast schedule, which had initially planned for four months to freeze the aerodynamic configuration. The tests were completed less than three months after the first flight and the decision made just a few days later. The good news was that the aerodynamics at low speed were remarkable. The only modification (which had been contemplated right from the start, anyway) was a slight increase in the deflection of the flaps for one of the take-off configurations. But there was some bad news too as we realised that we would have to reinforce the fin.

THE FLIGHT TEST SCHEDULE

First of all, I just want to say that the sole purpose of what follows is to describe how the flight tests schedule and certification evolved, without going into the reasons that led to delays in delivery. The flight tests team never had an overview of the programme's progress, especially that relating to production schedules. So here, we will simply focus on what happened between initial planning, some years before the start of tests, and certification.

* * *

Before explaining how the A380 development schedule evolved over the years, let's take a look at how Airbus was organised at that time.

The 'Big Boss' was Noël Forgeard until he took on management of EADS about two months after the first flight of the A380. Noël had three main deputies: John Leahy for the sales side, Gérard Blanc, initially responsible for programmes, and Gustav Humbert who headed production, design office and product support operations. In February 2004, Gérard Blanc and Gustav Humbert switched jobs for reasons unknown to me.

Gérard Blanc had one great quality that he demonstrated on a number of occasions. He was ready to go out and get first-hand information when there was a problem. Thus, in the early 2000s, he managed to sort out the A340 and A330 production lines that were not being run effectively as deliveries were regularly late. However, he refused to be contradicted and, lacking training as an engineer, he sometimes flew in the face of obvious technical issues.

Seen from our position in flight tests, Gustav Humbert seemed relatively inactive, even when he replaced Noël Forgeard to head Airbus. But, clearly, we are not aware of all that happened at that level.

The A380 Programme Manager was Charles Champion for whom we always had great esteem. Unfortunately, given the strong personalities of Noël Forgeard and Gérard Blanc, he had very little leeway. Two key figures under his orders were Robert Lafontan, Chief Engineer, and Alain Ramier in charge of planning and co-ordinating the work of all the teams working on detailed definition of the aircraft. Put simply, the Chief Engineer with his team designed the aircraft and was there to resolve all the main technical problems. Alain Ramier and his department looked after actual implementation in liaison with the design offices.

Having taken part in many aeronautics programmes, Robert had all the qualities needed to pull off the huge task involved in designing such an innovative aircraft. He also had considerable experience as a pilot on a number of different airplanes. In addition, he qualified as test pilot in our school as we considered that, with that experience, his credibility as Chief Engineer could only be enhanced in the eyes of some head pilots in the customer airlines.

Engineering was ensured by Alain Garcia, firstly under the orders of Gustav Humbert, then under Gérard Blanc. Flight tests were assigned to him. Unfortunately, Alain Garcia was hostile to crew members and he publicly stated this on a number of occasions.

* * *

How many development aircraft do you need to certify a new machine in the best conditions?

Here, a compromise has to be found. Pushing things to a logical conclusion, development would take a very long time with a single aircraft, as would return on investment. Also, this would hardly be compatible with work on the cabin.

Conversely, having too many development aircraft means the cycle for certification can be reduced, but in any case there is always a minimum period required to install and validate modifications made due to problems that are inevitably encountered and there would be no way of using those extra aircraft to the full. They also entail added costs as they are brought to comply with the definition desired by the customer. All this leads to time being lost and often a lower sale price as some airlines are not too keen on buying aircraft that have been used for development.

At Airbus, the trade-off generally adopted for a completely new programme is to have three or four airplanes, with two for the purely technical side and one or two for the cabin. The two aircraft used for technical development and certification are absolutely vital considering the scale of testing to be conducted. Also, even if there is a major incident on one of them, the programme can still go on as you have a backup.

As far as the A380 programme was concerned, with the Rolls-Royce engine, we thus had two aircraft for technical development, # 1 and # 4 (I don't know how these numbers were assigned!). The cabin was mainly to be developed on # 2 and aircraft # 7 came on the scene just before certification to perform endurance flights. A fifth airplane, # 9 was used for development and certification of the version equipped with the Engine Alliance engine.

* * *

Discussions on the schedule started in 2000, but the first big meetings took place in 2001, four years before the first flight. As from 2000, we reckoned that at least eighteen months would be needed between the first flight and entry into service. Remaining reasonably optimistic and assuming there were no major difficulties, we ended up planning for a period of fifteen to sixteen months in the flight tests schedule between the first flight and certification, with one to two additional months up to delivery and the first commercial flight. This was surrounded by a certain number of conditions and especially the fact that we needed to have the first two test aircraft ready in a very short time that is about two months, as on previous programmes. In addition, the aircraft equipped with the cabin should also come on line very quickly.

However, there were a number of critical cycles up to delivery. The most important of these cycles related to development of the cabin and this led to a minimum period of fourteen months between the first flight and certification, and sixteen months for delivery of the first aircraft. Indeed, we had planned for a series of flights with the full complement of passengers known as Early Long Flights (ELF) with the aim of reviewing the state of the cabin to define all the modifications needed before being brought into service. Obviously, to take passengers on board, you had to have total confidence in the aircraft's capabilities and thus have conducted enough tests beforehand. We reckoned that these flights could only take place some eight months after the first flight. This period had been calculated on the basis of our experience on the A340-600, which had only been a derivative aircraft, but which had taken seven months. The last stage before certification was route proving or endurance flights that involved at least 150 hours of flying in operational conditions, usually with the customer airlines. At that time, some weeks away from delivery of the first aircraft, we had to show a near perfect aircraft. Again, based on prior experience, we reckoned that at least four months were needed to finalise the cabin between ELF flights and route proving. In addition to that, there would be almost one month of endurance flights and then a further month for the certification authorities to handle the administrative side. That is how we came up with the period of fourteen months between the first flight and certification. Traditionally, there is a period of two months between certification and entry into service, the time it takes to install all the last minute modifications, hence a total of sixteen months before the first customer receives their aircraft.

There were two major conditions to respect this schedule. First of all, the aircraft with its equipped cabin had to be ready at least two months before the ELF flights so as to do the minimum of development work and secondly, and above all, all the safety equipment had to be validated. So the countdown planned for the cabin to be fully installed six months after the first flight of the first aircraft. This was unrealistic. Due to the large number of sophisticated items of equipment it contained, the first installation could not possibly be accomplished as fast as some people had imagined and this is something we made abundantly clear. In addition, this work was to be done in Hamburg, meaning the aircraft had to be made ready for flight and be flown there before starting work on it. The other condition was that no major problem be encountered in development likely to postpone one or other of these deadlines. But there was no point being anything other than optimistic!

This critical cycle, added to the many technical uncertainties you may encounter on this type of programme and the likelihood of delay in the aircraft being equipped with its cabin, convinced us that an additional margin of two months was needed. Generally, Top Management is averse to granting such margins, hence their dissatisfaction at that time. But facts are stubborn, so we announced sixteen months while hinting that, with a huge amount of luck, we could maybe get things done quicker, though refusing to commit to that.

Without going into the detail, many meetings were held to try to find a way of reducing the test cycle. My personal experience at that time of more than thirty years in flight tests (since 1970) had led me to estimate that we could hardly do better but, unfortunately, I hadn't managed to persuade the decision-makers. Or maybe we had already started to live by the tenets of what was later to be known as the green culture: all indicators are on green and above all you avoid admitting something's wrong or that you are running behind. But green culture has no place in flight tests. In our world, we need to square up to the truth.

So Flight Tests were in a fairly strange situation. We were practically being accused of holding things up as we refused to reduce the duration of tests unrealistically. We hardly benefited from overt support from the Head of Engineering. The Chief Engineer, Robert Lafontan, shared our view but he had some difficulty supporting us when facing the Top Managers. Thus, even though there were three technical entities on the same wavelength, only one dared openly state its convictions: Flight Tests. We assumed full responsibility with Fernando and I taking it in turns to argue our point of view at all the planning meetings.

I won't recount all the planning meetings on flight test scheduling but will just summarise some of them to give an impression of the general context.

* * *

The first meeting, in the presence of Noël Forgeard, was held on 26 November 2002 (although we were not invited despite the fact that the flight test schedule issue was brought up). By pure chance, we recovered an e-mail from Noël's Chief of Staff with his notes taken at meetings. At that date, the delay was estimated to come to thirteen weeks to which a potential risk of ten weeks had to be added due to the cabin. Noël's comments were as follows: first of all we stick to the schedule of bringing into service on 1 March 2006. He then declared (according to the notes): "...If we go to Airlines to shift the planning, the programme is half-dead". He also said that, in all likelihood, Boeing would have its new B747, the B747-800 ready and mature in 2006, while our aircraft would not be ready. You need to bear in mind that Boeing delayed bringing out its passenger version to 2012 due to a host of technical issues! That gives a picture of the state of mind that prevailed at that time with the idea that we were no good and Boeing was much stronger than us technically. What ensued showed that this view of things was way off the mark to say the least!

A few days later, on 11 December 2002, at a side meeting with Top Management, I expressed serious reserves as to the possibility of reducing the time between the first flight and delivery of the first aircraft from sixteen months, as initially planned, to thirteen and a half. My main argument was that it was impossible to conjecture what difficulties we were likely to encounter during flight tests. I was sermonised on the idea that we all had to improve, be more open and change our mindset "...including flight tests". The underscoring of the word including in those minutes is authentic! But a crystal ball is not one of the things a test pilot carries around in his kit!

In spring 2003 one fine morning, Gérard Blanc came into my office with its commanding view out onto the aircraft apron. After making a few general remarks, he then went straight to the heart of the matter: he wanted the aircraft to be certified twelve months after the first flight. I explained to him that this was unrealistic as we were bound to encounter problems, even if they were minor, as from the first flights and that they couldn't be settled by waving a magic wand. What I said was based on my experience of flight tests. One of his arguments was that the B777 had been certified in less than twelve months. That was true, but with a certain number of reserves of which he was unaware, including a major one in that transport of passengers was not allowed! The certification we were talking about for the A380 was for an aircraft status allowing for operational flights run by an airline. Certification in an even shorter time could also have been envisaged, but without passengers, no pressurisation, no autopilot and with a limited flight envelope. But that was not really what we were looking for. So the conversation ended there.

* * *

Another major meeting was held with Noël Forgeard on the afternoon of 15 July 2003 to review the test programme. It was hot and, on the flight tests side, we reckoned there was no need to wear a tie – but we were the odd ones out! A number of presentations were made, including one from Alain Ramier and another prepared by Fernando Alonso and myself. Alain showed that, considering the various delays on the assembly line and installation of the cabin, the date for entry into service would probably be towards the end of June 2006. The estimation of flight tests was not far off, depending on the assumptions made, but we got to near the end of August 2006. We had to come to a conclusion and, surprisingly, Alain Ramier decreed that we keep to the date for entry into service on 1 March 2006. I shook my head ostensibly and Noël declared in essence the following: "Of course we retain the date of 1st March 2006 despite Claude's denials". This seemed to me to be a refusal to face up to reality.

Shortly after this, on 21 October 2003, a small informal meeting was held with Gérard Blanc and Gustav Humbert. We were used to alternating presentations with Fernando and it was his turn to officiate. He thus again introduced a programme that, obviously, failed to respect the 1 March 2006 deadline for entry into service. That date we set rather towards June-August, and even so, on condition that all went well. Gérard Blanc worked himself up into a fury against Fernando. Obviously, I supported Fernando, and Gérard Blanc found nothing better to say to me than: "You make no effort to show you're in the right". I was so astounded by such an accusation of 'sabotage' that I can't even remember exactly how I responded but certainly there was something of a shouting match. The meeting came to an end quickly and, to our surprise, Gustav Humbert came over to console Fernando and myself, telling us that things weren't that serious! But finally, we had no other choice than to put forward a programme that led to entry into service on 1 March 2006.

On 3 December 2003, there was a new presentation of the schedule to Noël Forgeard and an assembly of about 40 people. After that disastrous October meeting we had to 'lie' with some skill. It was my turn to do the presentation and I showed an ideal test programme that ended on the right date. Obviously, I insisted strongly a number of times on the implausibility of such a schedule: no engine change, no working party lasting more than two days to modify the aircraft, meaning a near perfect airplane right from the start and a cabin practically without rework as from the first flight, etc. All the aircraft development specialists knew that this was completely unrealistic, but as the flight tests group was somewhat isolated, at least in appearance, we were unfortunate in having no other choice than to give a version that was really a half truth: we showed a schedule responding to the expectations of the management, but they were quick to forget all the major conditions that were tied in with it. In other words, they were simply burying their heads in the sand!

Here are a few of the comments made by our Big Chief at this meeting: "I wonder if we have taken a challenge above our capacities?" "Are we better than Boeing technically? Not convinced". "We are spending a large percentage of our skills in audits: I will forbid new audits". In my view, at least that last resolution was a good one!

Noël came over to congratulate Fernando and myself for the efforts we had made to respect the schedule! But one morning a few days later as I was on my way to another meeting, I bumped into him in front of the lift and he invited me to have a coffee in his office. I took advantage of the opportunity to tell him: "I hope you didn't believe a word of what I said at the schedule briefing". In response to his look of amazement I added: "You'll be very lucky if that aircraft comes into service in September 2006". He was stunned! I explained to him how I had been led to make such a biased presentation. The completely unrealistic conditions tied up in it had failed to make an impression on him and despite my insistence he continued to see the world through rose-tinted spectacles!

* * *

In 2004, everything calmed down for flight tests as the problem was no longer really our schedule but rather to know how we were going to catch up on the delay on the assembly line. Usually, on other programmes, the various elements like fuselage sections, wings, etc. came in almost fully equipped with all their wiring, and hydraulic and fuel pipes so that assembly itself was extremely quick. This was not the case on the A380 as it had been decided that the parts were to be routed on given dates, regardless of their current condition. Equipping them therefore had to take place at the same time as assembly and a great many Germans, British, Spanish and French from various plants had to converge on Toulouse where assembly was under way. One might reasonably surmise that things would have been more efficient and time gained without this 'green milestones' culture. By way of an example, on 6 May 2004, the assembly line situation was described to Noël Forgeard. Outstanding work in relation to the initial schedule, which included among other things the work that should already have been done by the various manufacturers of elements in their own plants, then came to some 23,000 man hours on the first aircraft. It was reckoned a further 30,000 to 40,000 hours would be needed for each of the following aircraft and this figure might well have been higher as it included only what had been admitted to. Obviously, catching up on such a delay represented a real challenge and for anyone realistic, this was not feasible in the required lead-times.

In 2004, considering likely delays, the date of the first flight could no longer be set at 17 January 2005, but the date for entry into service was maintained against all the odds. At this stage, there was no point going on about the schedule but rather let the entire team responsible get on with completing the first aircraft and prepare it for flights. It was urgent now to get it flying as this would mark a major stage from the technical and media perspectives. Oddly enough, the 'Reveal' date for the airplane to be presented to various public figures and the media was set for 18 January 2005, that is after the date planned for the maiden flight!

At the end of 2004, even if the official dates remained unchanged, the forecast schedule showed a delay of two to three months on the maiden flight and six months on certification. Despite this, we still had to hang on to an improbable timeline even though (and this was new) a major delay was already foreseen for the other test aircraft.

* * *

Over the first quarter of 2005, aircraft # 1 was prepared for its maiden flight with the difficulties inherent to such an exercise. Finally, being just over three months late for the flight of a completely new aircraft shows that the Design Office engineers were far from as inept as some people would have had us believe! Due credit should also be given to the teams in charge of preparing the aircraft who put in remarkable efforts to make sure we were not overwhelmed by the backlog of outstanding work. Although certainly not an example to be followed, one should also bear in mind that Boeing had its new B787 first fly more than three years late.

The situation hadn't improved for the following aircraft that were well behind schedule. There was more and more work to do after assembly as much of it had not been done in the various plants producing the components. Thus, it soon transpired that we would only be seeing aircraft # 4 towards autumn. Similarly, aircraft # 2 that was to have been equipped with the cabin was going to be significantly delayed, thus impacting the schedule because the critical path cycle related to the cabin.

A very discreet memo from Alain Ramier on 27 March 2005 announced the decision to postpone certification to the end of June 2006. I am not aware when and how our customers were informed.

Three months later, on 20 June 2005, in light of the delays on aircraft # 2 and # 4, a new memo informed us of a second postponement of the date of certification to 2 October 2006, and it was also decided that there would only be two A380s delivered in 2006.

* * *

In the spring of 2005, the battle was on for the position of Co-Chairman of EADS. According to the information given by the press (we weren't in on the secret!), Noël Forgeard wanted to replace Philippe Camus at any price and even become the sole Boss. Finally, he was appointed early in May and started as Co-Chairman in June. For Airbus, a number of weeks were to elapse before we got to know the name of his successor, Gustav Humbert, who was finally named CEO in July.

On 19 July, we learnt that Gérard Blanc was going to be leaving the company, probably disappointed at not having succeeded Forgeard. The speech he made at his farewell drinks party at the end of October certainly had a lot of needle in it.

A new organisation was published, but the Technical Manager still didn't form part of the Executive Committee, Airbus's decision-making body. However, in companies where technical issues play such a vital role as in the aeronautics industry, this position should certainly not be pushed into the background. All the managers under Alain Garcia's orders therefore jointly signed a letter to Gustav Humbert to express their disapproval. This wasn't a personal matter, but one of principle, as we reckoned that the Design Offices were not accorded the esteem they deserved at Airbus. On Friday 29 July, Gustav Humbert received us to explain that Technical Management was under heavy criticism due to problems on entry into service of the A340-600, delays on the A380 and A400M, systems development issues, etc. He wanted to review all of this at a later stage if the situation improved. I first of all argued that our goal was not specifically to defend anyone, but that we didn't agree with his analysis: there was a problem of principle in not having the Head of Engineering included in the small circle of decision-makers. Despite it being during the summer break, there were still quite a few of us and the other participants pursued the discussion. What shocked me most was that when I put a straightforward and perfectly natural question to him as to what he intended to do, he flew into a rage and started shouting at me. But surely, when a Boss finds things aren't working the way they should, it is up to him to make decisions to improve the situation!

Subsequently, after Gustav Humbert had left, in mid-2006, to be replaced by Christian Streiff, Alain Garcia was appointed to the Executive Committee. Finally, well deserved recognition for the Design Offices and those working around them.

* * *

Over a number of meetings towards the end of 2005, we again expressed our scepticism as to the schedule. There were two main reasons for this. Firstly, the period between the ELF flights and endurance flights had been reduced from four months to two and a half months, and above all certain modifications demanded for the aircraft were not decided on or put into practice quickly enough. This was the case, for example, for the reinforcement of the fin as covered in the next section. At that time, it was urgent not to argue about the dates for certification, but rather to make progress on development.

In an e-mail dated 27 December 2005, Fernando argued that the picture was clear: the first aircraft equipped with its cabin, # 2, would be coming off the Hamburg cabin assembly line more than two months later than was planned in June. Similarly, aircraft # 7, that was to be used for route proving, had not yet made its first flight in Toulouse, meaning it too was more than two months late.

* * *

As from the end of 2005 and in January 2006, the pressure on flight tests mounted, as the Design Office stated that it would urgently need all the test results. This was fairly logical in itself as the engineers working in all disciplines wanted to move forward with development and prepare their certification dossiers. But, of course, not everything could be settled at once and priorities had to be set. The atmosphere was somewhat electric during that period, with rows flaring up from time to time between the protagonists.

For us, one of the options would have been to keep quiet and let the programme run its course with certification finally taking place when everything was finished. But setting an unattainable goal had a perverse effect: we were sometimes led to ask for a modification to certify or improve the aircraft and the manager responsible would reply that this was impossible considering the planned date for certification, while it would have been quite readily achieved on time with a reasonable objective. All this led to delaying improvements on the aircraft, disrupting certain tests and, finally, a greater cost burden.

On Thursday 2 February 2006, without consulting flight tests, external noise measurements were brought forward by the Management, while cabin tests on aircraft # 2 were delayed. Fernando was furious with these changes that disrupted his flight schedule. The afternoon of Friday 3rd, a memo was sent to us. It summarised how things stood with testing and the documents to be forwarded to EASA for certification on 2 October 2006: dossiers on engineering calculations of stresses on the structure, external noise level measurements, engine test results, etc. Practically everyone was running late and unable to respect the lead-times. Obviously, at flight tests we were much in the same boat, with nearly a further 1,700 hours of trials planned, which would have meant us doing more than 200 hours a month per airplane. This value should be compared with the 90 hours on average since the start of flights (outside working parties), which is a good performance to be credited to the airplane itself and the first line team. It was impossible to do more and so that date of certification had to be ruled out.

The same evening, furious and exhausted by this lack of coherence, I decided to send an e-mail to the Management. Fernando and I prepared it together. The gist of our message was that development of the aircraft was far from completed; and that we had also experienced a loss of control during tests in January (see the chapter LOSSES OF

CONTROL). I expressed my conviction that certification and delivery of two aircraft by the end of the year was impossible and that the schedule would have to be reviewed. Indeed, as we shall see later, certification did take place in December 2006, but with restrictions that prevented operational flight.

The reaction was immediate. That very evening, all the recipients were told by telephone and e-mail that they were forbidden from answering my message or sending it on to anyone else.

On Monday, I was called in to see Gustav Humbert early in the afternoon. There were five of us around the table: Gustav, Alain Garcia, Charles Champion, Alain Ramier and myself. Just before going into the office Alain Garcia told me: "Your mail was spot-on". So I was optimistic about receiving support during the meeting. I started by telling Gustav about the loss of control in January, as he had not been informed of the incident. I then explained to him why the schedule had to be reviewed and why the delivery of two aircraft at the end of the year was impossible. I informed him that on each aircraft we always had at least four pages of limitations and that had remained constant since the first flight. I made it clear to him that it was impossible to do all the flying hours needed for a delivery by the end of the year. Once I had finished, Alain Garcia whispered something in Gustav's ear, though I didn't catch it. But what did happen immediately was that I got a true dressing down. I was even accused of sabotage. The discussion went on as a duo between Gustav and me! To try and find a consensus, I proposed we certify on 15 December with certain restrictions and make a symbolic delivery before the end of the year, knowing that entry into service could take place later, early in 2007, after those limitations had been raised, but this suggestion was firmly rejected. Finally, I came to the conclusion that Fernando, who had told me my intervention would be of no avail, was right, but, with hindsight, I do not regret having penned that e-mail as we saw later that some of the Top Managers made claims to not having been informed of delays in production. At least on the development planning side, this was clearly not the case.

In fact, this incident was not completely ineffective as I realised a few days later that Alain Garcia had received the order to spend at least half his time on the issue. Dedicated progress meetings in small groups were then organised every Monday morning, which no doubt improved efficiency.

* * *

On the flight tests side, we decided to follow the progress of tests more closely as, hitherto, on all programmes, we had been somewhat lax on that issue. We had methods for evaluation based on the judgement of the most experienced flight engineers which, in fact, gave very good results. However, from outside, they lacked credibility as they were not based on numbers. We therefore started to count domain by domain, the number of flights to make to get through to certification.

Meanwhile, we had an estimation of the number of aircraft slots available. For each test aircraft, apart from the one equipped with the cabin, there were two slots a day corresponding to two flights or test series on the ground. Knowing that tests were possible six days out of seven, with the seventh being reserved for maintenance, that gave in theory twelve slots a week per aircraft. The statistics showed that about 30% of the slots were lost for various reasons: aircraft problem, weather conditions, unsatisfactory result, etc., thus leaving eight to nine useful slots per aircraft per week.

Every week, the number of successful flights corresponding to the test programme was checked and compared with that scheduled and, by way of a joke, we called this the peanut count.

Clearly, there was a problem of principle as the number of slots available changed in a fairly linear fashion over time, with only a few variations relating to working parties or airshows. Conversely, demand for testing was much stronger at the start since all departments wanted test results rapidly to prepare certification dossiers. We then had to make some tough decisions to optimise the organisation of flights.

Every week, the various curves on our charts were readjusted and, finally, they were globally respected: the number of successful flights was sometimes less than forecast but, meanwhile, the positive results obtained meant we could often reduce our estimates as to the number of flights remaining.

Right from the start, examining these curves, and taking into account the availability of aircraft not equipped with the cabin, showed that practically all the tests leading to a certification with minimal restrictions could be completed by December. Certification with the lifting of restrictions, that prevented entry into service, was clearly impossible in 2006.

* * *

On the Technical Management side, we hardly benefited from much support. As soon as Fernando and I put forward a realistic schedule, it was contested because not in line with the demands of Top Management. This led to regular disputes as we refused to kow-tow to the 'green culture' philosophy.

At the end of February 2006, during the Singapore Airshow, Fernando had the opportunity to again discuss the situation directly with Noël Forgeard. I assume Noël recounted what was said there to Gustav since he called Fernando in for a face-to-face interview.

On 6 March, we learnt that Management had decided to postpone certification to 15 November 2006 and that two aircraft were to be delivered before the end of the year, even if entry into service was delayed until the following year.

On 14 March, Gustav summoned me to give a clear message: for media-related reasons, we absolutely had to deliver those two aircraft to Singapore Airlines before the end of the year. Everything else was to take second place.

Time went by, but the decision-makers still refused to face the truth. Thus, in April, while Fernando was getting ready to discuss participation in endurance flights with Singapore Airlines, we held a meeting with the Management. The aircraft equipped with their cabins were far from being ready to fly and we suggested postponing those flights. But we were talking to a brick wall.

During an EADS cocktail at the Berlin Airshow on 16 May 2006, I had a one-to-one discussion with Noël Forgeard. As I was unaware of the problems in production, I told him that entry into service would not come about before April 2007, but he simply replied that I was always being pessimistic. This seemed to be the refrain that came back whenever I opened my mouth to talk!

On Friday 9 June, we learnt over the grapevine that a delay of six months on the programme was likely to be announced, with a single aircraft delivered in 2006. The official version was that all was going well on the development side, but that industrialisation had suffered delays.

On 13 June, there was a bolt from the blue for those outside with the announcement of a global delay on the schedule of six to seven months still with a certification for the end of 2006 and delivery of a single aircraft that year. For 2007, just 9 airplanes were to be delivered instead of 25. On the flight tests side, this was only half a surprise as, firstly, we always thought that a delivery in 2006 was illusory but, secondly, we had no reliable information from Hamburg concerning cabin installation. We knew that there was some delay in light of the fact that we were running late on our two cabin test aircraft, # 2 and # 7, but that could be ascribed to assembly line start-up.

The next day, the headlines blasted Airbus and its leaders and the French stock exchange was shaken as EADS shares plunged by 35% within 24 hours.

By pure chance, a technical press briefing had been organised for 14 June 2006. I was urged to pass the bitter pill by showing how well flight tests had progressed. That was a fairly easy task for me as even though we were running behind schedule, I was totally convinced that we had a great aircraft. The journalists forgot the bad news for a moment but, unfortunately, that failed to see a change in the headlines the next day!

Following on from the controversy on stock options that gradually amplified in the press in late June, on 2 July 2006 Noël Forgeard resigned from his position as Co-President of EADS. Meanwhile, Gustav Humbert stepped down from the Airbus executive. They were immediately replaced by Louis Gallois for EADS and Christian Streiff for Airbus.

On Monday 3 July, a meeting of EADS's French shareholders was held at the CNIT conference hall in the Paris business quarter of La Défense and obviously, considering recent events, there were more people present than usual. I was to give a short presentation on A380 testing to put a brighter complexion on matters. I was just getting to the end but a few members of the audience, who were barely listening to what I had to say, wanted to get straight to the hard talk and interrupted me.

Meanwhile, others, who were listening, tried to get them to pipe down. Finally, as Forgeard was not there, having resigned the day before, it was hard to put tough questions on the stock options issue.

* * *

A few days later, on the eve of the Farnborough Airshow, I had the opportunity to take Christian Streiff on a visit of the aircraft. As we both travelled back to the chalet in the back of a golf buggy, he gave me his conclusions: "We have to organise a meeting with all the stakeholders and have a man-to-man discussion". He also asked me whether certification was possible by the end of the year. As usual, I replied: "Yes, on condition we accept certain limitations that will mean the aircraft cannot be operated by an airline."

On 25 July, despite the summer holidays, Christian Streiff summoned us to attend a general meeting on the programme. There were about 25 of us around the table. He spoke for twenty minutes with a few bitter-sweet comments on the aircraft design, for example the weight and complexity of the electrically operated doors. But his message was very clear: the first priority was to certify the aircraft before the end of the year; he then wanted to start delivery rapidly to Singapore Airlines and he targeted nine aircraft in 2007. He demanded a calendar, even an imperfect one, for the end of August as he admitted to being astonished to see such a complex project lacking a reasonable schedule. The discussion that followed on was very open and we all now felt much more at ease in making our ideas known.

* * *

A few days later an audit was kicked off by Louis Gallois to determine the possibilities for certification and delivery of one aircraft by the end of the year. The McKinsey consultancy was entrusted with this. None of the people who came in to see me knew about aeronautics. So apparently, you can become a consultant without having studied the area you work in! Luckily, they were accompanied by Gordon McConnell, a brilliant aircraft designer previously Chief Engineer of the A340-600 and now Chief Engineer for the A350. We thus had a frank and open discussion. He arrived at the same conclusion as us in that, yes, it would be possible to certify the aircraft for the end of the year, but with restrictions preventing the airplane being operated commercially.

* * *

Come 2 September 2006, Programme Manager Charles Champion was replaced by Mario Heinen.

On 21 September, we were delighted to have Christian Streiff fly with us on the A380 and this provided a good opportunity to get a few messages across. As he had been a pilot and even joint owner of a small single-engine private plane, I let him take my seat on the left after takeoff. He thus had a chance to pilot the A380 himself before Jacques Rosay brought the aircraft back to base (we didn't want to show him how easy it was by letting him land!). It was a great birthday present for him!

The EADS Executive Committee finally decided on 3 October that only one aircraft would be delivered in October of 2007, and 13 in 2008. At flight tests, far away from the cabin installation problems in Hamburg, we were amazed. We knew there were difficulties, but not to that extent. That same day, the Power8 economy drive was unveiled.

Friday 6 October saw Christian Streiff bring together all the Technical Managers to issue a straightforward message on his vision of the causes behind the development difficulties of the A380: "Not everyone is telling the truth and there is a problem of integration at Airbus." He wanted to streamline the organisation and have reliable indicators. Late afternoon three days later, news of his resignation came in, with him leaving to take over management of Peugeot.

Louis Gallois brought together all the leading Airbus managers in Toulouse on Tuesday 10 October 2006. He confirmed that he shared Streiff's analysis and that he understood his reaction. He announced that he was taking over as CEO of Airbus in addition to heading EADS: "I report to myself", he added with a smile. Other EADS and Airbus management communication meetings followed on quickly to explain the orientations adopted and more especially the new Power8 plan. In particular, on 16 October Louis Gallois again got the managers together at the Toulouse Convention Centre where he insisted on the two points already developed by Christian Streiff: tell the truth and ensure better integration at Airbus.

The next day, the media announced the appointment of Fabrice

Brégier as number two at Airbus and this was made official at the end of the month.

* * *

With the configuration retained for certification having been decided on (with a few operational limitations), we were able to make the last validation flight on 15 November 2006. Endurance flights still remained to be performed though (as discussed in a separate section) these being completed on 30 November. Alongside that, a number of preparatory meetings for final certification review were held over the last three months of the year (see the CERTIFICATION chapter).

The European and American administrations issued Airworthiness Certificates to us at a ceremony held late afternoon on Tuesday 12 December 2006.

As already planned in October 2006, the first aircraft intended for Singapore Airlines was handed over on 15 October 2007. Twelve airplanes were later delivered in 2008.

* * *

Did we take more time than planned for between the first flight and certification?

Remember that shortly before the programme was launched, the official duration was to be thirteen and a half to fourteen months, then reduced to twelve months later due to the delay for the first flight. At flight tests, our calculations led us to a total minimum duration of fourteen months. The real time it took came to a little less than twenty months, this mainly being due to the delay with two aircraft, # 4 then # 2. So clearly the lead-times were not respected.

But it is also interesting to take a look at another parameter, i.e. the number of months of flight for the aircraft planned for certification. In September 2003, the schedule adopted for flight tests was to use four aircraft, # 7 being specially dedicated to endurance flights. At that time, our forecast based on the assumptions explained at the start of this chapter was for 44.5 months of flights broken down as follows for the various aircraft: 16 months for # 1, 14.25 months for # 4,

7.75 months for # 2 and 6.5 months for # 7. As we got through to certification, we had pursued about 42 months of flying on all aircraft. But from that figure, you need to deduct everything that was not directly related to development and certification as with sales promotion campaigns, airshows and flights to provide information to customers or potential customers. It is difficult to calculate what this represents in the number of days to be taken away but a good reckoning would probably come to two and a half months. Thus, we took less than 40 months of flights to certify the aircraft, bearing in mind too that there were technical limitations to be raised so that the aircraft could be operated by an airline and, in particular, the need to certify automatic landing. All this upgrading took several additional flying months. You can thus globally estimate that the activity foreseen in 2003 of about 44 months was correct and was respected. We owe this in part to the fact that we only used three aircraft instead of four, # 7 arriving only after certification. But above all, there was no major technical problem and the A380's remarkable aerodynamic characteristics meant we could move forward relatively quickly.

In September 2003, we had reckoned that 2,300 flying hours would be needed, whereas the 'official' forecast was 1,800 hours. The final calculation of what would really happen was still extremely difficult as tests were performed during commercial trips or during ferrying to airshows, although these could not be taken advantage of to the full. But the evaluation we made was not far from reality considering the number of flying hours accumulated at the time of certification.

* * *

Flight tests are extremely expensive, costing about 40,000 Euros per flying hour including the major part represented by all the work devoted to analysing what is recorded. Incidentally, when I had engineers asking us to repeat tests to try and get absolutely perfect results and the impeccable curves they wanted to see on their graphs, I was used to telling them that it was superfluous as all we wanted were satisfactory and unchallengeable results. The magic figure to convince them was the cost per second of testing of 10 Euros. This is obviously an artificial kind of number as the marginal second isn't that much, but it does give an order of magnitude and helped convince some of them not to overdo things.

One day lost can sometimes lead to major financial losses, especially if lead-times accumulate, and that day can finally cost more than the additional flying hours needed to finalise a result. That's why the message we always passed on to crews was that the priority, apart from the safety aspect of course, was to gather as much information as possible relating to the planned test points, even if that meant spending more time in flight.

* * *

Drawing up a development schedule for a new aircraft is a delicate exercise as there are so many uncertainties. As soon as flights start, you are sure to get your share of good or bad surprises. Airbus went through this on all its programmes, and all you have to do is take a look at the specialised reviews to see that Boeing faced the same problems. The idea that you can get an aircraft right first time is illusory. The managers need to beware of the temptation to believe that with plenty of work upstream, development will be a cakewalk. Clearly, it is essential to be well prepared before the start of flights to minimise unknown factors. But, on a completely new aircraft, it would be naive to imagine that development on such a scale will be free of setbacks, and these are bound to have an impact on the schedule. Apart from production problems not covered by the present testimony, technical development of the A380 went pretty well, with no major difficulty in the end. A truly well designed aircraft, although the modifications that had to be made on all its systems took a lot of time.

The conclusion must be that any complacency should be ruled out when you devise the development schedule for a new aircraft.

STRENGTHENING THE FIN

Reinforcement of the fin was the only major modification made during the entire period of flight tests on the A380. Thousands of other improvements were made but either these were minor or they involved the software.

The need to strengthen this part of the aircraft was guessed at as from flight # 9 on 20 May 2005, and confirmed during subsequent flights. During the first climb tests with one engine inoperative and the others on maximum thrust, loads on the fin were relatively higher than on the previous aircraft as we went into a turn (see the chapter devoted to OPTIMISING THE AERODYNAMIC CONFIGURATION). In these first months of flight tests, considering all our various occupations, we failed to follow up this matter seriously enough, as we awaited decisions from the engineers responsible for the structure.

* * *

On 18 August 2005, the engineer in charge of the fin's design told us that, on 13 July, the decision had been made not to modify it. We had not been informed. The reason forwarded for this was that, thanks to the size of the aircraft and its fin, we could take advantage of its 'flexibility' to increase the acceptable loads. This was all Greek to us as the measurements made clearly illustrated the problem and we asked for clarifications from the Design Office. At the same time, considering the available information, we established a limitation on the take-off weight for each configuration so as not to go over these limit loads, flying at the recommended speed in the event of an outboard engine failure on take-off. We thus started to bear with a certain number of restrictions to move forward with testing.

Early in September 2005, there were a whole series of meetings and discussions on the matter. A lot of in-flight manoeuvres led to detections of abnormal loads and we insisted a modification be made rapidly. But no decision was forthcoming.

Then in mid-September, Fernando and I decided to bring things to a head to show the need to react. On 21 September, during flight # 93 of aircraft # 1, I stabilised the aircraft at a speed of 153 kt corresponding to minimum speed with one engine shut down at an overall weight of 470 tonnes. I set three engines to maximum thrust and one outboard engine to idle. We were wings level, the exact conditions for engine failure on take-off. I then went into a series of starts of turns, first slowly, then faster. Fernando announced to me the level of loads in relation to the limit loads after each manoeuvre: 90%, 95% and then 98%. I just wanted to reach 101%, but the value measured on the last test was 103% even though the turn was not so sharp. Clearly, this was an untenable situation. The Head of Engineering and Chief Engineer were immediately warned.

This test result triggered a new series of meetings, including one on 23 September that lasted all day, but nothing positive came out of it. A few days later, during a further interview with the Head of Engineering, the principle of a detailed calibration of loads in flight was retained. However, we first needed to reinforce ribs at the top of the fin.

* * *

A new dramatic development arose on 30 September when some specialists decided to reduce the acceptable load limit for the fin. I then immediately stopped all flights until we had the assurance that normal handling manoeuvres with a failed engine would not risk heading for a disaster.

After a few days interruption in flights, we got together with the Chief Engineer, Robert Lafontan. During the static test, failure of the fin had occurred at 1.49 times the limit loads, but having applied a wide safety margin to cater for the negative effects of temperature, humidity and fatigue. We were far from risks of failure. However, there were doubts as to the quality of the riveting and an inspection was deemed necessary. After this inspection where nothing untoward was found, we again authorised flights on the evening of 4 October.

* * *

Two series of load measurements on the fin were conducted during flights # 96 of 5 October 2005 and # 101 of 11 October. The technique used was to initiate sideslip, with or without simulation of engine failure, and play on the relative deflections of the upper and lower rudders (unlike other Airbus models, there are two rudders on the A380) thanks to ADIS.

Next Monday morning (17 October), at the traditional technical meeting, I discovered that a decision on reinforcing the vertical stabiliser had been made behind the back of flight tests and without addressing a certain number of problems. What's more, it had been decided to install a rudder travel limiter or RTL. This was a system that limited deflection of the rudders according to sideslip. Basically, the idea was a good one and had already been seen on the simulator. But there were many implementation and development difficulties. Some reactions in crosswind gusts were unknown and we also had to adapt the certification rules. In all, we reckoned that installation of this RTL, which we considered to be too complicated a contraption with too many potential problems, was a big mistake and that we needed to keep things simple by just getting on with reinforcing the fin.

9 December saw a meeting take place to report back on development of the RTL. It proved that on the ground, it was only marginally of interest, since maximum rudder deflection was possible in complete safety up to 165kt. Its considerable complexity also argued against it, with the need to take gusts into account, filter data, cater for delays between the detection of crosswind gusts to the front of the aircraft and deflection, etc. The RTL was finally abandoned without us ever finding out who had made this sensible decision!

* * *

Nothing much more happened for a number of weeks. We were waiting for limitations to be raised and above all the reinforced fin. On 24 February 2006, during a review meeting with the Head of Engineering, Alain Garcia, I again insisted on having a date for the removal of the many limitations and especially those relating to the fin.

Then, just over two weeks later (13 March), drama again with the announcement that, on each rudder, two servo-controls could, in certain cases, exert antagonistic forces and thus generate very strong loads.

It was even suggested that failure by buckling (the compression phenomenon that occurs due to forces in a material meeting in opposite directions) could occur at 82% of limit loads on aircraft # 4 that had no reinforcement. Stopping the flight was envisaged, but all this seemed somehow unlikely and we decided to run tests on the simulator first.

The next day, a new announcement was made informing us that there would be a further two weeks delay while the fins were fully reinforced. A whole number of tests had to be postponed and the entire schedule reconsidered.

On 27 March, nothing clear had emerged on this possible failure under buckling. Flight test engineer Didier Ronceray therefore prepared a take-off procedure with reduced thrust on the outboard engines. After reviewing the matter together, an e-mail was sent to the load specialists and to the Chief Engineer as this procedure, which we were happy with, led to reducing the safety margins. While awaiting an answer, the decision was made to ground # 4 (the only aircraft concerned), whereas # 1 already had a partially reinforced fin.

The next day, the Chief Engineer signed an approval note, but with a new value on limitations. In addition, the procedure was tested on the simulator, but failed to work! A new technique was thus defined and the verdict to accept it fell on 29 March around 7:00 p.m. However, this was conditional on limiting thrust of the outboard engines to 70% of the maximum value at low speed and low altitude. This was obviously extremely penalising for the pursuit of tests.

On 30 March, a further meeting took place with the Chief Engineer with the aim of determining whether we were to install the vertical stabiliser currently mounted on aircraft # 1 (that of # 9) and that had been partially reinforced, on aircraft # 4 to be able to pursue certain tests. The other solution to avoid this change and with no risk of buckling was to systematically work the servo-controls of each rudder section in parallel. This was the option that was retained, although on the flight tests side, we had expressed reserves as we felt that not all failure situations had been envisaged. So it was decided not to do anything before the situation had been clarified.

On Monday 3 April, much as we had guessed, it emerged that after a simple failure, operation with the 2 servo-controls in parallel could be lost and, in this case, limit loads were likely to be exceeded. So we all backtracked and the decision was made to mount the vertical stabiliser of aircraft # 1 on # 4. But that very evening, further computations showed that even with this new stabiliser, it was still possible to reach limit loads. What an imbroglio!

At the same time, we learnt that 800 hours of work remained to complete reinforcement of the original fin of aircraft # 1, meaning the schedule had to be reviewed again. This held up work on the aircraft for a week and stopped us moving forward on certain tests. If we had known about this beforehand, we would have organised works and flights on the two development aircraft differently. We were particularly disgruntled with this lack of transparency that disrupted our activity.

*   *   *

Then on 11 April 2006, we learnt, by pure chance, during a meeting with EASA, that the fin of aircraft # 9 that had been on aircraft # 1 for a number of flights and was to be mounted on aircraft # 4 was limited to 60% of limit loads. Why? Quite simply because a restriction relating to manufacturing had initially been forgotten about, but had then reemerged in the documentation. The situation was the same for aircraft # 3. For precautionary reasons, it was decided to check things out on aircraft # 4 to see if it too was affected. In fact, no difficulty was encountered.

Finally, on 2 May, both our development aircraft had reinforced fins. That from # 1 had come back to its original aircraft, while aircraft # 4 had the fin of aircraft # 9 (previously on aircraft # 1!). This eliminated the limitations that were holding us back from pursuing certain tests.

*   *   *

Unfortunately, that was not to be the end of our worries. On Monday 22 May, during a lateral handling flight with one and then two engines powered down, throughout the range of speeds, we exceeded the limit loads with two engines on idle on the same side and the other two at maximum thrust. The team responsible for load measurements in the telemetry room asked us to interrupt the flight. However, it was precisely on this aircraft # 1 that we were supposed to be able to conduct preparatory tests for VMCL-2 (see the section on MINIMUM CONTROL

SPEEDS) where it was accepted we slightly exceed limit loads, this not being a situation you currently encounter in operation.

The next day, the Chief Engineer announced that the following week-end, new reinforcements were to be installed on the vertical stabiliser of aircraft # 4 and new limitations were to be adopted. While waiting to have the reasons for these reinforcements and new limitations explained to me, I decided, once more, to stop flights on aircraft # 4. That evening, during a long meeting, we got to the truth of the matter. It turned out that a plus or minus sign had been reversed during the calculation of certain loads. In these conditions, rivets would become deformed as this fin, first installed on aircraft # 1, had already exceeded the limit loads.

Next morning, I discovered that, as from Monday, there was to be a meeting on this matter with the Head of Engineering, but that we had not been informed of the problem. I was furious and therefore shot off an e-mail to the Management recalling that, for flight safety reasons, we wanted to be kept abreast of such difficulties immediately. The aim wasn't to block the aircraft on the ground but to adapt the flight programmes according to the discrepancies encountered. After a further three-hour meeting, the decision was made to proceed with reinforcement by installing rivets. By how much were we to reinforce the structure? With Fernando, we favoured the strict minimum, to cover a single engine failure. The Head of Engineering agreed with us. The Chief Engineer wanted an additional margin of 20%. Further haggling led to a trade-off at 10%!

Work was completed on 5 June 2006 (after an extension due to a shortage of rivets!) and we could finally continue all our tests with no further alert, with reinforced fins that responded to all safety requirements.

* * *

This fraught history of the fin reinforcement mainly provides an illustration of the conditions we had to work in, with cabals, the green culture ethos, infighting, but also national rivalries. Airbus integration was still a long way away and that is exactly what Christian Streiff discovered as soon as he joined the Company.

A few years after the event, poring through my notes to write this account, I can better understand why throughout this period of development and certification of the A380, I was more stressed than I had ever been before during my professional life of more than 40 years in flight tests. Reinforcement of the fin was just one of the many problems we had to cope with on a daily basis. And on top of that, there was the pressure exerted on us to keep up with a hopelessly over-optimistic schedule!

FLUTTER

So what is flutter?

It is basically a combination of different vibration modes in a system. Looking at an aircraft in flight, the ailerons (the surfaces on the wings that control roll) may vibrate at a certain frequency, which varies according to the airspeed. The wing too has its own modes of oscillation in bending and torsion that also depend on the speed. If, at a certain speed, the two frequencies are identical, they may combine with the vibration of the aileron amplifying that of the wing and vice versa. If the amplitude of these movements becomes too great, this can lead extremely rapidly to failure.

This phenomenon is similar to playing on a swing. The swing has its own rhythm of movement and by kicking out the legs at the same frequency, the swinging movement can be amplified.

It is not only aircraft that can have problems with flutter. Suspension bridges too must be sized to resist when there are high winds. Indeed, there is a striking film of the Tacoma Bridge (Washington State) twisting and vibrating before collapse in 1940 just a few months after it was built.

* * *

Flutter leads to oscillations that diverge extremely rapidly. It is imperative to avoid them within the flight envelope, and even a little outside it, to keep a safety margin. On big transport aircraft, the number of vibration modes is high. It concerns various elements: the wings, engines, empennages, control surfaces, etc. Furthermore, these vibrations depend on the quantity of fuel in the wings or the horizontal tail unit, the airspeed and many other parameters.

Theoretical calculations are now extremely reliable and normally give you a clear picture of critical conditions in advance. Such computations are based on mathematical models of the aircraft structure and are reset using tests on the ground, subjecting the aircraft to vibrations at varying frequencies. These tests are performed on one of the development aircraft before the first flight and last several days. Despite the good level of analysis, in flight, attaining increased airspeed and higher Mach must always be achieved with caution.

Four-engined airplanes often pose more problems than other aircraft when it comes to flutter. The vibration modes relating to the outboard engines are one of the main reasons for this. Thus on the A340, modifications had to be made.

* * *

Flutter test flights are obviously flights that involve a certain amount of risk. Until the flight envelope opens up entirely, they are always carried out with the whole safety paraphernalia, including safety helmet, parachute, life jacket and so on. The emergency evacuation door is prepared.

A number of parameters need to be taken into account, including speed, Mach, weight of the aircraft and fuel distribution. Thus, for example, for the same flight conditions, you often have to do flights with the horizontal tail plane tank either full, half full and empty, as this difference in filling changes the resonance frequencies and thus the risks of coupling.

For non-specialists, a few introductory words on the speeds we shall be talking about in what follows. The speeds taken into account for these tests are known as conventional speeds or calibrated speeds. These are what the pilot sees in front of him on his instrument display. They are calculated simply by applying a formula to the difference between total (pitot) pressure (that obtained when the flow of air is suddenly stopped - you will feel it if you put your hand out of the window of a fast car) and outside pressure. These are different to real speeds and lower, all the more so as the aircraft flies higher.

The maximum 'normal' speed of the aircraft is called VMO or Maximum Operating speed. The pilot can fly up to this speed and that is where the red line is to be seen on his screen. However, due to unforeseen circumstances, for example turbulence or strong variations in wind, this speed may be accidentally exceeded and therefore safety has to be ensured beyond the VMO. Thus, in flight, it has to be demonstrated that there is no risk of flutter up to the speed known as 'Dive' speed (VD). The difference in speed between VMO and VD is quite high, about 50 kt for a conventional aircraft. But on Airbus fly-by-wire aircraft, flight envelope protections prevent such a high dive speed being reached. On the A380, the value for VMO is 340 kt and VD equals 375 kt.

All that has been said above for speeds applies to the Mach value with the same definitions and for the same reasons. Thus, for the A380, you have MMO = 0.89 and MD = 0.96. Remember that the Mach is the ratio between the speed of the aircraft and that travelled by sound in the same conditions.

Note also that beyond VD and MD, the ruling for certification requires that it be demonstrated by calculation that there is an additional margin in speed and Mach where the aircraft is free from any risk of flutter.

* * *

For each test point in flight, certain control surfaces are activated by a dedicated computer that sends oscillations at a variable frequency. A test lasts three minutes. The frequency increases for the first part then decreases during the second and returns to the initial value. The crew clearly feels the coupling modes as there is an amplification of the oscillations. If they consider them to really be too strong, they can stop the test at any time thanks to a switch in the cockpit. As it is difficult to know in advance what the amplitude of excitation adapted to each speed is, to have an adequate structural response for analysis, the flight engineers generally dispose of several levels of excitation for each mode. Sometimes a test is repeated in flight so as to obtain sufficient amplitude for validation, but not as strong as to cause damage.

The control surfaces are activated in different ways. With the ailerons, there are symmetric modes where the surfaces of each wing are deflected simultaneously upwards and then downwards, and antisymmetric modes where they are deflected in phase opposition. Most tests are performed by activating the ailerons, but some are also conducted using the rudder and the elevators.

All parameters are transmitted by telemetry. The specialists on the ground analyse each test point as soon as the measurement is gathered in. The specialist in whom we had utmost confidence was Detlef Schierenbeck from Germany. He was in charge of validating the flutter tests and followed all the flights. The time it takes for this examination varies according to the degree of confidence and consistency with the models. Sometimes, we had authorisation to pursue testing immediately. On other occasions, we had to wait several minutes. Finally, on some other programmes we ended the flight for a more thorough study, but this never occurred for the A380.

* * *

The flying technique was a delicate matter on Airbus models up to the A380. Indeed, such tests had to be conducted with a direct law so as not to introduce flight controls deflections commanded by a complex outside control law. On previous aircraft, we were even deprived of touching the stick. We had to hold altitude with trim, meaning that we deflected the whole horizontal tail plane using a wheel close to the throttle. The bank was adjusted using the rudder with tiny impulses applied to the pedals and, when that proved insufficient, with small dissymmetrical variations in engine thrust. Speed had to be maintained with great accuracy using the engines. The main difficulty arose from the fact that each variation in thrust led to a nose-up/nose-down and, as a result, correction had to be made to trim. One of the pilots looked after the flight path while the other devoted his energy to maintaining speed, all of which required perfect co-ordination.

For the A380, considering the aircraft's considerable inertia, we were a little worried about having to maintain flight conditions so precisely for the three minutes each test point lasted. Finally, the Design Office engineers managed to fix it for us so that we could use the stick without disrupting the test thanks to a special piloting law. Running test points then became a piece of cake.

Picking up speed is fairly slow, generally by 15 kt increments. This increase is achieved keeping the same Mach, and thus having to change altitude, as the relationship between Mach and speed depends on altitude. Similarly, gradual increase in Mach is applied while retaining the same speed.

For high Mach values, you cannot always maintain level flight, so tests are pursued in descent. Beyond MMO, and depending on the type of aircraft, you may encounter more or less severe buffeting. Also, drag increases rapidly at the same time as velocity (Mach) and you then have to dive significantly to accelerate. For high Mach values where flight conditions cannot be stabilised, it proves impossible to conduct analyses by oscillations at varying frequencies. These are replaced by very short pulses on the control surfaces. Moreover, as buffeting normally prevents the coupling of modes by disorganising oscillations, the risks of flutter are then reduced.

The final test is diving at VD/MD. You need to start at maximum altitude, dive at full throttle, stabilise and then hold MD in descent. The indicated airspeed increases gradually. The test comes to an end once you reach VD so you can reduce power and pull up. Throughout the test, pulses are sent onto the control surfaces on the different axes. The engineers have to work quickly as these test points are made pretty much instantaneously considering the high rate of descent.

* * *

On the A380, the first flutter test flight took place on 9 June 2005 on flight # 21. The weight on take-off was 533 tonnes and 485 tonnes on landing. During this flight with all fuel tanks full, the speed envelope was opened out to VD, i.e. 375 kt at fairly low Mach, with no particular difficulty.

The next flutter test flight took place early in August. This was flight # 54. Indeed, after the first flight, we had to move on to other activities, with the Paris Airshow, then optimization of the aerodynamic configuration. During this second flight, the envelope was opened up to the conjunction of VD, 375 kt and MMO, at Mach 0.89. No particular discrepancy was observed during the flight, except for the last test point where a coupling giving some lateral acceleration was encountered. Later analysis showed there was no real problem there. However, as we came down from the aircraft, we saw that there was considerable damage to the belly-fairing between the fuselage and the wings. Clearly, reinforcements were needed before the next flutter tests.

Tests were only resumed on 25 August (flight # 69) once the fairings had been modified. The envelope was opened up to VMO and Mach 0.95. We were to go up to MD, i.e. Mach 0.96, but between 0.95 and 0.96, there was a sudden jump to Mach 0.98 as a small shock wave (a small zone where a supersonic flow of air becomes subsonic) must have passed in front of the static pressure probes. It was clearly impossible to stabilise 0.96 due to this rapid variation.

As from 13 September, we were able to run one to two flutter test flights a week. There was therefore enough time available between flights firstly to analyse the measurements made, and secondly to make inspections on the aircraft as each series of tests led to the aircraft being 'shaken up'. During the first flight in that series performed with the horizontal tail tank empty, the envelope was opened up to VD and M 0.89.

Tests were pursued with different quantities of fuel in the empennage and the wings. They were then interrupted during the tour in Asia and Australia and its preparation (see the SALES TOURS chapter).

On 1 December 2005, Jacques and I went through a dive exercise at VD/MD. Due to the Mach measurement problem, calibration was only known up to Mach 0.957 and it had been decided to stick to a maximum of Mach 0.955. But, as from the first test, it jumped to 0.988. On the second dive, aiming at just Mach 0.950, the indicated value reached was 0.983. For old hands in aviation, this jump in Mach indication may well recall supersonic dives in fighter aircraft designed early in the 1950s! Some other dives were performed keeping down to a lower Mach value. According to the experts, for certification, demonstration at VD and Mach 0.943 was enough.

A last flight at VD and Mach 0.89 was made the next day to make additional measurements. This was to mark the end of flutter tests with the Rolls-Royce engine and it finally went off without any problems.

Another series of tests was conducted later with the second Engine Alliance power plants, again with positive results.

MINIMUM CONTROL SPEEDS

When the aircraft has one engine shut down and the other three on maximum thrust, it is subjected to a torque effect that brings it to turn to the side of the inoperative engine. This torque can be countered by applying the rudder to keep the heading. However, when speed diminishes, there comes a moment when the aerodynamic forces on the fin and the rudder, that are a direct function of the speed, are no longer sufficient. As the rudder reaches its stop position and balances the engine torque, a minimum control speed is reached.

On any aircraft, below minimum control speeds (VMC), there is a risk of losing control when one of the engines is shut down (outboard engine for a four-engined aircraft) and the other(s) are at maximum thrust. There are several VMCs: for take-off configurations, you have VMCA (with the A standing for Airborne) and, for approach, VMCL (with L for Landing). On a four-engined airplane, VMCL-2 corresponds to the failure of two engines on the same side. Determining this value is particularly important when the aircraft takes off for a ferry flight on three engines, without passengers, and has the misfortune to encounter a failure of the second engine on the same side (see the chapter on FLIGHT CONTROLS DEVELOPMENT). Finally, there is the minimum control speed in the case of acceleration on take-off known as VMCG (with G for Ground).

However, nothing is that clear cut and just going slightly under the VMC does not mean an accident is bound to occur. But what is sure is that at VMC, the pilot is at the limit of manoeuvrability and he cannot do as he likes with the aircraft. The rules to determine these VMC speeds may appear pretty strange and it sometimes seems difficult to see the underlying logic. Nevertheless, they have been around for a long time and experience has shown that they work. For all VMCs associated with flight itself, determination is achieved statically and a dynamic check is then made. For VMCG, measurement is done dynamically as will be seen.

VMC determinations are by their very nature tests that involve risks as the engine is cut off at low altitude and low speed. On a twin-engined aircraft, the failure of the single engine still operating leaves just enough time to relight the one cut off. On a four-engined airplane, the loss of the engine next to the one already inoperative leads to a loss of control if the other engines are not immediately reduced. In that situation, you also have to rapidly relight the engine shut down.

These risks of another engine failing are not what is most to be feared, given their very low level of probability. However, performing dynamic manoeuvres can very quickly lead to loss of control. Here, extreme vigilance is demanded and these tests are only performed by very experienced test pilots.

* * *

VMC measurements are not considered the main priority during the first months of development of a jumbo airliner. Indeed, they are low enough not to affect take-off and landing performances, except in the very special cases of low aircraft weight operations. Nevertheless, these measurements have to be made fairly early as you can never rule out a nasty surprise that could have an effect on performances at higher aircraft weights than predicted or that could require a modification, for example to the flight controls.

For the A380, we were unable to get on with these tests at an early stage since, as already mentioned, the fin had to be reinforced first.

* * *

How do you determine the VMCA on an A380?

Once the engines have been set (with one cut off and the others on maximum thrust), the test starts at about 20 kt above the forecast value with a slow deceleration, keeping a constant heading. The pilot gradually deflects the rudder to the full. He then gently rolls the aircraft to the opposite side of the shut down engine, continuing to reduce speed and still keeping the same heading, with the rudder fully deflected. VMCA is obtained in these conditions with a bank angle of 5 degrees (figure 1). This bank angle has a major effect as it means you gain about 5 to 10 kt as compared with the same manoeuvre performed wings level. How was this strange rule discovered? In fact, it remains a mystery! Maybe, in the earlier days of aviation, when there were no reliable measuring instruments, someone thought they could gain a little thanks to the tolerance on the bank angle. In practice, precise measurement is a delicate matter as the bank angle is very difficult to stabilise when the rudder is at its stop and therefore the engineers continue to pick up stabilised values at different bank angles around 5 degrees to determine the final result. This way of maintaining control with a few knots less speed available has become a part of the tradition and is now officially recognised.

Figure 1: VMCA – © Airbus

The principle for determination of VMCL and VMCL-2 is similar for landing configurations.

But that's not all there is to it! You also need to ensure that manoeuvrability in roll is adequate. The rules are slightly different for VMCA and VMCL, but here we shall simply take the example of the demonstration relating to VMCL. This involves checking that, on the side where handling characteristics in roll are more limited, that is opposite the shut down outboard engine, there is still an adequate roll rate. Thus, with the aircraft flying at VMCL with three engines running at maximum thrust, you have to be able to go from 5 to 25 degrees bank opposite the cut-off engine in less than five seconds. This test is hazardous as sideslip gradually increases to become extremely significant at the end. It cannot be reduced as the rudder usually used to control it is already fully deflected. As soon as you exceed 25 degrees of bank, recovery has to be extremely rapid and smooth with a gradual reduction in the engines and an increase in speed. Note that, many years ago, on another programme, we experienced a transient loss of control in this type of test. Since that event, in order to limit risks, the roll rate on all aircraft at low speed has been reduced by a half when the computers detect a significant dissymmetry in thrust. This avoids excessive and hazardous sideslip in the event of too sudden a manoeuvre.

Other dynamic manoeuvres have to be performed in flight for VMCs. Some are pretty similar to what has just been explained to check the roll rates. There are also other dynamic tests but they involve no special problems for the development of an aircraft like the A380.

* * *

The first VMCA/VMCL flight took place on 26 May 2006 on aircraft # 1. I accompanied Ed Strongman, who was later to be appointed Chief Pilot for military aircraft. The briefing had started off badly as some specialists wanted to prevent us from performing VMCL-2s under the pretext that not all the load calculations on the fin had been completed. We expressed our astonishment as this fin had been reinforced and installed a month ago and VMCL-2 tests were just one of a series of structural computations (see chapter STRENGTHENING THE FIN). The flight was made in poor weather conditions, but we were able to settle a few issues. The results complied with what had been forecast. The roll rates were readily demonstrated, but we did, nevertheless, reach a sideslip of 19 degrees in recovery on a roll rate demonstration.

A few days later, I made a second flight to bring in some more results. With Jacques Rosay, we had decided to take advantage of this opportunity to give some training to a young test pilot. This time, we were able to complete all the tests in good conditions, including the VMCL-2 tests.

The final certification flight for all VMCs in flight was made on aircraft # 1 by François Fabre of EASA and Jacques Rosay. The positive results were confirmed.

Finally, setting aside the load problems on the fin, no particular difficulty was encountered with VMCA and VMCL that turned out as forecast.

* * *

The VMCG is determined during a dynamic test. The aircraft is accelerated with all engines at maximum thrust and the nose wheel orientation control disconnected to simulate a wet and slippery runway. At the planned speed, an outboard engine is suddenly cut off by closing the fuel supply valve. The pilot has to counter this effect using the rudder. If this induced failure is introduced at high speed, as for a VMCA, he can control the aircraft with a minimal deviation. However, if the speed is low, even with the rudder fully deflected, he will be unable to avoid a significant deviation. The VMCG is, by definition, the engine cut-off speed such that this deviation is by 30 ft (Figure 2).

Figure 2: VMCG test © Airbus

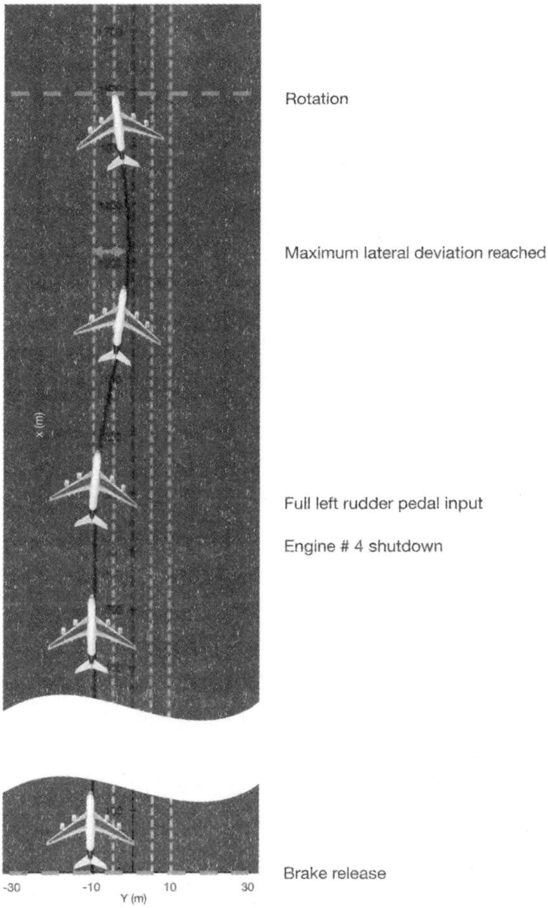

This test has to be conducted in calm weather as the slightest crosswind or turbulence can disrupt measurements. Generally, the first takeoff takes place at dawn. Obviously, we aren't going to tackle the critical situation directly and the first test is made at a speed about 10 kt higher than the planned VMCG. We then go on to make the cut-off at lower and lower speeds going down by increments of 2 to 3 kt, or even 1 kt. Generally, after six tests, the targeted deviation is exceeded, which means that the VMCG can be calculated by interpolation. However, we have to be cautious because as you get close to VMCG, this lateral deviation varies extremely rapidly with the shutdown speed.

For these tests, the pilot seated left is responsible for path control. He has to minimise lateral deviation and once the maximum value has been reached, he takes off. The other pilot is responsible for cutting off the engine at the desired value. It is important that the same pilot performs the same action as, if there is a bias on the cut-off speed, it will in principle be kept for other tests and therefore decrease in speed will be as gradual as planned. Analysis of the recorded data allows the real value to be determined. In the cockpit, after take-off, the test flight engineer applies engine relight procedures provided by the manufacturer to minimise damage caused by a cut-off at maximum thrust.

According to aircraft types, you may perform certification tests directly with a pilot from the authorities on board and an Airbus pilot entrusted with engine cut-off. The idea is to reduce damage to the engine when cut off at full power and avoid having to conduct the tests twice, once with the manufacturer's pilots and then again for certification. Nevertheless, for the A380, considering the uncertainties surrounding what had been predicted and the new features on the various systems, we wanted to rehearse things. It will be seen in what follows that we had pretty good intuition.

* * *

We were finally authorised to perform these VMCG tests on aircraft # 1 after partial reinforcement of the fin. This test is not difficult to carry out, but opportunities to go ahead with it are extremely rare. That's why, as for VMCA and VMCL, we had decided to bring Hugues Van der Stichel, a test pilot responsible for certification, along with us. So he was seated on the left seat, responsible for minimising lateral deviation and I was on the right to cut off the engine at the chosen speed.

That was on 30 March 2006. The aircraft take-off weight was 450 tonnes. Planned VMCG was 122 kt. During the pre-flight briefing, I found that the first speed proposed for cut-off of 130 kt was too low, so I asked for it to be increased by 2 kt. The reason for this was that we had seen a whole load of numbers flying around on the VMCG with variations going up to 13 kt so I was wary for this first test.

Hugues lined up the aircraft a little over 30 ft to the left of the centre line, bearing in mind we were going to cut off the starboard outboard engine. All along runway # 2 at Toulouse-Blagnac Airport, there are blue lines at 16.5 and 33 ft (5 and 10 metres) from the central axis. These are used for various types of tests and, in this instance, they helped keep a steady path exactly 33 ft from the central axis. They also directly give the pilot an estimation of any deviation. So I cut off the engine at the planned speed of 132 kt. Consternation followed as what should have been a deviation of about 6 ft turned out to be well over the 30 ft limit at about 50 ft. VMCG had to be at least 12 kt higher than predicted. That was a lot and risked impacting take-off performances.

We decided to perform a second test at a slightly higher speed of 134 kt. This brought another surprise as the deviation was hardly less pronounced, at around 40 ft. The videos clearly showed the main landing gear tyres skidding.

A third test was made at 136 kt. The result was 59 ft. The deviation was no less at higher speeds and this seemed crazy!

The next day, to try and understand what was going on, we repeated a test with a lower deflection of the flaps (configuration 1+F instead of 3). With this new position, we hoped to have greater loads on the landing gear and the wheels and thus improve adherence and reduce skidding. We also planned to push the stick less than we had been doing with the aim of forcing the nose wheels down onto the ground (standard take-off procedure). This new technique also allowed the vertical wheel forces of the main landing gears to be increased. The cut-off was made at 135 kt and deviation was by almost 60 ft. So all our efforts had come to nought! On top of which we also discovered another major problem. The rudder deflections were fairly slow and on one of them, there was a hidden failure that had led to having just one servo-control active instead of two, which thus slowed down deflection. This failed to explain the difference in relation to the predicted speed, but the system was decidedly not robust enough. The end result was that various modifications would have to be made to be able to pursue VMCG tests.

Thus, the Design Office initiated changes to the flight controls computers to improve lateral control of the aircraft. These involved deflecting, on the opposite side to the failed engine, one of the seven spoilers and two of the three ailerons upwards, and the central aileron downwards as the rudders came up to their fully deflected position (figure 3). The deflection of these four surfaces created a drag force contributing to 'turning' the aircraft to the opposite side of the failed engine. Thus, this effect added on to the one given by the rudder. Having the three ailerons deflected in different directions allowed the roll effect to be minimised. In addition, the operating logic of the rudder servo-controls was also re-considered to obtain faster deflection.

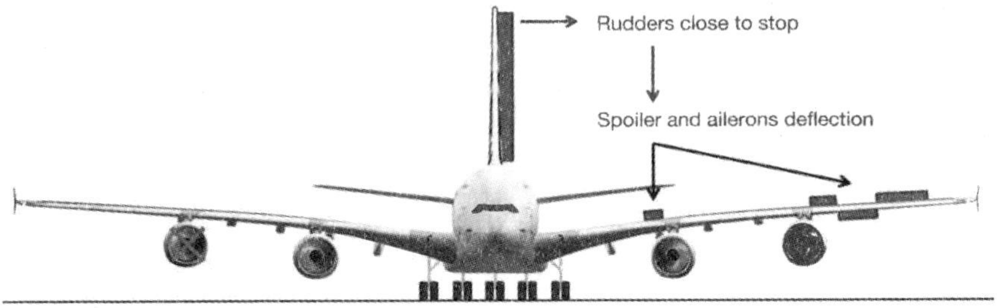

Figure n°3: Improving yaw control on the ground – © Airbus

The VMCG tests were performed at Istres on aircraft # 1 with these modifications during flight # 258 on 14 June 2006. Jacques Rosay came back to Toulouse with a big smile on his face just as I was reporting back on the tests to a group of journalists. He was thus able to give them the very latest news – a further obstacle had been overcome!

But at Istres, we'd created a bit of damage. Indeed, as we had to line up 30 ft from the central axis, one of the outboard engines was practically over the gravel. This threw up projections that had damaged a number of panels alongside the runway. To make matters worse, we had forgotten to warn the sweeping services, so we got a full dressing down! But what did that matter? – We'd resolved the VMCG problem.

The VMCG certification flight took place on 7 September 2006 with EASA pilot François Fabre. I should add something here. VMCG is extremely sensitive to the pilots' reaction time, which is roughly 0.6 seconds. Now, a very slight variation can significantly alter the VMCG, with differences of up to 2 kt from one pilot to another. The reaction time in itself has nothing to do with the skills of the pilots themselves. It isn't the best pilot who reacts quickest or vice versa. Now, at EASA there were two excellent pilots, an Englishman (Dan) and a Frenchman (François). After all the tests conducted on other programmes, we knew that François had a slightly shorter reaction time. So, now it's over and past history, we can own up to having tweaked the schedules to make sure it was François who performed the certification tests!

I accompanied François. The last test was made with a cut-off at 119 kt. Finally we obtained values very slightly lower than our forecasts: 119 kt or 121 kt, according to the maximum thrust of the engine and the option chosen by the customer (much like the engine power you choose for your car).

The take-offs were made at Istres, from the start of the runway facing south, which is normally only used by Airbus. There is no shoulder in this part and we were aligned 30 ft from the axis. So when we opened up the throttle we threw up a load of stones onto the runway and the parking apron we were using for turnaround. On return to Toulouse, 12 tyres were damaged, including one that had deflated on arrival. But the tests were completed and the results highly satisfactory.

CROSSWIND

Right from the start of the programme, at flight tests, we were worried about the aircraft's possible behaviour on landings and above all take-offs with crosswind. Some time later, those who saw the spectacular film made by the local Icelandic TV Channel at Keflavik International Airport, where we landed in gusts of wind reaching more than 55 kt, must have thought it was as easy as pie! But let's go back before that.

On take-off of the A340-600's first flight in April 2001, with a very light crosswind coming from the left, I had been surprised by the left wing's tendency to lift up a little – something spectators must have noted. This phenomenon is due to the wing's dihedral as the wind rushes under the wing and banks the aircraft. The A340-600 has three main landing gears, including one under the fuselage. Later, we gradually explored capabilities in crosswinds and I had been reassured, in March 2002, by a flight including a number of take-offs at Basel-Mulhouse aerodrome. The crosswind component was then greater than 25 kt and we had noticed that the bank of the aircraft did not increase significantly with the wind. Indeed, beyond a certain value, the force bearing on the shock absorber of the landing gear opposite the incoming direction of the wind is stronger, thus limiting bank. We were therefore reassured and, later, we did A340-600 demonstrations with a crosswind of 40 kt.

* * *

On the A380, the problem was rather more complicated as you have four main landing gears, including two under the fuselage. On take-off, there was still the same tendency as on the A340-600 for the wing to rise slightly with the wind, until you get to about 2 degrees of bank in regular situations. This is something we had all seen during the first flights. On landing, the situation is not critical because, as soon as the wheels touch down, the spoilers (control surfaces located on the upper part of the wing) deflect upwards to push the aircraft down onto the ground. The bearing force on the landing gears is then such that there is hardly any residual bank. On take-off, some tests had led us to fear that, at a fairly high speed, the main part of the weight of the aircraft would bear on the two fuselage landing gears and, being so close together, the bank angle might not be contained satisfactorily. Many discussions ensued with the Design Office. Of course, the pilot could always be asked to counter this bank. But this solution would have brought two drawbacks with it. First of all, there would have to have been a change in procedure compared with other Airbus airplanes as pilots are asked not to move the stick laterally during the take-off run. Also, when deflecting it on the left, for example, the spoilers on the left wing come out, thus pushing the wing down towards the ground; but unfortunately they also considerably increase drag, thus reducing the aircraft's acceleration and increasing the take-off distance (obviously something to be avoided). A host of preliminary tests were conducted as from the first flights. Finally, the Design Office offered us a dedicated flight controls law that, during the take-off run, automatically deflects the ailerons in relation to the bank to attempt to bring the latter back close to zero without deflecting the spoilers and with no need for the pilot to intervene. On that basis, we set off on our quest to seek out and take on crosswind.

Looking for crosswind sounds easy. But in reality, a whole number of conditions have to be brought together simultaneously. Firstly the aircraft must not be fitted out with specific systems that are incompatible with this type of test, with the constraints it always involved. But above all, the weather conditions have to be right and this is rare as runways are built precisely along the axis of the prevailing winds, thus making strong crosswind a relatively rare occurrence! The specialists thus scrutinise the forecasts some days in advance. Our favourite aerodromes are Brest in France, Shannon in Ireland and Keflavik in Iceland (although this is not an exhaustive list). It was at Keflavik that we always managed the best scores. Unfortunately, it lies at a full seven flying hours there and back. This aerodrome has the advantage of having two runways perpendicular to each other, meaning we could sometimes run crosswind tests on one runway before immediately pursuing a head wind test on the other. At Brest, you can also sometimes get a fairly strong southerly wind. However, it rapidly turns to the west to get back onto the axis of the runway once the spell of bad weather is over, meaning you have to plan forward and react quickly, abandoning the initially planned test flight and preparing a new flight plan. We sometimes got there too late, but every opportunity had to be seized on if we were to stick to the certification lead-times.

On the A380, we took some time before getting on with these crosswind tests as, firstly there were plenty of other things to do and, secondly, not all the conditions were brought together.

The first 'alert' was one Saturday morning in September 2006. It was about time as, with certification scheduled for December, a demonstration had to be made with at least a minimum of crosswind. This was planned for about 25 kt on the runway at Tarbes. To be honest, I remained unconvinced as I looked out at my garden in Toulouse (just 75 miles away) and listened to the TV weather forecast! But you can't miss out on an opportunity like that through a stroke of intuition. So we made the flight. The control tower announced 20 kt but that looked a lot less on our instruments so we soon gave up.

* * *

The next opportunity came a few days later on 21 September 2006. The whole day's events were difficult to organise as we only had one aircraft available for these tests, # 1. Moreover, it had been planned for Christian Streiff, our short-lasting CEO, to come over and visit us early in the afternoon and make a demonstration flight. At just that time, crosswind blew up in Brest. We were on the horns of a dilemma! We tried to make both flights, as it was hard to delay the flight of the CEO.

After landing from the demonstration flight, the aircraft was rapidly prepared to leave for Brest. We were still at the controls with Jacques Rosay. Just as we started our descent, we heard our friends from Dassault Aviation over the radio as they left having completed their certification tests on their brand new fly-by-wire Falcon 7X. Over a secondary frequency, they announced their score to us: 25 to 35 kt. Mission accomplished for them. But for us, the first two approaches gave about 20 to 25 kt crosswind. This was not enough for certification, but it still served a useful purpose as we were able to see that the aircraft was easy to control in such conditions and there would certainly be no real problems with stronger winds.

To summarise, it might be said jokingly that it was the CEO's fault if we had missed that flight! But that was not all. That 21 September was also the date of my sixtieth birthday. So the entire flight tests team had organised a small party towards 6:00 p.m. But this was not compatible with test flying and as it was difficult to postpone these tests, we postponed the party to the next day!

* * *

Validation for crosswind does not just include aircraft handling tests. Checks also have to be dedicated to one of the engines to look at fan blade vibrations, risks of stall (obstruction of the air flow that can lead to overheating and possible damage), etc. To do so, one of them is kitted out with dedicated sensors for that test series, mainly carried out at low speed, on the ground.

The opportunity arose late in September 2006, at Shannon, with aircraft # 4 as provided with an equipped engine. The crosswind was blowing at about 25 to 35 kt. Air traffic control was highly co-operative, allowing us to take off in one direction and land in the other, despite commercial traffic. With the crosswind capabilities demonstrated, we had the minimum required for certification, but this was still rather less than needed for entry into service - we wanted to be able to guarantee more than that.

Our departure from Shannon was accompanied by the shock of the landing gear not retracting. As there are two control computers, we tried using them individually, but to no avail. We then got into lengthy but inconclusive discussions by satellite telephone with the specialists in Toulouse to find out what we could do. The situation was as follows: the nose landing gear was up, but the four main landing gears had stayed down and worse than that, the main doors had remained open, meaning a huge increase in fuel consumption for the distance ahead.

To make matters still worse, ATC (Air Traffic Control) was unable to give us authorisation to head straight to Toulouse as there was also a flight plan problem. The air traffic services finally gave us their agreement after a quarter of an hour. Looking at our speed and consumption, it was clear we would not be able to get straight back to Toulouse unless we were willing to risk getting back with the tanks practically empty (just one tonne left was the estimate!). Obviously, taking such a risk was out of the question and the decision was made to land at Brest to refuel. We informed operations by telephone and everything was sorted out in record time with the Brest Airport managers. Even the Sub-Prefect was standing at the bottom of the steps to welcome us!

During that short halt at Brest, we hosted a lot of visitors for the usual look around the aircraft interior. That was the least we could do for the team that had welcomed us so kindly. But meanwhile, the nagging concern remained in our minds to determine the causes of our misadventure. It emerged that on one of the manoeuvre circuits, on each of the central landing gears, a sensor was broken. This was a weakness that had already been recognised on this system and it was just a pity that the new sensors had not yet been installed on that aircraft.

But that failed to explain why the landing gear did not retract on the other circuit. On closer inspection, it emerged that a connector had come loose on the left wing landing gear. So our DIY specialists went into action and, with a little aluminium tape, everything was repaired in just a few minutes.

On take-off, we managed to use one of the two computers to retract the landing gear as if nothing had happened. Finally, only a few people were in the know. The only time I was ever challenged about the matter was by a plane spotter from Brest who was used to seeing us conduct different types of tests and asked simply: "Why did you land with the landing gear doors open?" Spot on! As a matter of principle, I never lie, but in that instance I could hardly tell the truth unless I was willing to see the whole thing splattered across the Internet that very evening. So I just pretended that this was normal because we were doing a certification test! May God forgive that little white lie!

In all events, with three failures on the landing gear, works to upgrade in line with the latest standard became an urgent necessity and were immediately initiated by programme management.

* * *

What we had achieved at Shannon was still a bit short of the mark as we wanted to provide a demonstration with a crosswind between 35 and 40 kt. A further opportunity arose at Keflavik on 10 November 2006. The day before, the forecasts were good, but when we got in to our operations department early the next morning, Keflavik was announcing gusts of more than 60 kt, which was far too much for us. We delayed departure towards mid-morning as the wind was going to drop according to the forecasts and we had a three and a half hour flight to get there. On arrival, lateral gusts were still reaching 50 to 55 kt. Jacques Rosay was at the controls. He hesitated a little, but finally we agreed to make a go of the test, even if we had to go-around if things turned out bad. In the end, everything went fine. Take-off was on the same runway, in the opposite direction, with lateral gusts of around 50 kt. I made the second landing and we left the following four to the EASA test pilot, François Fabre. At the end of the day, for landing and take-off, we validated about 56 kt of crosswind in gusts. This was a record for Airbus, and maybe also for other manufacturers. What's more, the maximum deviation in relation to the central axis of the runway did not exceed 18 ft over the six landings, which was within the mean for all those performed on that aircraft, with and without crosswind. Admittedly, they were performed by three experienced test pilots, but that result shows that this big aircraft is perfectly manoeuvrable.

• See photo n°16 – page 236

The take-off was one of our concerns, but, thanks to the automatic control law described previously, the bank was limited on average to 4 degrees without the pilot activating the stick. We nevertheless managed to reach 6 degrees on a take-off with gusts at 55 kt. All this was satisfactory, considering that the aircraft was finally limited to 35 kt on take-off, for reasons relating to the behaviour of the engines.

At the end of testing, snow started to fall and as the only runway in the axis of the wind used for commercial flights was going to be cleared, we just refuelled and set off back home, mission accomplished.

Icelandic TV produced a great film on these landings and take-offs that was broadcast that same evening on the news programme of one of our French channels. Those images remain unforgettable.

FLIGHT CONTROLS DEVELOPMENT

I would like first of all to point out that what follows in some of the following paragraphs is more or less intended for people who are familiar with systems installed on Airbus aircraft or operation of airplanes. Between choosing to evade the issue of explaining flight controls development (given its complexity) and going through some of the tests we did to obtain present-day results, I've preferred the latter approach. Indeed, some readers may well be interested by development methods and the reasons behind our choices.

A number of other topics relating to flight controls are covered in the appropriate chapters.

* * *

Development of Fly-by-Wire (FBW) is a complex process. It starts with computations based on estimated aerodynamic models of the aircraft that are then checked and corrected using wind tunnel tests. This allows you to prepare a first version of the flight controls computers. The next stage is to use a development simulator where the computers are coupled with the very latest aerodynamic models. First evaluations can then start with the simulator pilots specialised in that type of testing then, when necessary, with test pilots assigned to follow that particular programme. Obviously many problems arise and so the computers have to be gradually improved on. The true test is when it comes to the actual flight. The models developed are generally reliable, but they never really represent how the aircraft behaves at very low or very high speeds, not to forget the ground effect. The start of flight testing also exposes the pilots for the first time to the way the aircraft responds to their commands. Thus, the airplane's flexibility can play tricks, obviously in terms of comfort, but can also have a resulting effect on handling.

When the models used for calculations or on the simulator give good results (as is generally the case) there is nothing much to be done after in flight validation. But sometimes the aircraft's behaviour does not respond to expectations and in-flight identification may then allow the aerodynamic model to be reworked so the Design Office can define new settings for the computers. This can sometimes prove difficult, either because simulation of the ground effect is not satisfactory, or because the aircraft's flexibility prevents correct simulation. So development has to be done in two stages: first using mathematical models and then directly in flight, where the flight test engineers and pilots decide in real time on the modifications to be made. It is really common sense and feeling ('seat-of-the-pants' flying as it used to be called) that guide changes. It is important to note this since many non-specialists think that in flight we simply validate the results obtained on the simulator. This does not hold true for a significant number of tests: development methods have hardly evolved since the last century, apart from the extra help provided by computerisation. Often enough, sensations and impressions show the way forward and, to save time, as explained in the CREWS chapter, the flight test engineers are in a position to modify some flight controls settings in flight.

This development is a tremendously exciting challenge. Practically each day sees its batch of pleasant or unpleasant surprises (although the A380 wasn't the aircraft that gave us the most worries). At the time, Dominique Chatrenet, a highly experienced engineer who had taken part in all programmes since the A320, ran the Design Office department entrusted with defining and developing the flight controls. He was backed up by an extremely high level team that did remarkable work. A special mention should be made here of the group responsible for drawing up flight controls laws headed by Sophie Lambeaux. On the flight tests side, Didier Ronceray ensured co-ordination of development. He was assisted by Stéphane Vaux, Laurent Lapierre and Matthieu Rouzeval, all flight test engineers and a number of other engineers including Jean-François Grandjean.

It would take too long to go through all the tests conducted, so only a sample of the most interesting ones will be covered here. A number of other campaigns, related to flight controls development, are also mentioned in other chapters.

* * *

Working far upstream of the actual launch of the programme, the Design Office has to make decisions that are vital for the future: how many computers, flight controls surfaces, hydraulic circuits and so on. All this is dictated by the analysis of failures, combined with a first estimation of the flight characteristics. Even in the event of multiple failures, the aircraft must remain controllable. One of the failures that can have the most serious consequences and that guides a fair number of decisions is the non-contained explosion of an engine rotor disc. Here we assume that a third of that engine part will be ejected and will penetrate the fuselage with very high energy. The engine is normally designed so that a failure does not expel debris outwards. But in terms of aircraft design, additional precautions are taken. The path that part will take is computed according to precise rules. For example, you need to ensure that all power sources (electricity and hydraulics especially) will not be affected simultaneously. This could eventually have catastrophic consequences. Obviously, this design work is much more complex on a four-engined than on a two-engined airplane due to the number of discs that could shatter on the inboard and outboard engines. Note that this scenario pretty much corresponds to that of the Qantas A380 whose engine exploded just after take-off from Singapore a few years after entry into service. Even though the aircraft was in a degraded situation no doubt beyond what is envisaged conventionally in design, the crew disposed of all the systems needed to land safely.

Many other factors are taken into account to choose the general architecture. Potential weight saving (obviously without reducing safety levels) is no doubt the most important criterion.

* * *

Fly-by-wire has greatly contributed to aviation. Easier handling and protections to avoid loss of control of the aircraft have already been mentioned. But there is more to it than that.

Before, flight controls were designed firstly to 'please' the pilot and, secondly, to answer to certification criteria. With fly-by-wire, three possibilities have been added on: improving safety through restrictions on manoeuvrability for the crew so as to limit risks of losing control, reducing the weight of the aircraft by prohibiting some actions that could increase loads on the structure (these protections dispensing with the need for additional reinforcements), and finally enhancing passenger comfort. All this inevitably leads to making development of the computers more complex.

* * *

A few remarks now to help make what follows clearer, without going into too much technical detail.

The A380 has six basic flight control computers: three primary computers (PRIM), each of which can take over all the aircraft handling functions, and three secondary computers (SEC). The SECs are not able to provide stabilised laws like the PRIMs, and they have very different software, meaning that a bug affecting all the PRIMs will not 'contaminate' the SECs. They are also designed to be more robust if there is a lack of the data needed to draw up flight controls laws. Each type of computer has both a control channel and a monitoring channel. A seventh computer known as the Back up Control Module (or BCM) can be resorted to in order to control the aircraft if the six others are lost. How it works is described later.

The A380 has only two hydraulic circuits instead of three on Airbus airplanes of previous generations. The third circuit was done away with to be replaced by relocated hydraulic generation systems. For some servo-controls, a small electric motor built into the actuator generates the hydraulic power needed to operate it. These systems go by the name of Electro Hydraulic Actuators (EHA) or Electrical Backed up Hydraulic Actuators (EBHA). This new type of architecture with two circuits represents a weight saving of several hundred kilogrammes, mainly through a reduction in the number of pipes.

Some control surfaces are separated into a number of parts with different electrical and hydraulic controls. Thus, there are two rudders instead of one as on other Airbus airplanes, and four elevators instead of two. There are three ailerons on each side instead of just one as in the A320 family and two on the A330 and A340. There are two main reasons for this: easy reconfiguration to resolve different cases of failures and weight saving. Each control surface (apart from the spoilers) is operated by two servo-controls which are powered by different hydraulic circuits, or by EHAs or EBHAs. Each of these systems can itself be controlled by two different computers, PRIM or SEC (and sometimes three with the BCM). A multitude of failures would thus have to occur for any control surface to be lost.

Spoilers are surfaces located on the upper part of the wing. When extended, they destroy part of the lift effect of the wing. They are used dissymmetrically (using spoilers on one side only) to roll the aircraft, and symmetrically, in flight, to boost the rate of descent or reduce speed (airbrakes), and on landing to push the aircraft down against the ground and improve braking.

As a last electrical standby (that is if there is no further electrical power provided by the four engines or by the APU), electricity comes from a generator driven by the Ram Air Turbine, a small propeller that comes out automatically under the wing in an emergency.

* * *

Even though certain settings can only be adjusted and validated in flight, the Design Office always needs to have an updated aerodynamic model as soon as possible as that is the only way to use simulation to effectively prepare adjustments to the flight controls laws in line with the aircraft characteristics. As from the first weeks, this identification of aerodynamics in flight is one of the general test programme priorities and so, as from 31 May 2005, flight # 16 was devoted to identification in pitch. During July and August, no less than fifteen flights were dedicated to tests of the same nature (roll, pitch, effect of the engines, etc.), and further flights ensued over the following months.

Note that these identification flights are unrelated to those that have to be made as from the end of development to prepare the models that will be installed in the training simulators. These last flights have to be conducted according to a very specific protocol. Moreover, the training simulators do not have the role of representing handling characteristics in extreme situations, while conversely, to develop computers, the Design Office must have identified the aerodynamics in marginal cases.

* * *

On the A340-600, tuning of the control law for take-off led to a few difficulties. It is worth recounting what happened here to give an idea of the obstacles you are likely to run up against in testing.

To take off, you have to pull back on the stick so as to bring up the nose of the aircraft fairly quickly and so minimise the take-off distance.

However, this action should not be too aggressive so as to avoid the rear of the fuselage touching the runway. On the A340-300, the pilots were unanimous in recognising that the aircraft's reaction was too 'soft'. A few years before the first flight of the A340-600, weighing some hundred tonnes more than the A340-300 and above all longer by nearly 40 ft, we therefore initiated a study to improve settings. A large number of tests were conducted on the simulator, then we tested the improvements made on an A340-300. We were pretty happy with the results! On return from the second flight of the A340-600, whereas the first take-off had been made under direct law, we decided to make a second take-off to check our new law. The captain for that flight, appropriately enough was Ed Strongman, who had ensured its development. Right from the start of the manoeuvre, the aircraft went into a series of violent oscillations in pitch, referred to as Pilot Induced Oscillations (PIO), where the pilot tends to react by instinct to an unexpected movement of the aircraft. This calmed down after about six trips. I was flying alongside Ed and wanted to check this out so I tested it myself, with the same result even though I was forewarned.

Why did this happen? As the front part of the A340-600 is longer, there is a kind of leverage effect where you are projected upwards astonishingly quickly, leading to the pilot naturally seeking to counter the movement and causing the PIO effect. So it was back to the drawing board. Then, after some work on the simulator to obtain a new control law, tuning took place at Istres through a series of fifteen take-offs with an experienced test crew.

The principle is simple: on each take-off we try, using the ADIS system (or its equivalent at the time), to improve on what the pilots felt during the previous take-off and what the engineers saw on their plots. For example, the law can be made more or less effective on initial deflection of the stick. You can also slow down the nose-up movement as you approach the take-off attitude, but you need to make sure it doesn't occur too early or too late. There is also the need to restrict nose-up where there is a danger of the rear fuselage coming down too close to the runway. Finally, you have to juggle between all the setting options, especially gain, damping and filtering and many other parameters so as to reduce the take-off distance and ensure safety in the most critical cases, as with engine failure, a pilot pulling the stick prematurely, and so on. This means the flight engineers have to have perfect control over all the influential parameters.

This example shows the limits to what you can do either with models or with simulation for some phases in flight, especially for manoeuvres close to the ground. In all events, thorough preparation is essential to make a good start and give the engineers the tools they need to make any necessary changes on the ADIS system.

With our considerable experience on the A340-600, we chose to adopt the same methodology for the A380, meaning good preparation of a basic law thanks to models and prior work on the simulator, followed by tuning during flight tests.

For all these tests, whether for flight controls development or, later on, measuring take-off distances, there is always a risk of touching the runway with the rear part of the fuselage as we were flirting with the limits of the aircraft's manoeuvrability. Also, to avoid any risk of damage to the structure, the aircraft is equipped with the same protection skid as that used for the VMU tests.

A highly experienced test pilot, Guy Magrin, was entrusted with this take-off law tuning with the team of flight test engineers working on the flight controls. The first flight took place on 29 December 2005 and, after 15 take-offs, the results obtained were satisfactory. We made an additional flight in February 2006 in order to finalise the setting of the protection to avoid touching with the fuselage, with one very slight impact on the skid during one of the tests, which proves to what extent we were seeking the minimal margin while ensuring safety. Later calculations showed that there would have been no impact if the skid hadn't been installed! Finally, a last flight on 8 March allowed the law to be validated for very high aircraft weights. Note that the first take-off was made at 596.5 tonnes, more than 30 tonnes above the maximum weight for certification. We first had to have enough fuel to reach the Istres base to pursue testing after departure from Toulouse. But we had also learnt from experience that it is always preferable to be a little heavier for this type of flight as customers fairly frequently ask for an increase in take-off weight shortly after entry into service. Therefore we sought to preempt this and so avoid having to repeat flight tests to that purpose.

Immediately after validation of this law, measurements of take-off distances needed for certification commenced with EASA crews.

* * *

Development of the longitudinal landing control law was settled quickly. Right from the start of flights, we knew that landing was particularly easy on this aircraft, but we still had to perform some settings for the various conditions for use in terms of the weight and centre of gravity position. For the flight part, we merely made the aircraft's response a little less abrupt that it had been initially.

The main modification was in doing away with what we called the derotation law on the A330 and A340. On these aircraft, as soon as the main landing gear touches the ground, the law comes into effect and helps the pilot control the aircraft's attitude until the nose wheels touch the ground. This law, which doesn't exist on the A320 family, was installed during development of the A340-300 as, in this landing phase, during a demonstration flight, an airline pilot had 'pumped', meaning he had encountered PIO (this was a Monarch's personal pilot and I was accompanying him!). Indeed, the A340 lands with a fairly pronounced nose-up attitude on the rear wheels of a bogie landing gear, which is itself tilted upwards. In addition, nose wheel touchdown occurs with the aircraft's attitude being slightly negative. The nose wheels must therefore come down from a quite significant height on landing. This derotation law allows the authority of the stick and the elevator to be substantially reduced so these wheels can be led gently down towards the ground.

A similar law had been carried over onto the A380, as a precaution, even though that aircraft had none of the characteristics of the A340 mentioned above. Once, when making a landing on what was a relatively short runway for the weight of the aircraft, at Morón in Spain, the pilot had a few difficulties in this derotation procedure. In all events, this law only comes into effect for two to three seconds. Now, in May 2006, during flight # 221 of aircraft # 1, when pursuing tests on settings for the longitudinal approach and flare laws, we decided to do away with it on an experimental basis, just keeping the flare law. After a number of landings, we were convinced that this suppression was the right decision. Since then, the pilots regularly used this modified law to ensure it didn't have any drawbacks. Following on from that, a few minor adjustments were made for the approach and flare. We needed to respond to what the majority of pilots wanted, after all! A wider authority to control nose-up and reduce the risk of a hard landing on an emergency return at high aircraft weight was also granted.

* * *

On flight # 51 that took place on 27 July 2005 in fairly strong and turbulent wind conditions (the vent d'Autan south-easterly characteristic of the Toulouse region), we made several approaches and realised that the control laws in roll had to be reworked. Pilots were exerting too much activity on the stick, the ailerons were moving too much and introduced unpleasant lateral accelerations, especially to the rear of the airplane. None of the settings available on that day (gain, damping, etc.) allowed these problems to be resolved. This fault absolutely had to be corrected but it looked like being quite a tall order.

In mid-October, new flight computers (PRIM) arrived. The engineers had installed an aileron based controlling system they had called 'VDA' for Valse Des Ailerons (Waltz of the Ailerons). When commanding left roll, for example, on the left wing, the inboard aileron moved upwards, with the outboard aileron also moving up, although to a different position, and finally the central aileron deflected either initially downwards in the direction opposite to the others before moving up, or directly up but after slight delay. A number of settings were possible, with different ratios between deflection of the inboard and outboard ailerons, and the central aileron following a different logic. This strange kinematics was intended to put a stop to oscillations of the wing as there were two of these at close frequencies that could combine. Looking at the display facing the crew representing the position of the control surfaces during a turn, it was easy to understand why this movement of all the ailerons, that appeared completely disorderly, had been given the dance epithet. A similar differential deflection was also installed on the two rudders, taking a similar name of VDR for Waltz of the Rudders. The improvement to comfort was spectacular. However, a few settings still had to be made.

On 18 January 2006, a new computer standard for flight controls came into effect. Improvements had been made on the VDA laws with, in particular, a reduction in the activity of the ailerons and progress was quite substantial. But a number of decisions still remained to be made as there were differing opinions within the flight test community. Thus, for example, an option remained open to have what we call a pre-command, meaning that for each action on the stick, there was a partial but immediate direct order to deflect the ailerons. A pre-command makes an aircraft more responsive, but a bit less comfortable. Without pre-command, the aircraft is a bit slower to react. In all events, the airplane's manoeuvrability is always adequate.

Final setting of the VDA law, for speeds lower than 300 kt, involved a deflection of the inboard aileron 2.5 times greater than the outboard aileron; and the median one with the same deflection as the inboard but with a delay of 350 milliseconds. Some improvements were still needed at high altitude because of effects due to Mach.

Another issue still to be resolved was the setting of the spoilers. They initially came out as soon as there was a roll command, leading to some buffeting. Then in mid-February 2006, new settings were proposed where there was no longer any deflection of the spoilers for a small movement of the stick to start a roll. In addition, in the absence of any special demand from the pilot, their extension was limited to 3 degrees so as to reduce vibrations for normal manoeuvres. Without this solution, one of our British test pilots told us he had the impression he was being 'punished' when deflecting the stick too much in roll due to the vibrations that ensued. Where there was a need to make more aggressive manoeuvres, it was decided to have a greater extension of the outboard spoilers than for the inboard ones as they create less buffeting.

For those flights where we had to judge controllability but also comfort, it was important to have a qualitative judgement at different places in the aircraft. At the front, the pilots were in a position to give their dual opinion on both handling and comfort. The flight test engineers, approximately in the centre of the aircraft, gave their views according to their sensations and the various plots available to them. Right at the back, we had installed a seat that was connected up to the pilots and flight engineers by interphone where a young engineer spent the whole flight to give his opinion on comfort. Considering the number of sharp turns we made on each flight, he needed a good stomach for it! Note too that right at the start of the programme, we conjectured a possible difference in comfort between the two decks. The first flights showed us there was no real difference.

On your next flight on the A380, whether you travel in the business or economy class, I recommend you book a seat close to the windows, up against the wing or slightly further to the rear where you can see the ailerons work (tough luck if you are in first class as you will be placed too far forward!). Then, just after take-off, you will see that around the median position (as they are deflected downwards with the flaps) as we start a simple turn, and taking the inboard aileron closest to you as a mark, the outboard aileron deflects at the same time, but just slightly less; while the central one is just a little later in joining the inboard aileron. If alternating roll corrections are made, considering the different deflections of the outboard and inboard ailerons and the delay on the centre aileron, you will see some ailerons move up while others move down. The term 'Waltz of the ailerons' as found by the Design Office is truly apposite – and the whole system works!

* * *

Fine tuning of the protections at high angle of attack, allowing loss of control to be avoided for all types of manoeuvres at low speed, must be performed systematically for all our aircraft. Test pilots are well familiar with the test techniques. This first of all involves setting the engines to idle, accompanied by slow and then more rapid decelerations until the stick is full against the stop, and then checking that stalling is not reached. These tests are then repeated in turns and then with maximum thrust. Going into turns followed by changes in heading are performed at maximum angle of attack obtained with the stick up against the stop and at different thrusts. The collision avoidance manoeuvre is one of the most spectacular to be performed, with the pilot pulling the stick into the corner for maximum roll and full nose-up. These tests are performed for all slats and flaps deflections but they also have to be made for extreme positions of the centre of gravity as well as for maximum and minimum weights, as the aircraft's reaction will depend on all these parameters.

As from the first flights, we evaluated these protections, but almost statically, meaning with very slow deceleration so as to be sure not to reach stalling because, as explained previously, we had doubts relating to the loads on the horizontal stabiliser. The engines were held on idle with the intention of getting an initial idea of the setting. On flight # 7 in particular, with aft centre of gravity position, the first results were satisfactory.

Real testing started as soon as the slats and flaps configurations were finalised. So, on 28 July 2005, during a flight with middle and then rear centre of gravity (ballasts and fuel transfers allowed us to adjust it), a first fairly comprehensive evaluation was made, still however avoiding too dynamic manoeuvring as uncertainties remained as to the loads on certain parts of the aircraft. The results were good overall, with the exception of configurations 3 and Full where the angle of attack was not contained with a consequent risk of stalling. For such flight conditions, we therefore decided not to go into turns at full thrust, with the stick pulled fully back.

Improvements were pursued with the arrival of different flight controls computer standards and, very quickly, towards October 2005, the settings of the protections were satisfactory. This was effectively demonstrated by in flight presentations at the Dubai airshow in November.

However, the final setting for these protections has to be made with 'ice shapes' and sometimes the maximum authorised angle of attack has to be slightly reduced. Detailed explanations on the tests conducted on the A380 in this domain are given in the ICING section.

* * *

The aircraft must also be protected against losses of control in deceleration or in tight turns at high and middle altitude, with slats and flaps retracted. When you pull on the stick, there is no visible stalling as at low altitude with the slats deflected, but what does gradually occur is strong buffeting that very quickly becomes deterrent if the pilot insists on pursuing the manoeuvre. These vibrations appear earlier as Mach increases, meaning that the setting has to be adjusted in relation to the latter. In practice, for each Mach, the test crew chooses an altitude such that tight turns can be made without the risk of eventually damaging the structure, bearing in mind that day after day our development aircraft are subject to really harsh conditions. A first exploration of turns, without the protections, allows the angle of attack for onset of buffeting to be identified. Then, based on these results, tests are resumed with a first setting of the protections, starting with slower turns and then moving on to more rapid ones, while seeking to stabilise Mach precisely. If you fail to encounter buffeting with the initial limit, the flight test engineers set a new, slightly higher angle of attack value to repeat the tests. If, conversely, buffeting occurs too early, the maximum angle of attack has to be reduced and manoeuvres repeated. Thus, for each Mach value, the setting is determined by trial and error but, by the end of the flight, you usually get the result you want. In practice, tests have to be performed for extreme positions of the centre of gravity.

We started these tests on 31 August 2005 with flight # 78. However, the results turned out to be unsatisfactory. Firstly, it proved impossible to set the maximum angle of attack between Mach 0.6 and 0.7. Then there were handling difficulties in roll during turns between Mach 0.80 and 0.84. Finally, between Mach 0.80 and 0.89, the aircraft tended to pitch up by itself, without intervention from the pilot, which led very rapidly to buffeting.

These various problems were gradually resolved. Note that the pitch up phenomenon requires extremely delicate adjustment of the flight controls laws. Indeed, when pitch up occurs, the law must automatically deflect the elevator down so as to gently counter the effect, whereas, without this automatic deflection, it would give a very pronounced movement of the aircraft. Finally, on 10 May 2006, a general review of the settings was conducted, with excellent results.

Some additional flights were however needed to validate the behaviour of this protection with the airbrakes extended. Indeed, pitch up varies considerably with the position of these surfaces as their extension reduces the lift of the wing. The first tests quickly showed us that the estimations from models and the wind tunnel were incorrect and we therefore proceeded with identification flights with various deflections of the airbrakes so as to define the automatic compensations to be introduced in the flight controls.

* * *

Just as there are protections towards low speeds and high angles of attack, there are others at high speeds and Mach. The characteristics of these latter protections have seen a few changes introduced since the first A320s, as it emerged that, initially, they came in too early. Indeed, they result from a complex trade-off. They have to be effective in not exceeding the maximum speeds demonstrated during flutter tests and be reasonably comfortable so as not to worry the passengers if they come into operation unexpectedly. They must not activate too early, otherwise, during a significant variation in the wind leading to VMO being exceeded, the aircraft would immediately start to climb, with a potential and unnecessary risk of collision with another airplane flying a little higher. When acceleration of the aircraft is rapid, they must come in early to prevent the maximum speed being exceeded. Whence the need to conduct many tests with different flight conditions to set all the parameters and define a good middle way: start of activation, anticipation in relation to acceleration and slope, load factor for the recovery in relation to Mach and speed. Remember that the load factor is the coefficient representing the forces sustained in relation to weight and that it is equal to 1 in level flight and greater than 1 in a turn or during a pull up.

As very high speeds will be reached during these tests, tuning can only start once all flutter flights have been completed. Furthermore, these latter flights allow the validation of aerodynamic models at high speed to be performed. There is therefore everything to be gained from waiting for analysis by the Design Office to be terminated and application of settings for the protections implemented. The protection must be active in Mach (to limit MMO overspeed), and in speed (VMO) for the same reason.

A first flight to explore protection at high Mach was made on 10 May 2006. Then, on 30 May, a whole flight was devoted to test protections on speed and Mach. Globally, the settings were unsatisfactory. Firstly, the protections activated too early and, secondly, at the end of recovery, Mach or speed remained in the red zone and thus too high with a failure to return to more normal values. Additional adjustments were needed.

On 17 August 2006, we made the certification flight for these protections. The imposed manoeuvres had to be adapted for the new fly-by-wire aircraft approach. The best-known test is called the FAA dive. For this, the aircraft is stabilised at the desired altitude with a thrust from the engines that just allows it to hold MMO. Then the pilot performs a slight nose-up manoeuvre keeping the thrust constant, and goes into a dive to go over the MMO with a descent attitude that is exactly equal to -7.5 degrees. The stop-watch is triggered as MMO is crossed. The pilot must try to maintain attitude even if this means pushing the stick fully forward, without affecting the thrust of the engines. After twenty seconds, recovery ensues, with engines reduced and the airbrakes extended. The stick has to be released if the load factor commanded by the protection is greater than 1.5. Otherwise, the pilot can pull a little on the stick to obtain that value. Of course, VD and MD speeds must not be reached in any manoeuvre for the result to be satisfactory. This procedure has to be repeated at several altitudes to validate Mach and speed protections, but especially at the one where VMO and MMO are encountered simultaneously. During these tests on the A380, recoveries were always made with the stick released and all the results were satisfactory.

However, that 17 August the flight continued and a test similar to the FAA dive was performed with the landing gear extended. In this configuration, there is no protection in the flight controls to slow down acceleration in dive. The maximum speed with the landing gear out is 250 kt but the airplane is designed to resist a speed of at least 300 kt.

I was with François Fabre, a pilot from the European Aviation Safety Agency (EASA). The first test was unsatisfactory as we had encountered a bit of turbulence close to the maximum speed of 300 kt. The second was good and, as far as the crew was concerned, the manoeuvre was a success. However, on arrival at the apron, the mechanic informed us that we had lost the lower half of the wing landing gear doors on both sides. I asked him to make sure nobody came close to the aircraft or took photos and then telephoned security for reinforcements to keep people away. As I came down from the aircraft, even before the security personnel arrived, I saw someone rushing over to take photos with his cell-phone. You can imagine the reception he got! The event could have been extremely damaging if it had got out onto the Internet. As it turned out, the incident went unnoticed and the only photographs taken were the official ones intended for the Design Office. If pictures had got out it is easy to imagine the sort of headlines certain newspapers would have indulged in, like "A380 narrowly misses crash". The parts must have been lost to the north-east of the Montagne Noire, in a remote region with very few inhabitants. We avoided launching search parties for the same media-related reasons. So if you now find parts of the doors during a walk or a run out there, you're welcome to keep them as souvenirs. Airbus won't claim them back from you now!

• See photo n°17 – page 237

* * *

The study of failures is a permanent process as from the start of design. The teams that define the aircraft have an essential preoccupation of covering failure cases and this represents a huge amount of work before the first flight. The goal is, of course, to ensure the safety of test flights, but also to detect risk situations as soon as possible. A huge number of configurations are reviewed, including some which we can be practically certain will never be encountered in flight. For all the anomalies that are reckoned to be possible in the life of all aircraft that will be brought into service, we need to ensure the behaviour of the airplane and its systems and check the procedures that will have to be applied by the crew.

Failures are essentially validated on the simulator, especially on the Iron Bird. But some are also reviewed in flight, where this is risk-free and also when the pilots' appreciation is especially important and there is a doubt as to the results obtained on the simulator. A number of flights were performed for these tests. There is no point going through all the failures analysed in flight here, so I will just mention a couple of them.

The first test flight concerning flight controls failures took place on 29 August 2005. During this flight, the eight pumps providing energy to the two hydraulic circuits were cut off and all manoeuvres demonstrating that the aircraft can be brought back readily, thanks to EHA and EBHA, were performed. In practice, I'm not sure that all pilots would see the difference in manual flight handling if they did not benefit from a whole series of warnings and procedures!

On the same day, another major failure was seen in flight. The test involved flying with just two flight control computers available, a PRIM and a SEC, corresponding to the case of the emergency electrical configuration. All this flight's results were good.

* * *

Simple failures on systems relating to flight controls have no consequence on handling and many of them are not even displayed to the crew during flight. They merely feature on the maintenance report edited on arrival and are also forwarded by radio or satellite to the airline's flight monitoring centre.

A dual failure may have no effect whatsoever on handling or, conversely, generate a flight controls law change. On an A380, there are few failures of the latter type due to the redundancies implemented.

If the computers are no longer capable of maintaining a normal flying law, they switch to one of the two so-called 'degraded' laws, alternate and direct. To schematise things, the alternate law retains complete stabilisation in pitch (maintaining the aircraft's longitudinal attitude without pilot command), but the various protections are lost. The principles of direct law were explained in previous sections. Here, an effort to simplify matters has been made as there are many variants to these laws according to the data available for the computers that work to develop them.

When flying with direct law, the pilot has to trim the aircraft, adjusting the position of the horizontal tailplane. Otherwise, permanent deflection of the elevator would be necessary, whence a deflection of the stick and an associated force. This is what the pilot will cancel out thanks to trim. This is the flying mode for all aircraft without electrical flight controls. On the A380, this is applied using switches, whereas on all the other recent Airbus planes, trim is controlled by a wheel located close to the throttles. Also, in direct law on the A380, a further system intended to damp pitch movements has been added (absent on Airbus planes of previous generations) due to the more rearward position of the centre of gravity.

The first take-off of the A380 was made with a direct law but that was completely unlike the one finally certified. Indeed, as there were so many unknowns for that flight, we didn't want to take the risk of being too constrained by the various stabilisation systems and the law was thus made as simple as possible.

Tuning of degraded laws is mostly done on the simulator before the first flights. We check that operation is satisfactory for all the weights, all centre of gravity positions and all configurations. There is no search for absolute perfection, such an outcome remaining extremely remote, but safety remains the byword. A certain number of settings can be worked on in flight, for example to adjust the sensitivity of the controls; this is basically routine work with no particular difficulty involved.

Controlling fly-by-wire aircraft at high altitude with degraded laws is much easier than on aircraft of the former generation where there is a mechanical link between the control column and the control surfaces. Electrical flight controls contribute to real improvements in safety as all the degraded laws on the A380 (as also on other Airbus planes of the new generation) include a full complement of stabilisation systems compared to the 'old' generation.

* * *

The A380 thus includes a seventh flight control computer, the Back up Control Module (or BCM). Failure of the six PRIM and SEC computers has practically no likelihood of occurring in the lifetime of any of the aircraft and, as a result, according to the applicable certification rules, the A380 was perfectly acceptable as it was (meaning without BCM installation), and certainly the BCM didn't need to be approved by the certification agency. Nevertheless, you can never be too cautious and we decided to install it to cover cases of wildly improbable failures. Many years ago a pilot did have the strange idea of cutting off all the flight controls computers on an A320. He then immediately re-connected them with no adverse effect (in any case there is a mechanical back-up system that makes the A320 controllable in that situation).

The BCM only uses a limited number of flight controls surfaces. However, the quality of handling with this system is of an extremely high level as it gives control over the three axes and includes stabilisation systems, with pitch and yaw damping, turn coordination (automatic deflection of the rudders as you go into the turn to avoid sideslip), etc.

We thus proceeded to test the BCM after cutting off the six PRIM and SEC flight controls computers. These tests had no urgency attached to them in so far as they were not required for certification, so they took place fairly late, on 8 September 2006. We had an aft centre of gravity. At low altitude, operation was satisfactory and it would have been possible to have landed with just this system available (although for safety concerns, we did not do so). Final verification was obviously conducted on the simulator. At high altitude, the situation was a little different as the aircraft went into pitch oscillations. These were not divergent, but correction nevertheless proved necessary.

In February 2007, further tests were performed with a few modifications. The results remained the same.

After investigation, it appeared that the problem stemmed from the fact that control was conducted using the inboard elevators. Indeed, with the BCM in action, the outboard elevators were not held firmly in the neutral position by the hydraulic systems and tended to go into small, non-divergent oscillations. This is what is called Limit Cycle Oscillation or LCO.

The design of the BCM was thus completely reviewed so that pitch could be controlled using the outboard elevators, while the inboard ones remained unaffected by LCO. Further tests took place in October 2009 to validate this modification. The oscillation had disappeared and the results were extremely good. Aircraft coming off the production line were rapidly equipped with this new version of the BCM.

* * *

The flight controls on the A380 are equipped with a number of devices intended to enhance passenger comfort and minimise loads on the structure when encountering turbulence or jerky movements induced by the pilot. This reduction in loads has nothing to do with preventing structural failure in flight of the aircraft since considerable margins are taken for sizing. However, the wing lifetime is a factor and more generally the fatigue effect on the structure. Such fatigue corresponds to the successive application of small loads that can gradually lead to failure. This is the effect that can be seen when you apply alternating loads on a piece of sheet metal until a crack appears to be followed quickly by breakage. A wing, an empennage, a fuselage, etc. must all be able to withstand an impressive number of cycles of severe turbulence or pressurisation with the absolute certainty that such a failure will not occur during the lifetime of all the aircraft in service, and this with a huge built-in safety margin.

The first device working to this purpose is the Passive Turbulence Alleviation (PTA) system. This in turn includes two functions, Manoeuvre Loads Alleviation (MLA) and Turbulence Loads Alleviation (TLA).

The MLA system applies a slight symmetrical deflection of certain spoilers and ailerons upwards when the load factor commanded by the crew is at least equal to 1.65 in order to reduce the loads on the tips of the wings.

When developing the MLA, we had to ensure that engagement of the system occurred without any jolting effect that could have created handling difficulties. Passenger comfort is not the first concern when the pilot performs such a manoeuvre but, as there is inevitably a pitch induced motion, it has to be countered correctly, thus calling on some settings. These adjustments are made when going into a turn until you reach a load factor of 1.7. The MLA must come in rapidly but without inducing too sudden a reaction from the aircraft.

The TLA is a dynamic system that is activated when the aircraft's longitudinal response does not correspond to the crew's command (this being due to turbulence). Beyond a certain threshold (load factor deviation by 0.15), the TLA will deflect the ailerons upwards and symmetrically thus reducing the loads on the wings. Validation was conducted flying through strong turbulence (not always easy to seek out!).

Active Turbulence Alleviation (ATA) is intended to minimise the effects of one of the symmetrical modes of wing oscillation. This is achieved thanks to small symmetrical deflections of the ailerons upwards or downwards.

Tuning of all these devices was not strictly essential for initial certification. Tests started in September 2006 and finalisation of the PTA part was completed for certification. Validation of the ATA was done early in 2007 to obtain its approval for entry into service of the aircraft.

* * *

Several years before the start of flights on the A380, we had decided to develop a system to facilitate handling even if speed and altitude data were lost. This went under the name BUSS, for Back Up Speed Scale. Tests started on the A340 and BUSS was certified on Airbus with fly-by-wire controls around the time of the first flights of the A380.

Development of this safety device was initiated because, in the 1990s, there had been two accidents on Boeings that, right from takeoff, had lost essential data, speed indications for one and speed and altitude for the other, without the crew becoming aware of the situation.

The principle is as follows: the speed indication is replaced by a simple scale with a green part in the middle, a red area at the top (FAST) and another red area at the bottom (SLOW). In practice, the speed is replaced by angle of attack data and the green area gives a band of angles of attacks that allows the aircraft to fly safely without getting too close to limits. The zones concerned are, of course, adapted to deflections of the flaps. The altitude scale, which is usually barometric, is replaced by altitudes given by the GPS, as on recent cars. Using this system is really easy for the pilot. He controls his new speed as previously, using the throttles when on level flight or the stick for climb and descent.

• See photo n°18 – page 238

Setting this system was an easy matter at landing speeds as the approach angle of attack with slats or flaps deflected is clearly identified, and a small variation in angle of attack does not lead to much change in speed. Conversely, in high altitude cruise conditions, a change in angle of attack of just one degree leads to a significant variation in speed. The system then has to be adjusted accurately to avoid going too fast or, on the contrary, taking the risk of stalling or inducing buffeting. Our experience acquired on tuning on the simulator and during the many development flights on various types of aircraft showed us that a satisfactory setting could be found.

It might be appropriate at this stage to make a comment relating to the Air France Rio-Paris accident that took place over the Atlantic on 1 June 2009. It should be recalled that this aircraft initially suffered from a loss of certain airspeed data. It was not equipped with the BUSS (not available when that airplane came off the production line some years previously). Some people may think the BUSS system could have saved that flight. That's far from obvious considering the dedicated procedure to bring it into operation. What's more, its engagement at high altitude has several drawbacks: when the BUSS is engaged, it is impossible to backtrack as there is no further way for the crew to know whether the speeds have become correct again or not. Moreover, all known situations of this type encountered at high altitude correspond to very short-lived speed data losses due to icing of the pitots. In this case, to simplify matters, the procedure is to maintain an attitude and thrust of the engines corresponding to the weight of the aircraft and the flight altitude. When, after a few seconds or minutes, the speed data return, a certain number of automated systems like the autopilot and autothrust can be re-engaged, which is not the case with the BUSS. After additional analysis, we came to the conclusion that using the BUSS at high altitude (above 25,000 ft) was not the optimum procedure and we thus restricted its use to low altitude. The reason for this is that a permanent anomaly on the speed sensors only normally occurs at low altitude. Known instances involved the sensors being blocked by their caches or by debris from birds having collided, insects in the pressure transmission pipes, or systems having disconnected. It was that type of problem that had caused both Boeing accidents mentioned previously.

As the procedure had not yet been modified at the time of the development of the A380, BUSS tests were performed at high and low altitude. The first flight took place on 22 March 2006 and was rounded off by a landing using this system, thus demonstrating how easy it was to use. Other flights were partially devoted to its final setting and certification.

* * *

Certification for a tailwind of 10 kt for arrival and departure manoeuvres does not require a specific test but beyond that, a certain number of rules apply. Thus, to certify for a wind of more than 10 kt, you need to demonstrate take-off and landing with 150% of the planned value. In so far as is possible, our objective was to authorise operations with 15 kt. As a result, we had to make some take-offs and landings with 22.5 kt of tail wind. It is well known that gusts increase with the force of the wind and while they are not very strong up to 10 kt, they pick up rapidly beyond that. That's how, in the past, we had been brought to land with a minimum of 22.5 kt of mean tail wind to validate 15 kt and this led to landing the aircraft with rear gusts of about 30 kt which was clearly unreasonable. The 'laws of the game' had to be modified.

Negotiation was pursued with the certification authorities to adapt the regulation in order to conduct tests in more realistic conditions while also ensuring operational safety for the airlines when making such landings and take-offs. Finally, it was accepted that the demonstration be made with at least 15 kt of mean tail wind and gusts higher than 22.5 kt. So that is how we proceeded with the A380. Note that the FAA refused this change in the regulation and so American certification includes a 10 kt tail wind limitation.

The demonstration flight coincided with the certification flight with the EASA pilot on board. It took place on 10 October 2006 at Istres. Indeed, the strong northerly Mistral wind regularly blows across this aerodrome and the runway is 16,400 ft long – perfect to make a success of testing when the weather conditions permit. The crew performed two take-offs and two landings with mean wind values equal to or greater than 15 kt and gusts very slightly higher than 22.5 kt.

* * *

Certification requires a demonstration of flight with a negative load factor. Remember that a negative load factor will throw the passengers, crew and more generally any object not tied down up to the ceiling. That's why you should always keep your seat belt fastened when seated in an aircraft.

Why such a requirement?

Of course, the idea wasn't to prove that a machine like the A380 can fly inverted like an aerobatic aircraft. The reason is to validate the operation of systems and especially those using fluids, including the engines, flight controls, hydraulic circuit, fuel system, etc., where the aircraft gets into a negative load factor situation due to strong turbulence. This means that if gusts bring the aircraft into such a situation for a few seconds, it should not be endangered. There are slight differences in the regulatory requirements between the European and American administrations. To cover all these rules, you basically have to go through passages under negative load factor during the same flight for an accumulated total of twenty seconds. One of them alone must last at least seven seconds.

From the flight controls point of view, at the time of the A380, we had to demonstrate that it was possible to reach -1 g, that is a value exactly opposite normal gravity, corresponding to the limitation by flight controls.

Before such tests, the aircraft has to be carefully prepared and everything has to be attached. The toilets are removed for obvious reasons. We board with the strict minimum of documentation so it doesn't fly around the cockpit and cabin as from the first test.

The demonstration is not performed flying inverted but over a rectilinear flight path. The aircraft is accelerated up to its maximum speed (VMO) then pitched up to the maximum attitude set by the flight controls limitations, i.e. 30 degrees. As soon as this attitude is reached, the pilot pushes the stick forward firmly to attain negative load factor. For the tests relating to the engines and systems, it is enough for the load factor to be slightly negative to obtain the sought result and the crew thus tries to stabilise a value between -0.2 and -0.5 g in accordance with the test protocol indicated above. Certification required several flights to be performed as they need to be done with different levels of fuel and oil (engines, generators, and hydraulic circuit, etc.) in order to ensure that everything works perfectly in extreme cases. This was done in May 2006. No difficulty was encountered on the systems.

From the flight controls perspective, the flight demonstration at -1 g posed a few problems. This requirement is an old one and aims to prove that the electrical flight controls still retain sufficient authority to dive. Unfortunately, it proved impossible to reach this value on the A380 even by reworking the flight controls settings and tests only validated -0.7 g on the flight certification on 4 August 2006. There were many discussions with EASA to convince them that, even if it had been accepted a long time ago, this rule was absurd. Indeed, the authority of the command to pitch up that has to be demonstrated is 1.5 g since you are on level flight at 1 g and the aircraft is limited by the structure to 2.5 g. Conversely, on dive it is 2 g since you need to go from 1 g to -1 g. Why demand more in negative than in positive g-force whereas the negative situation is much more dangerous for the passengers?

Note also that the fly-by-wire flight controls of Airbus airplanes have one special feature. Even putting the stick fully in the dive position, there is a very slight lead-time before the computers effectively command a negative load factor. These modifications were made, a long time ago, after a number of incidents where it was seen that aircraft could too readily switch to a negative load factor resulting from a sudden action of the pilot and thus risk injuring passengers.

Certification was finally accepted with that minor non-conformity that has no impact on safety. Immediately after this, discussions were entered into to modify the regulation by authorising the flight controls to limit the load factor to zero when everything is operating normally. In parallel with this, demonstration of system operation should always be made, but with a dedicated control law.

* * *

Although this does not strictly relate to flight controls, after certification, we have to look after the case of ferrying an aircraft with a failed engine. Indeed, for a four-engined aircraft, if a failure occurs at a remote airport, it may be preferable to bring it back to an aerodrome where engine replacement can be made in good conditions. You then have to be able to take-off on three engines and contemplate the idea of a second engine failing during the rest of the flight. Obviously, there are no passengers on board.

The most critical situation is on take-off with an outboard engine inoperative as that is what will give rise to the greatest dissymmetry of thrust. On take-off, the aircraft is initially aligned at about 15 ft from the axis, on the side of the two engines that are running. The inboard engines are powered up rapidly while the one symmetrically opposite the inoperative one has to be powered gradually so as to be able to counter the dissymmetry. Various thrust values are given to the crew in relation to the speed. Maximum thrust is reached at the minimum control speed on the ground with one engine failed. Take-off occurs at a speed slightly greater than that at which the aircraft remains controllable with two engines failed on the same side on the ground and in flight, i.e. about 160 kt on the A380.

For the approach, special procedures are also prepared. However, in this case, there are no real difficulties as the dissymmetry remains unpronounced due to the fact that the thrusts needed are less significant.

Development of this procedure begins on the simulator. Various situations have to be studied, with the engine running or blocked on purpose (fan blades removed), or an inboard or outboard engine failed. The acceptable limit for crosswind on take-off must also be determined, this being 10 kt for the A380. The procedures are then validated and often enough amended by flight tests. The great majority of cases of failure are validated on the simulator as there are a host of possible combinations according to the inoperative engine and the type of ensuing failure. The failure of a second engine may or may not affect the aircraft's symmetry (engine on same or opposite side). The number of control surfaces available can be modified, as also the condition of the electric and hydraulic circuits. There is a great deal of work to be done and the actual flights only represent the visible part of the iceberg.

Specific instructions are given to the crew. They will be confronted by a whole number of warnings as the aircraft computers are not designed for a normal take-off on three engines. A dedicated training course was developed and airline pilots appointed for this type of flight must have followed it recently.

The first test flight took place on 17 April 2007. In total we needed six flights to finalise procedures in the various scenarios. The last validation flight was towards the end of November 2007. The certification authorities obviously took part in a number of these tests.

* * *

Well after certification, additional tests were conducted to draw up the Configuration Deviation List (CDL) in order to facilitate operation of the aircraft with certain mechanical discrepancies. Let's illustrate the problem with an example.

Imagine that when towing the aircraft the tractor driver brings it into a minor collision against a hangar and the winglet of an A380 is damaged. What can the airline do to lose as little time as possible? The easiest solution is to remove the component, possibly also taking out its equivalent on the other wing, and fly again with passengers on board and no winglets. However, to do so, you need to be sure you can fly safely, even if this means imposing some limitations, as this would be considered to be a deviation in relation to the aircraft's certified configuration (whence the term CDL).

First of all, we check in flight whether there are limitations to flight handling or special restrictions to be applied. This may involve a change in the characteristics relating to stalling, vibrations at different speeds, etc. Having appraised these first results, we decide whether the aircraft can be authorised to fly like that with passengers on board. Next, there may be constraints that have no influence on safety but that have to be taken into account in operating the aircraft. Thus, it may prove that fuel consumption in cruise flight increases very slightly. The pilot must be informed of this to finalise the quantity of fuel to take on board. If the minimum flight speeds are affected, those adopted for take-off will also be changed so as to keep the usual safety margin. Whence the need sometimes for corrections to be made, which can also lead in turn to take-off weight limitations according to the length of the runway used. Similarly, penalties can also apply to landing speeds and distances.

This type of testing applies to many situations where components like the winglet, strake, small fairing, etc. are removed, but where the flight can still be pursued normally with some minor constraints or even none at all.

However, the CDL also covers more serious instances where, for example, the main landing gear door is damaged and has to be removed. In such cases, special procedures may be adopted as with flying with the landing gear down (but no passengers). This is fairly similar to an airplane ferry flight on three engines but this time with greatly simplified procedures readily accessible to any pilot and no need for special training. On a complex aircraft like the A380, a considerable number of flights have to be made to draw up the CDL.

LOSSES OF CONTROL

For us pilots, a loss of control represents any situation that could have led to an accident or very serious structural damage. We are of course used to going to the limits of the aircraft's flight envelope, and thus pitch and roll routines following a scheduled manoeuvre and that remain within reasonable limits do not come into this category. For example, when we stall, the aircraft goes into roll or pitch in an uncontrolled manner but this is not considered to be a loss of control as long as the bank angle remains limited, let's say at less than 90 degrees.

We have encountered such situations during development in practically every programme as we get close to the limits or when performing unusual manoeuvres specific to the tests. But passengers can rest assured that these extreme situations, deliberately sought after in testing for development and certification, are not to be encountered in commercial operation. After certification, such cases no longer arise during normal operations of the aircraft, even in the event of failure.

One might well wonder why, in the 21st century, with all the simulation and computation resources we have available, we can still encounter situations where we lose control of the aircraft? First of all, as already seen, simulation has its limits and what happens in real life is sometimes very different from mathematical models and wind tunnel results. But there are other reasons, one of the main ones relating to programming errors. Many discrepancies are detected before flight, on the simulator, and that's one of its main uses. However, on the Iron Bird, the results sometimes depend on assumptions made based on models or the wind tunnel. Suppose, for example, we decide to see how an aircraft behaves with a control surface that is no longer controlled but that just floats in the wind stream. We can simulate this situation on the Iron Bird to perform controllability tests, but will the deflection used be the right one? Only the flight can tell us with certainty. A similar situation led to the BCM design being modified (see the

FLIGHT CONTROLS DEVELOPMENT section). In the vast majority of cases, the forecasts prove to be correct, however, sometimes the result turns out to be unexpected. Luckily, this rarely leads to a loss of control since, where we believe there may be a discrepancy, we move forward tentatively. But as we shall see below, such loss of control comes as a complete surprise.

* * *

In an ideal world where difficulties are analysed with equanimity by real professionals able to explain any hitch and find a solution and where the media are not constantly seeking to latch onto the sensational, with the harm that can inflict, there would be no problem with specialists from outside the company being informed of such issues. Unfortunately, that's not the way it is and, to avoid the aircraft being unduly maligned, we have no other choice than to disseminate information on losses of control to just a very small circle of people.

Now all these anomalies have been corrected since certification, the aircraft's success is assured and the A380 cannot be harmed by the truth being given in full. Below, you will find an overview of losses of controls on that aircraft during development and certification. Not all the 'surprises' in flight are recounted in this book so as to avoid going into too much technical detail, but some other interesting ones are explained in the various chapters.

* * *

The first loss of control occurred on 12 January 2006.

The aim of the test was to finalise the airbrakes (spoilers that deflect partially and symmetrically). There are eight of them on each wing and we had to define the correct deflection for each of them. The problem lies in the fact that the spoilers closest to the fuselage are the most effective in slowing down the aircraft or assisting descent, but that they also generate vortices that exert major loads on the tail unit structure. Excessive vibrations that can make the flight uncomfortable for the passengers also have to be avoided. To resolve this latter difficulty, we provided the pilot with an intermediate notch for position of the airbrakes handle corresponding to maximum deflection in normal use, with an acceptable level of vibrations. Beyond this release level, the pilot can always deflect them to cover unusual situations, without this necessarily involving an emergency descent.

The main constraint for finalisation here was precisely for emergency descent. The certification requirements here are complex. Looking at things simply, they assume that there is a major calibrated air leakage leading the cabin altitude to rise rapidly. A rapid descent then has to be undertaken from the maximum flight altitude 43,000 ft to reach 15,000 ft within a time set by the regulation. To obtain the highest possible rate of descent, the engines are put to idle, the airbrakes set to maximum deflection and speed increased to MMO then VMO. At the end of descent, the altitude of the cabin must not have exceeded 15,000 ft.

We had conducted many tests, but not from the maximum altitude, to compare the rates of descent with various deflections of each of the spoilers. We thought we could achieve the best trade-off with maximum extension of each spoiler, where loads on the tail unit are limited, the level of vibrations acceptable and the rate of descent adequate. All we had to do was perform the test in real life and this was done on aircraft # 1. The two test pilots Ed Strongman and Francis Chapman climbed to the maximum altitude of 43,000 ft and stabilised cruise speed, Mach 0.85. The time signal was given and engine power was reduced, the airbrakes were extended at maximum deflection and a dive was initiated, initially with a slight turn, to accelerate to MMO. Once the aircraft got close to Mach 0.89 in acceleration, the pilot pulled slightly on the stick to nose up and stabilise the speed at MMO. But then nothing happened. The pilot pulled more and more, until the stick was fully to the rear. Seeing the aircraft accelerate in this uncontrolled manner, the other pilot, who thought there was a flight controls problem, also pulled his stick to full nose up but there was still no response. Finally, the nose came up just slightly and Mach stabilised at 0.96, which was maximum Mach that had been demonstrated in overspeed.

Everyone was taken aback. Why did this happen now, given that this type of dive had been made so many times? The explanation lay in the fact that previously we had not started off in the same conditions and above all not from such a high altitude. Why did the aircraft fail to respond, or respond so sluggishly, while the pilots were pulling fully back on the stick? The truth of the matter was that tests in the wind tunnel had shown aerodynamic characteristics that differed from what was really the case at very high speed and high altitude with the airbrakes out and, as a result, the flight controls had been set with an incorrect compensation for Mach related effects and thus with inadequate authority for the stick to pitch up. There was no way we could have encountered this phenomenon on the simulator as it used the wind tunnel model!

The test meant we were able to reset the wind tunnel results and, on that basis, the flight controls computers were reprogrammed to avoid any further problem. This perfectly illustrates the type of surprise flight tests can hold for you when you least expect it.

In our team, we all wondered the same thing: what would we have done if we had been there in the cockpit in the place of our friends to get out of such a sticky situation? Retract the airbrakes? Increase thrust? Switch to direct law? Thinking back on it, they had reacted as they should have done. As long as the nose came up very slowly, and despite acceleration, they simply had to continue pulling back on the stick and wait.

* * *

The second incident occurred on 7 August 2006 on aircraft # 1 with new software that had just been installed on the flight controls computers. After having obtained the usual test results for any new computer version, the crew ran a fairly ordinary certification test involving a touch-and-go (landing immediately followed by take-off) but this time without the nose gear touching the ground.

This first test went well but, on the next landing, the spoilers, that are supposed to push the aircraft down to the ground as the wheels touch down, came out completely while the aircraft was still 26 ft above the runway instead of waiting for the main landing gear wheels to reach the ground. The airplane plummeted at a rate of 16 ft per second whereas the normal limit is just 10 ft. One of the landing gears alone sustained a load of 360 tonnes and had to be replaced.

What had happened? The system controlling the spoilers manoeuvre in the flight controls computers had correctly seen that the aircraft was on the ground during the first landing but had failed to discover the conditions ascertaining renewed flight. So, logically, the spoilers were retracted as throttle was opened out and, as predicted, they remained in that position when, later, the pilot set the throttle to the Climb notch, the usual position during flight phases, except for take-off and landing. For the second landing, as soon as the pilot brought the thrust levers to idle while approaching the ground, the aircraft was detected as being on the ground by the computers, with the latter authorising extension of the spoilers, thus creating a huge loss in lift.

This incident related to an algorithm in the calculators and was subsequently highlighted on the simulator. We had not rehearsed this test previously but in any case we would not have been able to detect the shortcoming as the 'ground' and 'flight' conditions for the computers are partially established by simulation. We were only able to reproduce the problem on the simulator by tweaking the conditions.

* * *

A few days later on 17 August 2006 a loss of control could have had serious consequences. This involved testing relight of an engine on emergency electrical configuration. This is an extremely unlikely situation although it did occur at least once when a Boeing 747 went through a very dense volcanic cloud and all four engines cut out. Obviously, the main electrical power provided by the engine supply is lost, meaning the first engine to be relit has to rely on electrical backup. Once that's been done things become (relatively) easier since you will have recovered at least a part of the normal electrical network.

That 17 August we were going through the second flight of the day on aircraft # 4. The test was organised under Jacky Joye's leadership. On departure, there was a failure on the Air Data Computer (ADR1) that provides the pilots and the many other computers with speed and altitude data. Given that there are three of them on board and that the failure of just one of them is relatively inconsequential, the crew decided to pursue testing.

The flight started with a series of relights which were all successful and the crew came to the one in emergency electrical configuration at 30,000 ft. Clearly, it was out of the question to actually cut off all four engines for this manoeuvre, so the flight configuration was simulated by switching off the electrical generators of the four engines and only the actual engine being tested, number 1, was shut down. Electricity was then only provided by the ram air turbine. Once the entire crew was ready, relight was initiated. There was no problem with the engine that came back on perfectly and the test was apparently over.

All that then remained to be done was to get the aircraft back into the correct configuration by reconnecting all the electrical circuits. When reconnecting the generator of engine 1, the aircraft went into a series of extremely violent oscillations and the load factor varied between +3.2 and -1.3. This wasn't the first time we had gone through negative load factors as this is something that has to be demonstrated for certification (see the FLIGHT CONTROLS DEVELOPMENT chapter), but, for this type of test, everything is prepared in advance and with all objects being fastened down and here nothing of the sort had been done.

Towards the fourth oscillation, things all calmed down and everyone managed to get hold of their possessions again, including Jacky's precious stopwatch he was so attached to. It was slightly damaged, but over more than thirty years of test flights it had seen worse and it was repaired!

We had the impression that we had come close to a disaster on that day as the loads on certain points of the structure reached 150% of limit loads. The aircraft was therefore inspected for a whole day and the full array of load sensors over the structure enabled the specialists to more readily get an idea as to potential damage. Finally, nothing had been broken and the aircraft flew again two days later.

So what had happened?

There is no simple answer. The flight controls specialists' study, followed by tests on the simulator, enlightened us. There are three main flight controls computers and, when the generator comes back on line, the first one, that usually ensures overall control, had internally locked a direct law as there was a hidden failure (meaning one that was not indicated to the crew). Remember that, due to the multiple redundancies, a large number of simple failures are not indicated to the crew and merely feature on the maintenance report once the flight has been completed. The pilot's orders then went through the second flight controls computer that had a 'better' (alternate) law. But as a certain number of modules energised again, this computer received erroneous speed data due to the failure of ADR1. Mainly, it reckoned we were at low speed and therefore authorised greater deflections of the control surfaces than were necessary but it also saw that speed was increasing very quickly and gave a pitch up order to reduce it. By the time it took for the speeds seen by the computer to stabilise at the true value, around 270 kt, everything was over, that is in a matter of a few seconds.

Could this have been predicted?

On commencement of the flight, the crew couldn't have imagined a scenario of such complexity with respect to the flight controls as the incident was due to an accumulation of several elements: a test at the limits of system capabilities, a known ADR failure and another hidden sensor failure. The wisest course would have been not to have flown with that known failure. That's easily said from an armchair position but when you want to get on rapidly with tests, you try to evaluate the risks in real time and, obviously, if we reckon there's the slightest danger, we stay put. But no crystal ball gazing can resolve such issues!

This was the second time that François Barre had flown as captain on a Class A test flight as he had just finished further training to move on from test pilot (B) to experimental test pilot (A). A worthy first! After the flight debriefing, everything finished off with a round of champagne to boost morale.

THE AUTOPILOT

Some decades ago, in the early days of this type of automated system, the autopilot came to be known colloquially as 'George', being considered to be just another member of the crew! But nowadays, George works for more than 95% of the time on long-range aircraft.

What is asked of an autopilot in climb, cruise or descent is above all to ensure correct guidance of the aircraft in accordance with the flight plan decided on by the crew. It also has to fly the aircraft to ensure maximum comfort and the ideal situation is where passengers don't even notice its movements. This means it has to do its best to soften the effects of turbulence and if it can't manage to do this effectively, it should disconnect automatically to hand back authority to flight controls and the crew. It still has to be sufficiently robust as to avoid the pilot having to intervene in strong turbulence as much as possible.

On the A380, overall system design is such that the autopilot uses the flight controls computers to act on the control surfaces. Thus, all the aircraft's stabilisation systems are active permanently.

Outside the landing phase, development of the autopilot is reasonably simple, easier than that for the flight controls.

* * *

Development of the autopilot generally starts very early. Indeed, we got on with it as from the very first flight of the A380 and a few turns were made so as to get a first idea of the settings. A pretty substantial number of flying hours have to be devoted to all the adjustments as there are so many flight conditions to be covered. Thus, in cruise, low and high speeds have to be focused on, not to forget Mach related effects, the influence of altitude, the effect of the (fore or aft) centre of gravity position, the weight (heavy or light airplane), fuel distribution, etc. Evaluating all these interrelated combinations represents a huge work load. In addition, we also have to make all such settings for the various flap deflections.

Alongside the autopilot there is the associated autothrust system that constantly adjusts the thrust of the engines to regulate speed. Basically, this system gives maximum thrust when climbing and brings the engines towards idle in rapid descent. In level flight, it ensures speed is maintained and, when the pilot wishes to maintain a rate of climb or descent, it adjusts thrust to retain speed. Developing autothrust may appear an easy matter as it appears to be much the same thing as the cruise control in modern cars. In reality, setting is more complex than on a car since a gust of wind can suddenly change the speed in relation to the air and the system is called on to react immediately. However, too prompt a reaction will have an adverse effect on comfort and may cause problems maintaining speed. The ideal tradeoff to cover all situations and maintain speed precisely is hard to find.

Jean Muller, a brilliant engineer with considerable experience in developing autopilots, ran the development team at the Design Office. On the flight tests side, two men of experience were working on the front line: Bernard Masdoumier as flight test engineer, and Richard Monnoyer who was at the controls for almost all these flights.

Finally, no major difficulty was encountered in development although obviously adjustments were needed for some flight conditions.

Note that during endurance flights with the Engine Alliance engine in autumn 2007, that is well after basic certification, some of our passengers noticed an oscillation in cruise, at high Mach, in turbulence. This was not a problem per se, but its discovery in this type of flight was bothersome. This was a small pitch oscillation with a 5 seconds period. To stop this, the only solution was to disconnect the autopilot, then re-engage it. This discrepancy probably followed on from an upgrade that had not been validated correctly before the campaign started. Between two series of flights, on return to Toulouse, modified computers were installed a number of times but to no avail. Luckily, this defect did not affect the Singapore Airlines A380 that was coming into service at that time as it didn't have the same type of computer. The whole issue was quickly settled.

* * *

The availability of automatic landing is now fundamental on large passenger jets as airlines and passengers are unwilling to accept rerouting due to an aircraft not being able to land in fog with the outcome of five hundred people having to be provided with a hotel room and an aircraft that won't be ready for the next flight. Sometimes you also have to send out other pilots to fly the plane back once the fog has lifted or wait for the initial crew to have had their statutory rest period. A real kettle of fish for the operator.

At the time of the first Airbus programmes, it was the tradition to certify automatic landing six months after that of the aircraft, but later some of us considered that this situation should be improved as irksome for the first users in operational terms. So, in 1995, for the development of the A319, a year after I had taken over management of flight tests, we decided, with a small team, to do all we could to ensure that autoland was available as from entry into service of the aircraft. There was no objection from Management, on condition that certification was not delayed. In the Design Office, the engineer who accepted this challenge was Jean Muller, whom we were to find later on the A380, and at flight tests, at that time, Armand Jacob was the flight test engineer responsible for autopilots. Meanwhile, as a pilot, I was keen to follow this matter closely. We only had seven months to succeed in this as certification was scheduled for the end of March 1996. By mid-February, we had called a halt to tests and autoland was certifiable. Admittedly, touchdowns were sometimes a bit rough, but at least re-routing due to fog could be avoided. Over the following months, improvements were made to avoid these firm touchdowns. As from that date, we have tried to certify autoland for entry into service on all new programmes.

* * *

Developing autoland means you have to take an interest in many different aspects of flying. That's what the flesh and blood pilot always does without noticing it. Firstly, you need to be perfectly stabilised on an approach axis and a descent slope that are materialised by the Instrument Landing System (ILS). During this time, speed must be held precisely by autothrust. Then, you need to make a correct landing that is not too rough for the passengers' comfort and resistance of the structure, right within the zone planned for on the runway, neither too early nor too late. Guidance on the ground is also automatic and the axis of the runway has to be kept to accurately, even with an engine failure or with an anomaly on extension of a thrust reverser.

We have to validate all flight cases authorised for autoland, especially with extreme weight conditions (heavy or light aircraft) associated with the different positions of the centre of gravity. You also need to check for correct operation with the various types of wind that may be encountered: calm, head wind, tail wind, crosswind, and at various strengths up to the limit that will be certified. Two flaps deflections are also approved. Finally, we demonstrate landing with an engine failing during any phase in the approach.

All combinations of the various parameters could lead to a huge test programme, but common sense selectivity means the engineers can make progress intelligently without losing too much time. After several hundred autolands performed for development, at least 100 are demanded for certification and, of course, these have to cover all extreme situations.

Airbus decided to authorise autoland in the event of failure of an engine and this obviously leads to conducting additional tests to ensure good precision of lateral guidance in flight and then on the ground. These tests were sometimes impressive in strong turbulence or crosswind with the former versions of flight controls systems as the variations in thrust needed to keep speed precisely could lead to significant deviations in heading and roll movements due to the dissymmetry of thrust. Nowadays, more effective fly-by-wire flight controls considerably improve the aircraft's behaviour. We can sometimes be led to reduce wind limits, more especially for crosswind in approach with the failure of an engine, but this was not the case on the A380.

We also need to check that at any time during approach, the sudden failure of an engine does not lead to a hazardous situation. First of all, reductions in the thrust of an engine are made at various points in the approach, but the most spectacular exercise remains real and sudden cut-off of an outboard engine, as induced by closing the fuel cock. This manoeuvre has to be performed at different altitudes, and more especially close to the ground. Thus, on the A380, a number of tests were made at 700, 200 and 100 ft during development and then for certification. Clearly, this was to see the quality of guidance but also to ensure that the modifications caused to the electrical and hydraulic systems, and flight controls did not lead to anomalies, especially in the most critical phase, close to the ground, as transients, relating to reconfigurations of the electrical circuits (sometimes with very short cut-offs), are generally not very good for the computers!

Special attention is devoted to go-around at different altitudes. If this takes place at very low altitude, typically during flare, the aircraft may actually touch down with its wheels. During such a manoeuvre with the failure of an outboard engine, a slight transient lateral deviation is admissible, but it must not constitute a hazard.

The development of autoland requires a huge number of approaches to be performed and it is interesting comparing this with how a human pilot learns. The latter adapts much more quickly with it not being an infrequent occurrence for young pilots to be released to go solo on a light aircraft after some sixty landings. But the idea of proceeding so quickly for automatic landing cannot be entertained. However, once it has been settled, autoland works much more precisely and above all much more repetitively than the human being. Take the example of landing with crosswind. During the approach, the aircraft flies crabwise in relation to the ground, meaning that its longitudinal axis makes an angle with the runway that will increase as the wind gets stronger. On arriving close to the ground, the autopilot will 'decrab' so as to reduce this angle and again set the aircraft along the axis of the runway then let the wheels touch down immediately. Synchronisation of these manoeuvres is statistically not so good with human intervention, meaning that the aircraft can stray slightly from the axis.

One of the difficulties in development stems from the fact that the ground effect is poorly understood and cannot therefore be readily simulated. As we cannot count on the wind tunnel and models established before flight to ensure this development, we have to feel things out in flight to get to the desired results, which explains the relatively high number of test manoeuvres.

* * *

The development of autoland on the A380 hardly threw up any more difficulties than on any other Airbus programmes as, after all, the aircraft has two wings, engines under the wings and a tail unit just like all the others! However, to succeed within the lead-time, adequate priority had to be granted to this objective. Making development flights for the autopilot (often considered to be a simple form of assistance) right at the start of the programme may appear strange, but this was the only way to succeed. Thus, for the A380, it had been decided, from the beginning, to reserve a minimum number of flights each month for the development of automatic landing and this was especially important during the first weeks as these tests were used by the Design Office to draw up an aerodynamic model. We were able to resume tests a little later with the computers developed using this reference.

The first flight almost entirely devoted to testing the autopilot was flight # 17 and here what we were mainly interested in was identification throughout the flight envelope. The results were very good, suggesting that development would be pushed through quickly. That flight ended with the first autoland and, not only did the pilots not have to take over the controls again, but the touchdown was extremely gentle.

Considering all the other priorities, we could only resume development of autoland on 10 August 2005 with flight # 58. Here, what we wanted to do was make calibrated landings, with a command to nose up on the stick held for a few seconds, in order to determine what the autopilot should do by copying the human pilot. These landings were made in Toulouse and then nearby Tarbes where the runway climbs fairly steeply. It appeared that the autopilot had an initial command that was less clear cut than the pilot's.

Over that month of August, other flights were devoted to identifying the ground effect in order to establish a good working model. After work in the laboratory and elaboration of new control laws based on the initial identification, development tests on the aircraft resumed on 17 October 2005 and, as from that date, they were pursued actively. A great number of flights were thus made with different versions of the computers over the following months.

No major difficulty emerged from all these development flights. The goal was rather to optimise landings to ensure repetitiveness and comfort in all situations. Compared with our previous airplanes, a new feature was introduced for vertical guidance: in order to always follow an identical and pre-determined path, the Design Office cleverly conceived of the idea of controlling the lift of the wings by partial and variable extension of the spoilers. This guaranteed accuracy of the touch down point and the vertical speed at that instant. The advantage of that technique is that the action of the spoilers on lift is immediate and therefore adjustments of path guidance can be both precise and more rapid. The first flight with this device took place on 29 November 2006, following which progress was extremely quick.

Another new feature was also introduced, relating to landing with crosswind, as it had transpired during tests that, in some cases, the same wheel touchdown quality had not been encountered due to various aerodynamic effects. So, to soften impact, a small pre-command to pitch up was added on deflection of the elevator just before touch down when the computers detect a crosswind.

One difficulty that might seem trivial led to us losing a lot of time. This concerned interference on the signal of a glide reception antenna, a guide beam on the approach glide slope. Early on, it is an antenna mounted on the fuselage of the aircraft that ensures guidance, but when the aircraft descends, as it gets to 700 ft above the ground, there is a switch to an antenna secured to the nose landing gear. Indeed, the system is designed so the antenna follows the glide slope. On a big aircraft, if you leave the fuselage antenna in action right to the end on that slope, the landing gears would come down too low over the runway threshold and potential obstacles. As you get close to the ground, you therefore need to use an antenna placed lower down on the nose landing gear. Guidance cannot constantly be ensured on the landing gear antenna as, early in the approach, the landing gear has not yet been extended, whence the need for both these antennae. Unfortunately, the signals from this landing gear antenna proved to be subject to too much noise and a whole number of modifications to the insulation and the shielding of cables proved fruitless. As a last resort, landings were made in February 2007 without this switching, but this proved unsatisfactory as the start of flare came in a little too late. As from mid-April, various solutions were tried out. On 25 May, we finally managed to resolve the issue after two new flights: a simple extension of the antenna support allowed the problem to be settled. This is just another example of a difficulty that seems quite banal but that led to a huge loss of time and money. Thus, with final correction only coming in fairly late, we had to repeat a number of development and certification tests.

Meanwhile, a certain number of certification flights were conducted as from the end of 2006 to cover various wind conditions. A 43 kt head wind was demonstrated on 8 March 2007 during a development flight and similar values were reproduced later with an EASA crew. Crosswind, at 38 kt, was validated at Keflavik on 24 April 2007 and a tail wind of 16 kt on 26 April. The final values retained as wind limitations for automatic landing were slightly less than the values demonstrated in flight, with 38 kt for head wind, 10 kt tail wind and 30 kt for crosswind.

Late in 2007, it was decided to go to Bogota, at 8,300 ft in altitude, to proceed with endurance flights on the aircraft equipped with Engine Alliance power plants. A series of automatic landings was performed. This was rounded off by another series at the end of 2009, this time at Addis-Ababa (altitude 7,600 ft). These tests allowed the certified altitude to be taken to 9,200 ft, bearing in mind that an extension is accepted above the demonstrated value as it is established that variations in altitude have relatively little influence on the behaviour of the aircraft in such manoeuvres.

* * *

Further tests were run to cover specific runway configurations, one of which took place at Johannesburg aerodrome. Johannesburg is special in that the terrain descends before you reach it but that the runway itself then rises with a slope of about 1/100. Now, flare is performed using the radio-altimeter, a system that measures the distance between the aircraft and the ground using a sort of radar, a system fairly similar to the obstacle detector you find on the front and rear of some cars but with greater range. During the final part of the approach, the terrain descends at the same time as the aircraft, so the autopilot sees a fairly slow movement towards the ground. But then, suddenly, this rate accelerates as the runway goes up and the reaction has to be fairly decisive so as to avoid a hard touchdown. The flare law therefore has to be programmed to be capable of deflecting the elevator rapidly, when necessary, to reduce the descent rate.

This problem hadn't been brought to our attention until A320s were delivered to South African Airways. Modifications had to be made to the aircraft already built and, for new models, computers that take this kind of runway into account had to be provided. For obvious cost reasons, we wanted to avoid going to Johannesburg to develop this new 'all-terrain' guidance law. As the runway of Tarbes-Lourdes aerodrome has a similar rising profile, all we had to do was introduce a bias in relation to altitude into the radio-altimeter measurement during the approach. This simple trick meant we could simulate the Johannesburg runway and so get on with the job of development. Note that validation of a flat terrain before a rising runway was also done using the same technique, thus catering for the specific terrain features of certain aerodromes and check out the ruggedness of this landing law. Right from the start of development, we conducted tests at Tarbes as the control law was to cover all known cases and it would have been difficult to make a later modification just for the Johannesburg runway without impacting all the settings.

Endurance flights provided the opportunity to do a world tour by the poles and the first stage, on the way to the South Pole, was Johannesburg. This helped validate our simulation methods for the A380 using the bias introduced in the radio-altimeter. So, we made four automatic landings to demonstrate the effectiveness of our development technique.

* * *

Autoland certification tests are conducted through to the maximum landing weight. However, in the mid-1990s, during development of the A330-200, we decided to validate up to a weight close to the much heavier maximum take-off weight. The simple observation of a significant number of hard emergency landings immediately after take-off of long-haul aircraft of all types imposed this choice on us. On such airplanes, the pilots perform a very limited number of landings each month, and training for return to the aerodrome in heavy weight conditions is only done on the simulator. Autoland was capable of doing a better job than the pilots in the overwhelming majority of cases. By definition, certifying landing beyond the maximum landing weight was out of the question, but an emergency procedure had to be provided to be applied as the pilots chose.

Validation of this procedure is generally conducted in three or four stages. A limited number of landings are made, generally three for each weight, including one with a simulated engine failure (engine on idle), in good weather conditions, as what counts is above all to check that there is no significant degradation in the quality of landing with an increase in weight. Then one fine morning we are ready, the first landings are made and the aircraft returns to the apron to refuel. It then leaves for a new series of landings at a higher weight and so on. Tests come to an end once you get near to the maximum take-off weight.

On the A380, these tests took place during two series of flights in February and April 2007. At each stage the weight was increased by 20 to 30 tonnes and six short flights were needed considering the considerable difference between the maximum weight on take-off and that on landing. Finally a weight of 565 tonnes was demonstrated. This could feature in the Guinness Book of Records as the maximum weight for an automatic landing! This weight finally attained was actually slightly higher than the maximum take-off weight for certification.

* * *

Although this did not relate to operation of the A380 autopilot, we had another difficulty to manage relating to autolands with disturbance of the ILS lateral guidance system, the Localizer (or LOC), due to the very size of the aircraft. Indeed, the antennae are located just after the end of the runway and, when an aircraft taxis close to them after landing, it disturbs the data transmitted to those that are on the approach, with its fin causing parasite reflections. The bigger the fin is, the greater will be the disturbance, and this is exacerbated if the aircraft quits the runway close to these antennae. The A380's fin culminates at a height of over 80 ft while that of a B747 reaches 65 ft, which makes it logical for the potential deviations in axis created by an A380 to be more pronounced.

During an autoland, you obviously need perfect guidance and therefore disturbance has to be eliminated. Protected zones have thus been calculated according to the type of aircraft and none of these are to be found in what are called 'Category 3' operating conditions (landing in fog). Shortly before entry into service of the A380, two new methods to calculate these zones were prepared to cover all types of airplanes, one by the Ecole Nationale de l'Aviation Civile (France) and the other by Ohio University (United States). They take into account the local environment and especially parasite reflections of the radio waves from the hangars. Both methods have been approved by the ICAO.

However, many airports found them to be unnecessarily penalising, not only for the A380, but for all other aircraft, and they have therefore been rejected by a considerable number of operators. Another, much more refined method was later developed within EADS.

Special talks took place for operations at London Heathrow considering the extremely heavy traffic that only has two runways available to it. Singapore Airlines commenced operations to London in March 2008. Shortly after, a pilot flying a United B777 complained of major disturbance of the LOC as he approached behind an A380. The London airport authorities were worried by this and set the distance to be respected behind an A380 to 20 NM. The same B777 was subsequently to fall victim to the same misadventure in Munich, but this time there was no A380 to be incriminated! It turned out that there was simply a discrepancy in its guidance system. In July, things returned to normal at Heathrow with a distance spacing based on the regulation relating to the wake vortex, 6 NM for a B777 coming behind an A380 at the time.

THE AUTOPILOT TRAFFIC COLLISION AVOIDANCE SYSTEM (APTCAS)

Since the early 1990s, all airliners have been equipped with a system intended to avoid collisions known as the Traffic Collision Avoidance System (TCAS). This is based on an exchange of information and manoeuvring orders so each pilot can see the relative position and altitude of other aircraft in the vicinity.

TCAS data are given on a dedicated screen for older aircraft or integrated in that used for navigation on airplanes of the new generation. All aircraft, whether they be fixed wing or helicopters, flying close by and within a certain range of altitude are represented by a small diamond shape. Alongside this is a number giving the difference in altitude and an arrow showing whether the aircraft is climbing or descending. Where there is no danger, this lozenge is white and hollow. If the two aircraft start to get close, it fills out. If an avoidance manoeuvre is in the offing, the lozenge then turns into an amber coloured round shape and a voice announces "Traffic". The pilots may still make no move at this stage, this being what is known as the Traffic Advisory (TA) mode.

Where the situation becomes critical, the round shape turns into a red square. An avoidance manoeuvre is then coordinated between the TCASs of the two aircraft so as to ensure adequate distancing. Instructions are relayed to the pilots in two different fashions: indication on the variometer (the instrument indicating vertical climb or descent speed) and a spoken order. The variometer becomes red in colour to mark the forbidden zone and green to indicate the rate of climb or descent the pilot must hold. The pilot must then disconnect the autopilot and the flight director (manual flight control guidance system) and follow instructions from the TCAS. There is then an oral confirmation of orders, for example: "Climb, climb". This is known as the Resolution Advisory (RA) mode. If one of the aircraft fails to correctly follow the flight path demanded, the order may be modified, for example: "Increase climb, increase climb".

Many accidents have been avoided thanks to the TCAS and others could have been. One memorable accident was a collision that took place in 2002, close to the Swiss border, between a Russian Tu-154 and a German Boeing 757. In this accident, the pilot of the German aircraft followed TCAS orders, but the Russian pilot obeyed the air traffic controller whose instructions were in direct conflict with the TCAS. If both had followed the coordinated order of the TCAS systems, the aircraft would have come close but there would have been no collision. Following that accident, the ICAO made it compulsory to follow the orders from the TCAS. The Russians initially refused this directive but found themselves isolated.

* * *

In January 1996, we started to think about what the A3XX (as it was then called) would be like. How would it be for the pilots? I sent out an internal memo to the test crews asking them for their ideas. This was a good opportunity for them to exercise their imagination as we had plenty of time to sort through their responses. A few weeks later, I received a first list from Jacques Rosay, the pilot in charge of this new aircraft. A small group of us met a number of times to summarise the proposals coming in before commencing discussions with the Design Office.

It was a matter of urgency to meet the representatives of the airlines with whom the aircraft had been defined and start to take a look with them at the operational side to things. This meeting with the crews was delayed a number of times for various reasons and especially because the general definition of the aircraft's architecture had not been completed. This finalisation took place in December 1996 and we were able to launch the First A3XX Cockpit System Workshop, a first working meeting with the pilots of potential customers, at the end of April 1997. Even the Japanese, who were much more used to buying from Boeing, were there! On the Airbus side, a number of engineers from the Design Office also took part alongside three test pilots: Jacques, Etienne Tarnowski (aka ET), an old hand who had taken part in defining many airplanes, and myself.

We shared out the presentations between us. I started mine by announcing right away that I was going to raise issues that may well lead to polemic. Two proposals were put on the table. Firstly, automatic terrain avoidance. Since the ends of the 1980s, aircraft have had on board a system called the Ground Proximity Warning System (GPWS) that gives the pilot orders if there is an immediate danger of flying into the ground or an obstacle. The idea was to have a capability for the automated systems to take over if the pilot had failed to react after the first warning. Imagine your children were on board, would you prefer the system to intervene on the second warning, or would you continue to have faith in the pilot who didn't react first time round? There was, after all, a major technical difficulty in implementation: the mechanism coming into action in an untimely manner absolutely had to be avoided as the severity of the manoeuvre could lead to injury of passengers and cabin crew. In the event of a real danger, this was perfectly acceptable. But there was a real worry as to a false alert due to a possible malfunction. This system had to be robust and the matter was subject to lengthy discussions. We remained uncertain as to whether feasibility could be guaranteed by the date the first aircraft were to be delivered. Voices were heard in the room more or less equally divided between those in favour of the development and those against.

The second proposal was automatic aircraft avoidance, meaning the mechanised performance of TCAS orders. Here, the reaction was clear, with the idea being unanimously rejected by all participants. Why this was so is hard to say as there was nothing truly rational in this opinion. It was nevertheless decided to continue thinking about the matter although it was left somewhere in the background as a whole number of other tasks now took up the time of the Design Office on the A380.

* * *

Bringing the TCAS into service showed that even though it's a remarkable tool to reduce the risks of collision, a number of difficulties remained. Firstly, some pilots react extremely brusquely whereas the manoeuvre is designed to be performed immediately but gently. For example, on a Boeing, there was an instance of a pilot flying into a vertiginous descent at some 11,000 ft per minute. We have even seen cases of pilots reacting in precisely the opposite manner to the instruction, for reasons that remain a complete mystery.

You have to recognise that the TCAS guidance system, based on orders indicated on the variometer, is imperfect as those indications are filtered and you can't instantaneously display a value. For example, if you want to climb with a certain vertical speed, you have to pull on the stick, wait for the rate of climb to stabilise and check that the value reached corresponds to the one sought for. Otherwise, you need to make corrections.

Various improvements were made on how data is presented on all fly-by-wire Airbus aircraft, but these modifications were not adequate to prevent excessive or even wrong reactions.

Early in the 2000s, we decided to develop a guidance system based on the pitch attitude of the aircraft. Everything would then be easy for the pilot who would just have to maintain the computed attitude as displayed on his screen and the vertical speed obtained would be that required for avoidance. We had already made considerable progress when the Design Office discovered during the summer of 2004 that we had been beaten to the post by Boeing who had already patented this device.

Not to be outdone by our competitor, we decided to relaunch what came to be known as the Auto Pilot TCAS (or APTCAS) for avoidance by the autopilot or a variant, the Flight Director TCAS (FDTCAS). In the latter instance, the pilot uses the flight director in manual flight and just has to place his flight reference point on another one whose position is determined to follow the planned flight path.

* * *

Study of the APTCAS thus started late in 2004 and was directed by Isabelle Lacaze. Paule Botargues succeeded her for the development phase.

In order to evaluate and validate the guidance logic and principles, the Design Office had to invent a new simulation tool – an intruder generator – meaning a system that creates an artificial aircraft posing a threat. That's not as easy as it sounds as this 'aircraft' actually has to navigate in order to inject some realism into the scenario. What's more, it has to react by following its own TCAS orders as coordination of manoeuvres between the model or laboratory aircraft and the intruder have to match real conditions. The situation where the reaction is inappropriate also has to be covered, leading to changes in TCAS orders. In fact, two generation systems were used – a simplified one for the initial search phase and to validate the principles, and another more sophisticated one for development on the simulator and to be taken on board the test aircraft.

In June 2005, tests on the design simulator known as Epopée started with Airbus flight crews to pursue general definition. Everything worked well, even though much still remained to be refined. Demonstrations soon followed with a limited number of airlines to test the ground and, notwithstanding the fears we harboured after our presentation in April 1997, the system was very favourably received. People's attitudes had evolved and having seen the system work on the simulator contributed to convincing them as the system remains extremely intuitive and was in harmony with the general flight controls philosophy that prevailed on our aircraft.

Early in 2006, we were again attacked in the press and above all by some of the unions who reckoned that we were taking the pilot out of the loop. This was a truly odd way of reasoning when you consider that, on most flights, the pilots engage the autopilot just after take-off and only disconnect it on approach (and even sometimes on the runway when they make an autoland). So why not entrust the autopilot with this avoidance job? This recalls an incident where I was verbally aggressed from the back of the room by an airline captain at a briefing on this system in Moscow early in 2009. I seized the opportunity to make it clear that whatever happens, whether in ordinary flying conditions or during avoidance, he, as pilot-in-command, was ultimately responsible. I got a mild round of applause for that, as I'd flattered the egos of some of the pilots! But what I'd said was just the straightforward truth.

Finally, this incessant objection to automation of some of the pilot's tasks by a very small minority didn't take our minds off our objective and accident statistics bear us out on this. Over the last ten years, the new generation of more automated aircraft, whether from Airbus or Boeing, have had an accident rate three times less than previously. So all this gesticulation had no effect on us.

However, another difficulty arose as to whether the same airline could have some aircraft equipped with APTCAS while others weren't? If you take the example of Air France or Lufthansa, with more than a hundred aircraft from the A320 family between them, some of them, including those first brought into service, could not accommodate this system, which was incompatible with their original computers. If it was considered that operating the two versions side by side was unfeasible, then both airlines would have to dispense with using APTCAS. Many other operators faced the same dilemma. The idea of just reserving the system for new buyers was thought to be an option, but then airlines sometimes hire older versions on a temporary basis and these would not have been equipped with the system.

The initial view of the training pilots at the Airbus Training Centre was that having some aircraft with this system and others without it in the same fleet had to be avoided so that the pilots would not have operational procedures that differed from one airplane to another. I personally discovered this argument at a session of the Product Safety Board, the Airbus high level management body responsible for making collective decisions on flight safety issues. I disagreed as, in principle, this would amount to blocking any upgrade or progress in flight safety terms.

After internal discussions, especially with Jacques Rosay and Jean-Michel Roy, who was then Head of the Training Centre, we agreed that if such a situation was clearly not desirable, it still remained acceptable. All you needed was a placard in front of the pilot telling him or her whether the aircraft was equipped with APTCAS or not. Later, someone had a brainwave and it was decided to show TCAS in blue on the Flight Modes Annunciator in TA mode when there is a serious risk and it is ready to come in. This is much the same as what happens when the pilot is authorised to intercept his approach course and the system is shown in blue as being armed and ready to engage, switching to green as it becomes active with the aircraft in approach. With APTCAS, the pilot sees TCAS appear in blue on the vertical guidance mode indicator as it is armed in TA mode. If avoidance mode (RA) engages, the message turns green. If, as TA comes in, there is no blue display, this means the aircraft is not equipped with APTCAS. The pilot must then prepare to react manually as required, maintaining the requested vertical speed. That's how an elegant solution was found to a fleet management problem.

* * *

Let's summarise the operating principles of APTCAS.

When there is a potential risk of collision, in TA mode, the device is armed, with, as stated previously, a blue 'TCAS' message on the vertical flight control modes indicator, above the artificial horizon. If the warning disappears, it disarms.

If there is a RA, the two TCAS coordinate their orders. The TCAS guidance mode engages and the pilot sees it switch to green on his indicator. The autopilot follows orders: level flight, climb or descent at the required rate. Meanwhile, autothrust adjusts the engines' level of thrust to maintain speed.

At the end of the alert, the aircraft comes back of itself gradually to its level flight altitude or resumes its climb or descent according to the initial situation.

When the autopilot is not engaged, the pilot simply has to follow the guidance orders from his Flight Director (FD). The flight path will be exactly the same as that obtained with the autopilot.

Note that APTCAS is capable of following warnings sequenced by TCAS, for example with two successive aircraft. TCAS computes avoidance orders after coordination between the various aircraft and APTCAS ensures guidance for successive avoidance manoeuvres.

• See photo n°19 – page 238

* * *

The final decision to launch APTCAS, not only on the A380, but on all our fly-by-wire aircraft, was made, after all the preliminary studies, on 12 June 2006.

Once the underlying principles have been accepted, the development of such a system remains complex as you have to modify many different computers and sometimes get them to establish a dialogue between each other whereas that had not been initially planned. When the aircraft was designed, for example, it had never been envisaged that the TCAS was to transmit data to the autopilot. APTCAS thus had an impact on the autopilot, TCAS, flight controls, the Flight

Warning Computer, displays and even on the accident data recorder.

There are a very large number of variables to be considered for development and this starts with the 34 possible cases of RA, some of which have different rates of climb or descent. These RAs can also occur in many situations, either at high or low altitude, at any speed and with all flaps deflections. In addition to that, there are the different weights of the aircraft, from the highest on take-off to the lowest, with an aircraft with no payload at the end of its flight, to be considered. Not to forget the various centre of gravity positions between the forward and aft limits. The situation where two warnings occur one after another also has to be validated. All of which greatly increases the number of cases that have to be addressed. Finally, the study of failures, as ever, makes for a major share of development tests. Development of a system that initially appeared simple turns out to be a huge programme. In reality, with some common sense, a very large number of these tests can be conducted on the simulator considering the very high quality of the aerodynamic models available at the end of the aircraft development cycle. This means that once in flight, you can stick to checking those few issues that were considered to be critical.

So, having validated the basic principles on the design simulator, the next stage was to prepare a first version of the aircraft computers, then tests on the development simulator.

* * *

There was no standard set of regulations to certify such an entirely new system. Moreover, using APTCAS introduces substantial changes in the procedures the pilots have to learn. EASA therefore requested a human factors study in order to validate the concept formally and ensure that there were no negative consequences for the crew.

The first session was scheduled for October and November 2007. As we commenced, the A380 integration simulator was not yet available for use as one of the computers was not yet ready. So these tests were conducted on the A320 simulator, given that all the algorithms were the same on the two models. Five sessions were organised, in which Airbus and EASA test pilots, and instructors from our training centre took part. A number of recommendations were made and, overall, the results were considered to be extremely positive.

In September 2008, a second review session was conducted to validate the modifications requested and examine others that had been decided on following development tests. Everything was satisfactory and a few recommendations were made for pilot training.

Alongside these works, although this was not necessary for certification, airline pilots from Air France, British Airways, KLM, Lufthansa, Northwest Airlines (now Delta), Scandinavian Airlines System and United were invited between October 2006 and May 2007. The improvement made to flight safety was unanimously welcomed.

* * *

The first tests on the integration simulator with real computers and the actual aircraft environment took place in October 2007. Just for the flight controls computer (PRIM), no less than 17 versions were needed. That just gives an idea of the difficulties that lie in developing a system that initially seemed so simple.

Eleven flights led to certification, on top of numerous sessions on the simulator. The first test in flight took place on 21 March 2008. Apart from a few minor oddities and settings to be reworked, the results were already extremely satisfactory.

In May 2008, after six flights, most of the guidance problems seemed to have been settled and there remained just a few anomalies in logics on the systems. But never cry victory. In October 2008, we discovered that at low speed, when guidance orders led to too great an increase in vertical speed, the autopilot (respecting the planned logic of things) disconnected all by itself due to the excessive increase in angle of attack. Various modifications were considered. The defect was completely unacceptable and it was inconceivable that a safety system kick out in full swing! The setting for disconnection in angle of attack was reconsidered to set it at a higher value when the TCAS mode is activated. The last flight for adjustment thus took place in February 2009.

The two certification flights were made in May 2009. They were then followed by additional tests on the simulator to check situations that hadn't been encountered in flight.

In mid-August 2009, APTCAS was certified on the A380 and immediately installed on aircraft number 33, first delivered to Air France. Development continued meanwhile for the A320, A330 and A340 as the added safety advantage had clearly been proven.

MAXIMUM ENERGY BRAKING TESTS

Maximum energy braking tests are among the most expensive certification tests, costing three to four million Euros, mobilising some seventeen semi-trailer trucks and obviously a whole bunch of engineers and technicians. The aim is to validate the maximum energy level the brakes are capable of absorbing. It really is quite a spectacular test.

* * *

But before going into this episode, let's go back and take a look at the braking system.

To put it simply, on Airbus aircraft of the last generations, there are three main states. Normal braking is completely optimised. 'Alternate' is a degraded mode, but antiskid is retained. However, it fails to give best performance levels due to the loss of some systems. Finally, in the event of loss of the main part of the computer functions, there will be no antiskid. So, in order to avoid the tyres bursting when the pilot presses down too strongly on the pedals, pressure on the brakes is intentionally limited. The A380 systems are much more complex, bearing in mind that some brakes can remain in normal operation while others are in degraded mode. This makes development more complicated due to the large number of configurations to be tested.

Many of the tests to qualify the brakes are conducted on the test bed. The distribution of loads on each of the wheels (twenty in all on the four main landing gears) means that each brake can be applied differently and, as a result, development of the system can only be done on the aircraft itself. It requires many test flights as braking must be optimised for landing distances to be as short as possible. In addition, for a given weight, the runway length needed for take-off will depend on braking capacity as one of the sizing features is the ability to stop the aircraft at high speed in the event of a failure.

It should be noted that for the A380, braking efficiency is not as critical as it can be on other aircraft. The reason for this is that the landing and take-off speeds are considerably lower than on other jumbo jets. Now, as it uses the same runways, it does not need to have such a significant deceleration capability. That's why, a few years before the first flight, the brakes were removed from the two rear wheels of each of the two wing landing gears each equipped with six wheels. This made for a valuable reduction in the aircraft's weight. It was planned to install them again on heavier versions and in particular the A380F, in cargo version, that was to have a higher take-off weight (about 600 tonnes). However, this model was later abandoned.

There are various braking modes. Assuming the aircraft is extremely heavy, it may arise that the pressure of the pistons on the brake fails to block the wheel as it is being pressed firmly against the ground and driven to rotate by the forward movement of the aircraft. Braking capacity will then be limited by the torque applied to the wheel and antiskid will be inactive. Conversely, if the aircraft is very light, the wheel will not be held against the ground with the same force. The brake pistons actuated by the hydraulic pressure will then be capable of blocking the wheel and antiskid will be needed to avoid locking. Development of the braking system essentially involves work adjusting this antiskid function so it maintains an optimal ratio between the wheel sliding along the runway and its rotation.

Tests on take-off are conducted using acceleration stops, meaning that the aircraft is accelerated up to a certain speed, then the engines are reduced to minimum thrust and maximum braking is applied, either pressing the pedals, or using the automatic system that kicks in as soon as the throttles are brought to the idle position. To avoid any damage to the brakes and wheels, we perform these tests at weights and speeds such that the energy imparted to the brakes does not exceed half that which they are theoretically capable of absorbing. The maximum energy demonstration is made at the end of development as it is destructive.

For landing tests, maximum braking is applied as soon as the main wheels touch the ground. The nose gear will not yet be in contact and, due to braking, it will tend to descend very quickly. This movement has to be controlled with some skill as, firstly it must be brought down to the ground as soon as possible for braking to be more effective and secondly, contact must not occur at too high a vertical speed in order to avoid damage. Tests on the A330 and A340 family were delicate due to the nose landing gear being far from the ground on touchdown of the main wheels due to the nose-up attitude of the aircraft on landing and the geometry of the main landing gears. The A380 is different and the nose wheels can make contact very rapidly after the main landing gears.

To demonstrate optimal performances (the aim of any aircraft manufacturer), braking should be done on a part of the runway not contaminated by the rubber left by other aircraft. This meant we landed a good third of the way along the runway, on the part that is cleanest, as the point at which the wheels of commercial aircraft hit the ground is well before and the deposition of rubber comes from this moment of contact rather than from braking itself, which is generally fairly gentle. No safety issue arises here as, with maximum braking, you only need about 2,000 ft to stop the aircraft on a dry runway. As we soil the runway pretty quickly with our own rubber due to the slight slippage of the tyres on maximum braking, rubber clean-up operations have to be conducted at regular intervals, these being especially useful just after development, when we start measurements needed to determine the braking models used to compute performances on take-off and landing.

To be certain that performance will be optimal, the touchdown must be fairly firm in order to ensure immediate rotation of all the wheels. The aircraft must remain perfectly centred laterally to avoid a landing gear coming onto the white line going down the runway axis, which would reduce braking power (much the same as a car brakes less effectively when its wheels are on the markings on the road). Finally, the stick must be released as soon as the nose gear is on the ground to be sure that all the weight lies firmly on the landing gears and that there is no additional lift due to a residual deflection of the elevators. To avoid this, someone (generally the engineer) has to call out "stick" as the nose wheels touch, to remind the pilot to release it.

One of the difficulties with these tests was that each of them led to heating of the brakes, which can be detrimental to continuing flight. To quickly cool down the brakes and wheels, all our test aircraft are equipped with a fan on each wheel and brake. On long-haul series aircraft, this is an option that is very rarely chosen by our customers as it leads to big increases in weight and cost, while the flight stops are sufficiently long to ensure cooling. In order to avoid losing time in coming back to the apron to bring the temperature down, procedures were developed to sequence tests. In principle, if at the position we stopped at after a test (generally towards the middle of the runway), we can take-off again safely with the possibility of stopping the aircraft again in an emergency, then we start off again from that point with maximum thrust. Otherwise, we come back to the start of the runway. In all cases, after take-off, the landing gear is kept down to speed up cooling. We then return to the runway for the next landing when we reckon the maximum temperature of the brakes will not be higher than about 100°C when the wheels touch down.

All these tests have to be made when the runway is dry, but also when wet, as the setting can be different and you have to find a trade-off between the various situations. Finally, when it's pouring down with rain, we try to perform tests on what we call a 'contaminated' runway, meaning with some inches of water. A good deal of luck is involved in finding the conditions for this type of test: you just have to be at the right place at the right time.

Here again, developing a normal braking system is not all there is to it. You also need to cater for 'degraded' modes and determine, where necessary, the penalties on landing distances that will have to be applied when brought into use.

Generally, no major problem cropped up in developing braking for the A380, even if finding the best compromise for entry into service required a lot of work. After certification, other tests were conducted to further improve performances.

* * *

Let's now return to the maximum energy demonstration. On the A380, the brakes were calculated to absorb energy of at least 119 MJ (Mega Joules) and this theoretical value, as already validated in the laboratory, had to be confirmed.

This certification test involves going through a very precise scenario.

First of all, all the brakes have to be worn to the extent of 90% and so, during previous flights, braking was adjusted for this condition to be fulfilled. For the test itself, the aircraft has to taxi for three nautical miles making three stops, this being representative of taxiing before take-off. It is then accelerated up to a certain speed before the engines are reduced to idle and maximum braking applied, without using the thrust reversers. This speed is calculated to adjust the energy imparted to the brakes considering the wind, the residual thrust of the engines and the drag of the aircraft. Having come to a halt, even if there is a fire on the wheels, it must remain contained without spreading to the rest of the aircraft for five minutes. That marks the end of the test and the firefighters can then move in to cool down the wheels and brakes.

This test does not involve any particular difficulty in performance. The pilot responsible for speed reduction has a big screen on his knees that instantaneously gives him the ground speed (the on-board instruments always suffer from a very slight delay due to the time taken to compute and display the data). Indeed, there need to be no mistakes: one knot less and maximum energy will not have been demonstrated, and one knot too much may well be too much for the brakes! In fact, as there is a small safety margin, we prefer to be sure of making a success of the test and so the start of reduction commences by just exceeding the target speed, which leads to a slightly higher demonstrated energy.

* * *

Traditionally, this test was made a few weeks before certification. We had never encountered problems until that made on the A340-600 in February 2002. Let's come back to that interesting event. The emergency stop went well, but about three minutes after the stop, just as the tyres usually start to deflate, the wheels exploded, projecting debris around the aircraft. Luckily, no-one was hurt. What had happened? The sizing of the wheels was incorrect and the fuses mounted on the tyres that let them deflate when the temperature is too high hadn't had the time to react.

The A340-600 was seriously damaged as the fire spread to the main beams of the fuselage made of carbon. After a few days for repair on site, we ferried the aircraft back to Toulouse, at low speed with the landing gear down after which it took a month and a half to repair it.

We couldn't take the same risk with the A380. Obviously, we had taken what we had learnt for the sizing of the wheels and fuses into account, but another surprise was still possible. For certification, we initially validated 89 MJ, well below the maximum limit and therefore risk-free, and that was more than enough for the aircraft being used for our endurance flights and even for the first commercial flights, considering the routes planned by customer airlines over the first months.

One thing the A340-600 test did show was that there were risks for the firefighters in the event of explosion. So, after lengthy discussions to find a good solution, we modified the two firefighter trucks with hoses in a low position to spray the landing gears without the firemen having to get out of their cabin. These two trucks formed part of the convoy on a semi-trailer.

* * *

The only runway in Europe suited for such a test is that at Istres as the distance for emergency stop is a little greater than 13,000 ft. Airbus had financed the extension of this runway in the early 1990s for tests on the A340-300 taking it from 12,465 to 16,410 ft.

The weather conditions need to be good but the long term forecast also has to be satisfactory. Indeed, all the equipment has to be set up some time in advance. The 'launch window' became clear towards the end of February 2007. It appeared that the first week-end of March would be favourable with light wind along the runway axis, and there was to be no rain. So everything was got ready, bearing in mind that the decision to fly the aircraft out to Istres would be made at the last minute.

The chief organiser for the entire mission was Robert Lignée, who had been the flight engineer in charge of all braking tests for quite some time. It was up to him to set up and coordinate all the resources used. Jacques Rosay and I were to be the pilots. We hadn't planned for both of us to take part as, traditionally, we take advantage of this type of test (a relatively rare event) to make up a crew with one experienced pilot and a younger one who had not yet done it. But the communication people decided otherwise as they wanted the two pilots of the first flight to be filmed. So there was no refusing! Gérard Desbois, the test flight engineer and proud 'owner' of aircraft # 1 obviously didn't want to miss out.

On the morning of Saturday 3 March, the decision was made to go for a test on Sunday morning. This posed a few problems as some staff weren't entitled to work on Sunday, except with prior agreement from the labour inspectorate. But in our job, that depends so heavily on meteorological conditions, not everything can be planned in advance. We couldn't miss such a good opportunity that was likely not to come again before some days or even weeks to satisfy a regulation that was so unsuited to our way of work. Indeed, the next day, the weather took a turn for the worse and wouldn't have been suitable for this test. But everything turned out for the best and the authorisations finally arrived.

The aircraft was weighed before departure from Toulouse with full fuel tanks as needed for the ferry flight and to perform the test as there was no system to weigh such a heavy aircraft at Istres. We weighed in pretty heavily, with 590 tonnes on take-off, and the weight on landing at Istres 577 tonnes, above the normal maximum take-off weight and about 200 tonnes above the maximum landing weight, though this was not a record!

On arrival, we conducted a briefing session on the rejected take-off with the crew and EASA representatives only so as not to have to go into too many technical details during the major briefing the following day.

That evening it was free time for everyone and the entire crew dined at a restaurant, aux Heures Claires, at the small port of Istres on the Etang de Berre. After dinner, as we arrived back at the hotel there was an eclipse of the moon and all the photographers and cameramen were out there with their big telephoto lenses. It was a magnificent sight but we would have to be in fine form the next day so for us the show came to an end as the moon became almost completely obscured and we retired for the night.

* * *

Sunday morning, at 6:45 a.m., a major briefing session was held in one of the classrooms made available by EPNER. The main issue was to ensure good coordination with the teams outside the aircraft, including firefighters, photographers, technicians and mechanics.

The test thus started with taxiing over three nautical miles and three complete halts. In reality, we made four stops. This was because we also had to cover the case of aircraft equipped with Engine Alliance engines that had slightly greater residual thrust on idle, which heated up the brakes more before acceleration. It goes without saying that we were hardly keen to repeat the test! To avoid having too high a temperature on the brakes, the trick was to stop just after a turn as, with the bogies, the friction of the tyres in turning is such that it slows the aircraft down. This was not forbidden by the regulation, so why not take advantage? Just as we were lined up to start the acceleration, the temperature of the brakes was around 110°C.

Considering the adjustments due to the weight and the wind, the speed chosen for reduction was 166 kt. After a nominal acceleration, the engines were set to idle at 167 kt and the maximum speed reached was 168 kt as deceleration of the power plants is not instantaneous. We had chosen to use the autobrake function that automatically gives maximum braking rather than push down on the pedals. As from the start of deceleration, Robert announced to us that braking performance was very slightly better than expected. He regularly confirmed this throughout deceleration. Next, the stop was maintained for a few seconds to validate the performance.

Next came a phase that aroused some apprehension in me. We had to free the runway rapidly before the tyres deflated, otherwise the manoeuvre could become somewhat delicate and, for that reason, good initial thrust was desirable. But with too high a speed, as you again stop the aircraft, you will be adding energy to the brakes. Nobody could tell us exactly what was going to happen. So, we accelerated up to 30 kt, then decelerated using the thrust reversers (this was authorised as the braking test strictly speaking was over). About 90 seconds later, we were in position on the apron to the north of the runway. The tyres started to deflate one minute later thanks to their fuses. The brake temperatures shown to the pilots had been on their upper limit of 1000°C for some time. The suspense lay in seeing whether we could keep to the five fateful minutes after the end of emergency stopping.

Finally, Robert Lignée blew the final whistle for the end of regulation time and it was a winner. The firefighters could now intervene with the fire hoses and spray the wheels, with two units being devoted to each landing gear. This lasted about 1 hour 15 minutes, with eight fire-engines and three trucks to supply them with water. One of the issues that we were concerned about was the temperature of the wheel axles. If this exceeded 300°C, then the axle would have to be replaced. The hottest one did not exceed 290°C. This was pushing things to the limit, but at least we avoided having to go through the lengthy technical intervention of replacement. Following this test, the axles were modified to simplify and shorten maintenance work after high energy emergency stopping.

• See photo n°20 – page 239

* * *

We were coming up to midday and time for a snack alongside the aircraft on what little grass the parched landscape of the Plain of Crau had to offer!

For our return, I had Sophie Rosay, Jacques' daughter, who had just qualified as a pilot, fly my light aircraft, a TB20, over to us. As she was not qualified on aircraft with retractable landing gear, she was accompanied by an instructor. Having finished off our meal tray early in the afternoon, we flew back to Toulouse, with Jacques and I to the rear of the TB20. I think that was the first time I travelled in that position on that aircraft!

Meanwhile, the A380, weighing some 500 tonnes after some tens of tonnes of water ballast had been evacuated but leaving all its fuel inside, was jacked up that afternoon. Emptying fuel is a lengthy business and you also have to provide the tanks to store it in so it was best to try and avoid doing so. The sixteen wheels equipped with brakes were replaced the same afternoon and into early evening. Next morning, the aircraft was brought back to Toulouse.

Once all the calculations had been done, 121 Mega Joules had been demonstrated without damage to the aircraft. A terrific success!

THE NEW AUTOMATIC BRAKING SYSTEM

All Airbus aircraft allow the pilot to use automatic braking with a choice of several levels of deceleration. The maximum value is only used in a rejected take-off, with the system being armed before the aircraft goes onto the runway. If the pilot decides to stop during the take-off roll, maximum braking engages automatically as soon as he sets the throttles to idle once beyond a certain speed. Thus, the stopping distance is reduced to the strict minimum without taking the risk of having a delay due to the crew's reaction time or even inadequate braking as had already been seen.

On landing, the main advantage of automatic braking is in reducing wear on the brakes thanks to the regularity of deceleration. This is especially important when the discs are made of carbon as is nowadays the case on the great majority of aircraft. For the same reason, passenger comfort is enhanced. But there is a downside in so far as it is not optimised to vacate the runway on the desired taxiway as it works in 'open loop', with constant deceleration at the value set by the manufacturer (although the airline can still choose a slightly different rate, better adapted to its network). Thus, taking the aircraft slowly through almost to standstill, it can lead the crew to having to take a fairly remote runway turn-off. That said, the pilot can always disconnect the automated system to set speed and braking himself.

* * *

When these automated systems were designed, navigation wasn't as precise as it is today. GPS has come in and finally we changed everything as we moved from accuracy to within several hundred feet and sometimes more when there were no radio navigation beacons, to just a few feet. You can now determine your position on the runway exactly, just like in your car.

Fabrice Villaumé was a young engineer who between 1998 and 2001 worked on his thesis on possible improvements to control of the aircraft on the ground. Among his main conclusions was the idea that it was possible to develop smarter braking systems with the idea that we could certainly do better than what had been done in the 1970s and as repeated in practically the same way on all our models.

The decision to take matters further to demonstrate a concept with flight tests on the A340-600 was made in September 2001. At that time, no-one knew exactly how it could be done and there was no formal commitment to install a new braking system on any aircraft. Research and experimentation were the issues, and this phase required a fine level of judgement. Instead of launching a study as conventionally in a Design Office, with, let's admit it, the usual cumbersome administrative procedures of big companies, a small 'commando' task force of just three people was set up. This included Fabrice Villaumé of course, a flight test engineer, Robert Lignée, experienced engineer and specialist in braking and performances, and Armand Jacob, experimental test pilot who also had the advantage of considerable experience as a flight test engineer. This small team was there to guide all research.

The initial objective was to create a deceleration system that rapidly brings the aircraft down to a low speed close to the taxiway the crew wants to take. The time the runway is occupied is also reduced as strong braking only starts on approaching the turn-off, dispensing with several seconds of low speed taxiing on the runway. This means the pilot can either land close to the threshold of the runway and braking can be delayed, or touchdown of the wheels can take place well after the threshold and braking can be immediate and much sharper. If all aircraft were equipped with this system (wishful thinking maybe!), this would allow for greater runway landing capacity. An additional advantage is in reducing the energy absorbed by the brakes and having a positive impact on wear.

One meeting after another was held to come to a consensus as to how to install such a device on an aircraft. Alongside this, sessions were conducted on the design simulator to devise a system that can readily be used by crews, before fitting it out on our development A340-600.

* * *

This demonstrator's first flight took place in April 2004. Up until July 2007, the date of the last tests on this aircraft, five major iterations were made, in addition to which there were ten software versions. In total, 21 flights and 120 landings were performed with the A340-600.

A decision had to be made fairly quickly as to installation on the A380, without waiting for the end to experimentation. So, in March 2005, as development had moved on apace, a demonstration flight was organised in Toulouse and Paris-Charles-de-Gaulle. On board the A340-600, there were the A380 Programme Manager and a number of Chief Engineers from other Airbus programmes. The two test pilots were the Chief Engineer of the A380, Robert Lafontan, and Armand Jacob. In the cabin, the passengers could follow the landings on video screens that showed the runway together with the data shown the pilots. Three landings were made during this flight, two in Paris and one in Toulouse. In addition, those landings at Charles-de-Gaulle enabled us to test the high-speed taxiways approach, where you can quit the runway at higher speed, for the first time. For the tests, we had chosen a braking disconnection system speed of 50 kt, whereas it is just 10 kt when the taxiway is perpendicular to the runway. The decision to go ahead and install this automatic system on the A380 was made immediately once the flight was over.

The next stage was to convince our A380 customers as the system was an option and the best way was to get them to fly on it. Even if many things can be seen on the simulator, in flight operation is much more spectacular. The first demonstration flight took place in August 2005 and it goes without saying that the pilots were immediately enthusiastic.

This preliminary development was not always an easy matter. One of the first difficulties encountered was due to the fact that, at the time, there was no airport with adequately coded data on runway and taxiway coordinates in the databases available for navigation to answer to our test requirements. Furthermore, Google maps did not cover France when we started tests. We therefore had to do this encoding ourselves for the four aerodromes we were likely to use: Toulouse, Istres, Tarbes-Lourdes and Paris-Charles-de-Gaulle. This encoding was tricky as the slightest error or shift in the runway position could prevent the system from functioning normally, considering all the protections that were installed.

That reminds me of a story. We discovered a bug by pure chance. It turns out that the Greenwich Meridian runs exactly through the threshold of runway 20 (facing south) at Tarbes airport, this marking the 'border' between the east and west longitudes. Inside the computers, this leads to a change of sign of certain parameters as you cross that line. On landing, the system switched over to safe mode as the meridian was crossed since, having seen the sign reversed, it saw the aircraft moving away from the runway axis at great speed.

Sometimes, you have to turn into something of a DIY specialist to get things done. For example, we wanted to use the runway at Perpignan for some tests but, here too, there was no validated database. As the A340-600 was not available, Fabrice Villaumé, who was preparing his qualification on a TB20 (a single engine airplane with retractable landing gear) at the Airbus aeroclub, decided to go and do this calibration himself with that aircraft during a training flight. He therefore set off with his instructor for Perpignan and covered all the taxiways as also the entire length of the runway to validate all the geographical coordinates with some dedicated instruments. Everything was satisfactory and tests of a wet runway could start the following week with the A340-600 and no need for further verification.

Finally, with the tests on the A340-600, we laid down the basics for what we now call Brake To Vacate (BTV).

* * *

Clearly, such an automatic braking system cannot be installed without taking many precautions as you have to be absolutely certain that it will not lead to you coming off the runway. If it detects a risk while the aircraft is still in flight, the crew must be warned in time to perform a go-around. Once on the ground, if the computer detects the hazard's continuing presence, BTV must immediately warn the pilots and call on them to take appropriate actions. In some situations, BTV can act directly, for example by immediately applying maximum braking. That's how the team realised that what it had done in practice is create a system to give warning of risks of going off the runway, to prevent them and also reduce their consequences.

The causes of accidents have changed considerably over the last thirty years. Collisions with terrain have been well-nigh eliminated thanks to the GPWS and the EGPWS (Enhanced Ground Proximity Warning System). The number of losses of control has also been considerably reduced on aircraft of the last generation equipped with fly-by-wire and sophisticated protection systems. Nowadays, the main cause of accident is runway overrun, meaning where the pilot overestimates his aircraft's deceleration capabilities and lands anyway or his landing and braking technique is not up to scratch, or even both of these. Many accidents could be avoided thanks to the systems installed in BTV.

In practice, you don't need to have the entire BTV system to ensure protection of the aircraft against runway overrun. It is enough to just install the associated safety systems. This device was given the name Runway Overrun Prevention System or ROPS and includes a 'warning' part in flight and another 'protection' feature on the ground, with braking and warnings.

* * *

Let's look at how BTV works.

When selecting the runway and taxiway, the pilot displays the map of the airfield on his navigation screen and chooses the runway on which he intends to land. Two small bars, identified as dry and wet, appear along that runway image. They give the landing distances from the threshold of the runway for those two surface conditions (braking is obviously less effective on wet than on dry ground). The pilot can then select the exit taxiway according to those data. If the overall length of the runway is insufficient or marginal considering, for example, tailwind, the weight of the aircraft, etc., the crew grasp the situation immediately. Note too that calculation of the distance implies a touchdown point for the wheels that complies with usual trajectories and includes a small margin to take any scatter in handling into account.

With BTV, during standard operations where there is no safety issue, the start of braking is adapted in relation to the distance between the point where the wheels touch down and the chosen taxiway, with delayed braking if the distance is long or immediate and harder if landing took longer.

As you close in on the runway, the two bars come to life as they take into account the real energy of the aircraft. For example, if it is flying in faster than planned, or above the normal glide slope, they will start to shift up towards the end of the runway. If safe landing can no longer be guaranteed, there will be one or two warnings. The first will relate to the risks of landing on a wet runway, in which case the pilots will be able to read "if wet: runway too short" on their screens prompting them to perform a go-around if the runway is not dry. The second warning corresponds to the instance where landing is impossible, whatever the surface condition. In this case, "runway too short" will be displayed in red on the screen and a loud and authoritative voice will repeat the same message. When the crew hear this, there can be no hesitation and they will be bound to make a go-around. That's the warning part of the system.

• See photo n°21 – page 239

Once on the ground, the only solution is to stop the aircraft in the best conditions. So, if the airplane is still in an emergency situation and BTV has been armed, maximum braking will immediately and automatically be activated. Simultaneously, a voice announcement will ask the crew to bring the thrust reversers out into the position that will ensure maximum deceleration, and the warning will be repeated until this has been performed. It will also ask for these reversers to be kept at maximum thrust until the aircraft comes to a halt. This procedure is unusual as, generally, at a speed of around 70 kt, the thrust must be reduced on idle to avoid the risk of ingestion of the hot gases pushed forward by the reversers into the inlets of the engines which could then suffer damage. But in such an emergency situation, the priority is to safeguard passengers' lives rather than consider the material consequences.

* * *

Computing the required landing distance on the destination airport has never been problematic as the multiplier factors adopted in relation to demonstrated performances are realistic and have been proven over time, meaning there has never been a reason to change them.

However, until recently, crews also had the option of using the performances demonstrated by the manufacturer without adding the least margin for non scheduled diversion airfields. If airlines had authorised them to do this without taking precautions, this could have led to accidents where the chosen runway was just at the limit of the aircraft's possibilities. Indeed, if no serious accidents have occurred in these circumstances, it is because diversions are rare and also as the runways usually chosen are much longer than the minimum value.

Even so, the Airbus flight tests team has always been concerned by this potentially dangerous situation, even if the likelihood of an accident of this type occurring is slight. The general principle in determining minimum landing distance is to use tests and calculations done by the manufacturer to obtain the shortest length possible. Looking at the deceleration aspect on the ground, as explained in the previous section, for this demonstration, as soon as the wheels are on the ground, we apply full pressure to the brakes until the aircraft comes to a complete halt. This technique is far from natural given that airline pilots are not trained to brake while the nose wheels are not yet on the ground. You might be willing to admit, at a pinch, that an experienced pilot will be able to do it on the A320, but on a jumbo, this is completely unrealistic considering the difficulty inherent in the manoeuvre. To give an idea of the scale of magnitude involved, while about seven seconds elapse between the main landing gear and the nose wheels touching down, that lead-time has to be reduced to three seconds to obtain the best performance. This proves that the minimum landing distances demonstrated by manufacturers, in accordance with the regulations, must be used with precaution and that it remains wise to add a good safety margin.

So what to do? Increasing the landing distances on Airbus airplanes would have penalised our customers in daily operations and made them uncompetitive compared with the Boeing alternatives while, in reality, performances are similar. Imposing additional margins for such diversions could only be done through official regulations.

An accident that occurred at Chicago-Midway in 2005 provided a denouement for this situation. Following it, the FAA decided to set up a working group to study the problem. This took the name of TALPA ARC, standing for Take-off And Landing Performance Assessment - Aviation Rulemaking Committee. Airbus obviously played an active part in this team's work, with Robert Lignée as our representative. One of the final proposals was a simplified calculation method, including a certain number of margins, that were reasonable without being over-stringent, to calculate the landing distance at a diversion airfield.

Even before the FAA official documents came out, Airbus recommended its customers take this method on board to calculate landing distances while in flight. As far as the ROPS system is concerned, the algorithms used, initially based on common sense, are compatible with the working group's recommendations and, consequently, they take these margins into account and the crew is warned just before landing becomes abnormally risky. As for other warnings, for example ground proximity or anti-collision, the crew is informed slightly in advance, but its reaction needs to be immediate.

* * *

For entry into service of the A380, BTV development was not the priority. What we were interested in was firstly to certify the basic aircraft. Furthermore, development tests were being pursued on the A340-600 demonstrator with its cohort of minor problems and we wanted to avoid immobilising an A380 to refine the settings. The first BTV and ROPS flight on the A380 finally took place on 17 May 2008.

Whereas we had encountered a few difficulties with the databases for aerodromes at the start of tests on the A340-600, everything was settled at the time of A380 flights as the usual suppliers of such data had prepared coding for most airports. On the A380, the aerodrome map can be shown on the navigation screen with a zoom, just like on modern cars. The names of the taxiways just replace street names!

Setting BTV is a delicate matter. Indeed, on a dry runway, the ideal solution would involve using maximum braking at the last moment. You could thus benefit to the utmost from deceleration due to the aerodynamic forces at work and the thrust reversers and, as a result, reduce the energy applied to the brakes, their wear and the cooling time between two flights. But the pilots and passengers would hardly appreciate being subjected to too brutal braking at the last moment. Moreover, the system cannot know the state of the runway and, if it is wet, maximum deceleration will be significantly less and thus braking will have to start earlier. The system has to take precautions so as to be able to quit the runway at the planned location, whether dry or wet. I should add that, initially, it was decided to prohibit the use of BTV on runways qualified as being 'contaminated', i.e. flooded, snow-bound or slippery. To summarise, braking is applied gradually and the pressure on the brakes is adjusted constantly so that the deceleration profile is reasonable and also allows for the runway to be left quickly by the taxiway the pilot chooses.

One situation where it proved difficult to find the right setting was where the crew selects a turn-off right up at the end of the runway or close to it. Here, strong braking will kick in late and as they see the end of the runway loom up so fast, their instinct may tell them the aircraft won't stop in time. Right from the start, being aware of this difficulty, we had decided to bring the aircraft down to low speed 300 ft before the end of the runway, whatever the selected taxiway. Tests on the simulator showed that this distance was not enough to allay the fears of the crew and was finally taken to around 1,000 ft.

To conduct tests on automated systems and on all ROPS warnings on the ground, we obviously had to exclude any risk of coming off the runway. There was a simple solution for this as the system allows you to artificially modify the position of the threshold and of the end of the runway. Indeed, part of the runway may be out of operation during works or for other reasons and BTV must be provided with such information. Moreover, in the USA, where landing strips cross each other, there is a landing technique called 'Land And Hold Short' (LAHS). This requires you, after landing, not to overshoot a line located slightly before another runway crossing your own strip. That way, one aircraft can be landing on one runway while another takes off from the one crossing it. This technique is not used in Europe where it is considered to be too dangerous. To use BTV and ROPS, the system must be informed that there is a line not to be crossed before the physical end of the runway so the related data has to be injected into the aircraft's computers.

For ROPS tests, the pilots artificially reduce the length towards the end of the runway in their computers. In practice of course, the runway is still available. So, to operate automatic braking and trigger voice warnings, all you have to do is aim at a 'long' landing and then touch the wheels down in a zone fairly close to the end of the invented runway. The crew thus successively obtains "runway too short" warnings in flight and then, after the wheels touch down, automatic maximum braking and the voice request to deploy the thrust reversers. This technique was used not only for development and certification but also for demonstrations to potential customers. Triggering of maximum braking in these conditions always bowls them over!

We wanted to ensure that the whole set-up worked satisfactorily for all cases of landing and especially when the aircraft is extremely heavy. Indeed, as for any safety system, you need to be certain it will be effective in all situations. So, on 22 December 2008, we took off from Toulouse at 603.5 tonnes (a record for the A380) and the aircraft weighed 591.7 tonnes on landing at Istres (also a record, standing at about 30 tonnes over the maximum take-off weight) in order to check the settings. Resorting to the longer 16,400 ft Istres runway rather than Toulouse was, despite rehearsals on the simulator, the best option to reduce risks. One of the pilots on this flight belonged to EASA and was responsible for certification. All results were nominal.

Another story again shows the advantages of doing real tests on the aircraft rather than just relying on partial tests on the simulator. We sometimes flew to Nice for our tests as this airfield has the advantage of having high speed taxiways. During one test, as we were leaving the runway on one of these taxiways, automatic braking came in in 'High' mode, meaning an anomaly had been detected, and stopped us at the entry point to the taxiway. This was due to a new safety feature that had just been installed for future operations in poor visibility conditions that automatically triggers this 'failure' when deviation in relation to the runway axis exceeds 60 ft.

The first presentation on the A380 in flight made to airlines took place on 26 June 2008 for Air France and Lufthansa. Later, on 27 May 2009, we gave a demonstration for the specialised press. All the pilots taking part were highly impressed by the system.

Development and certification called on sixty flights and about 200 landings on the A380. EASA approved our BTV and ROPS systems on 15 October 2009 and they were installed for the first time on aircraft # 33 delivered to Air France at the end of October 2009.

* * *

This system was evaluated in relation to a number of accidents that occurred on Airbus aircraft, using recorded flight data. Let's take the example of the A340 that landed under torrential rain at Toronto and came off the runway at high speed (no injuries thankfully). If ROPS had been installed on the aircraft, there would have been a first "if wet: runway too short" warning shortly before landing and, considering the weather conditions, the crew would have made a go-around. Just after overflying the threshold, given that the speed was high and the aircraft had not yet touched down, there would have been a red "runway too short" alarm and the decision to call off the landing would have been imposed. When the aircraft actually landed, deployment of the thrust reversers took several seconds, whereas it should have been immediate. The repeat sound warning would have called for this action until it had been implemented. Maximum braking would have been applied automatically and immediately (in fact, the automatic system is only available with BTV, and if the ROPS function alone is installed, there will be an aural reminder to apply maximum braking). So, supposing the aircraft had been equipped with the ROPS system and in flight warnings had not been heeded, but that on the ground the crew had obeyed orders diligently, the speed of runway overrun would have been much lower. Clearly then, ROPS means this type of accident can be avoided or at least considerably reduces the consequences.

Having analysed other such accidents where we had all the technical detail available always led us to the same conclusion - they could have been avoided thanks to ROPS and, even if the crew had not followed all the orders, the consequences would have been less serious.

As the main cause of accidents nowadays is runway overruns, Airbus is gradually going to make the ROPS system a basic feature on all fly-by-wire aircraft. The decision to do so was made in 2010 but several years were needed for all Airbus airplanes coming out of the plant to be equipped, considering the lead-time for development on the many versions of aircraft in production. It will also be possible to install it on aircraft already in service.

Installation of BTV is more complex and, unlike the ROPS system, its impact on safety is limited, its role being more to provide assistance to the crew. Development for existing aircraft is therefore planned, but this will remain an option.

Other manufacturers have since stated their interest in the patents filed by Airbus and discussions are under way for adaptations on certain types of aircraft. You can well imagine such a system or a similar device developed by a competitor becoming compulsory on transport aircraft in the long term. This is what happened with the Ground Proximity Warning System, as first proposed by an equipment manufacturer, and that then became mandatory on transport aircraft at the end of the 1980s. The same applies for the TCAS anti-collision system that has been imposed for the last fifteen years. So why shouldn't having a ROPS system on board become mandatory within a few years time?

THE ROLLS-ROYCE FLYING TEST BED

The A380 can be equipped with Rolls-Royce or Engine Alliance engines. The latter are designed by a combined outfit bringing in two of the major transport aircraft jet engine manufacturers, General Electric and Pratt & Whitney.

The engine is certified independently of the aircraft after a whole series of tests on the ground. However, a host of in flight checks are required as the aircraft power plant interface plays a major role operationally. For example, the quality and optimisation of cooling will depend on the design of the nacelle. You really need to go through the validation process in an environment that corresponds exactly to that of installation on board the aircraft. Any special features of the aircraft can have a decisive influence for development, whether this relates to fuel supply, air intake, oil and fuel cooling, or hydraulic and electrical systems.

In order to detect any potential anomalies early enough, a flying test bed is frequently used to commence checks several months before the aircraft's first flight. This means corrections can be made, if possible before the airplane takes to the air and, otherwise, before the start of certification testing. Taking off for the first time with an aircraft whose engines have not already been tested in flight would require a host of precautions to be taken as the uncertainties relating to the aircraft would be compounded by those due to the engines and, above all, this could have an especially negative influence on the schedule for development due to additional problems mounting up. In the case of the A380, the Rolls-Royce Trent 900 engine was installed on the first development A340-300 that had kept the entire original test installation. Tests were conducted by Airbus, in co-operation with Rolls-Royce. As for the Engine Alliance power plant, it was tested in flight under the wing of a Boeing B747 at the General Electric test centre in Victorville in the Mojave Desert in California.

* * *

Before going into the detail on the tests performed, just a few words to explain how a jet engine works may help the less informed reader get used to the terms that appear throughout this section and in that devoted to development of the engine on the aircraft.

Modern jet engines comprise two or three rotating assemblies, each having a compressor and a turbine part. Rolls-Royce engines have three of them and Engine Alliance jet engines two. The combustion chamber is located between compressors and turbines. The fuel is injected into it and burnt with the air coming out of the compressors. This flow of hot accelerated gas drives the turbines and thus the compressors linked to them. It is then channelled into the exhaust jet pipe, thus producing part of the thrust.

Each turbine or compressor may include several stages of rotors (rotating blades), before which are to be found the stators (fixed blades). The very large diameter fan with its many splendidly curved blades is the first stage of the compressor (acting much like a propeller) and part of the air compressed by it avoids the combustion chamber to be channelled and accelerated to the rear, thus ensuring the main share of the thrust. This is the cold flow, while the hot flow is that going through the combustion chamber. Such modern power plants are thus referred to as bypass engines or turbofans.

One especially important parameter in how an engine works is the Exhaust Gas Temperature (EGT) as measured between two turbines.

All engines are now equipped with a system known as FADEC (Full Authority Digital Engine Control). This is a digital computer responsible for all engine operating and control functions throughout the flight envelope, as with adjustment of maximum thrust and idle, control over acceleration and deceleration laws, start-up on the ground and in flight and checking the various regulation valves, according to the altitude, speed and outside temperature. All FADECs incorporate two independent computers to ensure redundancy.

Development of the FADEC is initially conducted by the engine manufacturer on different test benches on the ground but, as soon as the engine is brought to fly, surprises are aplenty and the flying test bed allows a review of the various functions to be made and so prepare the necessary corrections.

The A380 can be equipped with engines of a maximum thrust going up to 80,000 lbs (36.5 metric tonnes). The customer may choose engine power from within a certain range between 70,000 and 80,000 lbs. This involves exactly the same engine in all instances but with a different FADEC setting. The engine obviously wears faster if the crew uses high thrust levels more often. Furthermore, the purchase price for the airline will depend directly on that value. In practice, our first customers who chose the Rolls-Royce engine kept to the minimum required thrust they needed, that is 70,000 or 72,000 lbs according to their routes (72,000 lbs for Qantas on the Australia-Los Angeles trip).

* * *

Installing a new engine on a flight test bed aircraft poses problems as the aircraft design does not take into account the special characteristics of the engine and adaptations therefore have to be made. For safety reasons, it is preferably installed on a four-engined airplane and placed in the inboard position so as facilitate handling when we make the inevitable sudden variations in thrust during tests.

• See photo n°22 – page 240

The Trent 900 has a much greater diameter than the original CFM56C on the A340-300 and, as a result, the nacelle was very close to the ground. To reduce risks of it touching the tarmac on take-off and landing, especially with crosswind, the shock absorber on the side of the engine being tested was overinflated. In addition, the pylon bearing that engine was tilted upwards by a few degrees.

Other precautions were also taken. Thus, as the Trent 900 was much heavier than the CFM56C, we had to fly with dissymmetrical fuel distribution in the wings (by about 6 tonnes) to compensate.

A number of dedicated systems were installed. Take, for example, the case of an electrical test bench (looking something like a big radiator) to which the new engine's generator was connected, as the latter's characteristics made it impossible to connect it easily to the aircraft network. Now, we wanted to be able to perform tests with high power demands.

The A340-300's original engine has a thrust of 34,000 lbs (about 15.5 tonnes) while that of the engine undergoing testing could have been taken to 80,000 lbs – quite a substantial difference. Indeed, the Trent 900 was voluntarily limited to 70,000 lbs (about 32 tonnes), the minimum value planned for certification of the aircraft, so as to limit handling difficulties. As tests were not made at maximum possible thrust, a certain number of them had to be resumed on the A380, but the verifications already made on the flight bench meant we saved a great deal of time.

The procedure to apply power on take-off was quite special. We wanted to be able to display full thrust of the Trent 900 as soon as possible for the tests, but obviously without taking any risks. That meant that we had to be at all times able to control the aircraft in the event of failure of one of the engines. Thus, as we came onto the runway, the throttles of the four engines were pushed forward so as to obtain a thrust greater than idle but still relatively low. Then, at the start of roll, towards 40 kt, the throttles of the three engines other than that of the CFM56C adjacent to the new engine were set to the front stop. This allowed full thrust of the Trent 900 to be obtained towards 60 kt. The CFM56C engine located under the same wing remained at low power to limit the dissymmetry. If the Trent had had a failure at low speed, all you would have had to do was reduce all the engines and stop the aircraft. Above V1 (which is the decision speed, as explained in the OPTIMISING THE AERODYNAMIC CONFIGURATION section), in the event of the Trent 900 stopping, the procedure was to immediately push the throttle commanding the CFM engine forwards to return to a configuration already known on an A340-300 with one inboard engine failed and the others on full thrust.

Unlike the usual flight test scenario, an additional workstation is installed for an engine manufacturer engineer (here Rolls-Royce) enabling a high level specialist to intervene directly on specific functions and, for example, make adjustments in flight on the characteristics of the FADEC. Thus, you can compare how the engine runs with different settings and choose the most appropriate regulation laws.

All the tests on the engine bench were run by Jacky Joye (see the chapter on MY FELLOW TEAM WORKERS) and by Emanuele Costanzo (aka "the Sicilian" due to his origins) a young engineer who had worked a lot in the engine field before becoming a flight crew member.

* * *

What can you do on an aircraft engine test bed?

Firstly, you cannot perform development on the entire power plant installation as integration on the aircraft is not identical on a number of counts. Interfaces like the fuel and electrical circuits, hydraulic system and bleed air can be different. You therefore need to concentrate on everything that is specific to the engine and its nacelle, and also ensure, as soon as you can, that there is no major problem that could prevent certification and entry into service within the planned lead-times.

The main tests we were able to conduct on this flight test bed are described below. The fact that validation was done on the flight test bed does not always mean there was nothing else left to do on the aircraft. The computers are subject to modifications between the two series of tests. Similarly, the engine's definition evolves between installation on the flight test bed and the start of flights on the aircraft, either because difficulties have been encountered on the bench, or quite simply because development is being pursued by its manufacturer for any number of reasons: improvement in performances, in reliability, operational characteristics, maintenance, etc.

* * *

Flight tests of the Trent 900 started on 17 May 2004 nearly one year before those of the A380. In total, 49 flights were made in two distinct campaigns. The first was completed on 10 August 2004 after 33 flights. The second, that took place from 25 April to 14 June 2005, was almost exclusively devoted to natural icing tests. Indeed, considering some mishaps on previous programmes, it had been decided to perform these engine tests on the engine test bed. These are described in the ICING chapter.

During the first two flights, the priority was not to try out the engine but to ensure there was no safety problem anywhere in the range of speeds and altitudes in which the engineers intended to pursue testing. We first checked that there was no risk of flutter and then that there was no particular difficulty in manoeuvrability. Obviously, as this flight test bed was only used by experienced test pilots, limitations that do not prevent you from performing tests would have been perfectly acceptable. The goal was not so much to have a great aircraft, but to be effective. Time-wise, five flying hours were needed to validate this useful flight domain.

* * *

Relights in flight are among the tests that can be conducted on the flight test bed. Indeed, the conditions for relight en route, meaning starting the engines again in flight at high altitude at different speeds, only depend to a minor extent on the aircraft, its sensors and systems and on-board installation. What mainly count are the engine characteristics. Thus, final validation can be made on the aircraft to be certified, without having to repeat all tests.

One of the tests needed to establish relight procedures involves determining values for autorotation of the high pressure compressor in relation to speed and altitude when the engine is shut down. Indeed, a jet engine can be relit in flight without mechanical assistance from its starter on condition that the high pressure compressor is 'windmilling' at a minimum rotation speed, this being dependent on the aircraft's speed being high enough. So, as a matter of priority, we have to determine the aircraft's speed in relation to the altitude at which autorotation is obtained. This engine's rotation speed, known as 'N3' (rotation of the Trent 900 engine high pressure compressor) was set at 7% (100% representing maximum speed). For this test, the engine is shut down at high altitude and very slow descent is commenced while adjusting the thrust on the three other engines. The speed must be held with considerable accuracy. Once the engine rpm have been stabilised at the target value, the speed is increased by 10 kt to seek another stabilisation point at a slightly lower altitude. You thus obtain altitude and speed couples where the N3 rotation speed is 7% with the engine shut down.

• See photo n°23 – page 240

As this relation between rotation speed, altitude and aircraft speed is known, a flight envelope can be validated where the engine restarts simply by letting the fuel enter the combustion chamber with the sparking plugs ignited. Verification is made using relights at different altitudes and speeds, bearing in mind that in general, the most critical point is that with the highest altitude and the lowest speed. The altitude targeted for the A380 was 30,000 ft, as on other programmes. The minimum speed was to be set from the autorotation characteristics.

For lower speeds, when the N3 autorotation speed is below the minimum value, relight in flight is performed thanks to the air starter that drives the engine's high pressure compressor in rotation as for start-up on the ground. The pressurised air is taken either from another engine or the APU. The sparking plugs are brought into operation and injection of fuel starts once a certain N3 rotation speed is reached so as to have adequate air flow in the chamber. Here, the minimum speed to relight the engine is generally low, being close to the landing speed.

To ensure re-light in good conditions, you can play on the various parameters involved, including the moment when the fuel is injected, its flow rate in relation to the engine rpm, speed and altitude, the rotation speed at which the start valve will close, etc. All of this is adjusted through the FADEC. The increase in rotation speed has to be reasonably fast and not lead to too high an EGT. Sometimes trial and error is involved in optimising these relight characteristics as you go through a number of tests with different settings. At the end of development, all these values are memorised in the computer that will then determine by itself whether relight must be performed in autorotation or using the starter according to the flight conditions.

The maximum altitude for relight in autorotation found after flight tests on the test bed was indeed 30,000 ft. However, considering a few failures encountered on the aircraft, the maximum altitude finally guaranteed was set at 28,000 ft with a minimum speed of 250 kt. For start-ups with assistance from the starter, the maximum altitude is 20,000 ft as on other Airbus airplanes and minimum speed 150 kt.

* * *

Validation of the engine acceleration time was also performed on the flight test bed. You have to demonstrate that, in landing conditions, and up to the maximum altitude at which airfields are to be found (that is 14,000 ft, a value also set by the regulations), the acceleration time between idle and a thrust allowing go-around to be ensured safely in all cases does not exceed eight seconds. That thrust is practically equal to that on take-off. This criterion results from a trade-off: firstly, the engine has to be able to rapidly give all its thrust and secondly, a high engine idle speed leading to a high residual thrust that would reduce the acceleration time, would prevent the aircraft from decelerating readily in approach. As a result, this regulation requirement is generally respected without a margin. If necessary, the engine rotation speed on idle is adjusted. At the end of acceleration, the rotation speed and EGT obviously have to remain within acceptable limits. Finally, the thrust must be stabilised rapidly, without engine rotation speed oscillations. If an anomaly arises, the FADEC settings have to be reworked.

For a number of years now, Airbus has had another criterion relating to the minimum thrust reached after four seconds. Indeed, on an aircraft certified in the last century (!), the thrust remained close to idle for four or five seconds before rising very rapidly. This long absence of thrust, followed by a sudden increase, disrupted the pilot's handling.

But these criteria are not necessarily the only ones that have to be respected. Sometimes such settings are reworked on the aircraft. During verifications on the A380, it transpired that when the pilot or the automated systems commanded an increase in thrust, the engine remained in conditions close to idle for too long. This disturbed operation of the autothrust as the delay could lead to too great a decay in speed. Following on from these difficulties with the settings on the aircraft, Jacky Joye decided, in June 2006, to increase the idle thrust by 20%, meaning a more rapid response could be obtained from the engine. Obviously, we checked that this new setting did not have a significant impact on rates of descent or deceleration with the engines on idle.

* * *

Tests referred to as Lapse Rate Take Off (LRTO) mean you can check the evolution of the engine parameters at maximum thrust for several minutes, just after take-off, especially the rotation speed of the low and high pressure shafts, temperatures in the various zones of the nacelle and above all EGT. These are important to ensure that limits are not reached during take-off with a failure on one engine, with the others being held on full thrust for quite some time.

Climbing takes place at a rate of about 500 ft per minute, which is approximately representative of aircraft performance when it is extremely heavy, with a failed engine. The tests may last from five to ten minutes and must be conducted with the engine being tested 'cold', which is the case on take-off after a long halt, and 'hot' to cover the case of go-around. To perform it, the engine is held at maximum thrust throughout the duration, and the pilot keeps a speed of 150 kt, representative of an operational speed in such a situation. The thrust of the other engines is adjusted constantly so as to keep to the required rate of climb.

Changes in the EGT and other temperatures during the test give you a first idea of the temperature margins in relation to limits. The results are extrapolated by the specialists in order to determine whether they are sufficient in all cases (especially with very high outside temperatures) while also taking into account the wear of the engines over time, leading the EGT to be increased. The aim is to have a margin so as to keep an engine under the wing to be operated even in extreme conditions for many thousands of flying hours.

These tests on the engine test bed were also intended to detect any anomaly as early as possible. Everything went as planned. However, some tests had to be repeated on the A380 as the engine installed on the bench was not in its final version and the modifications under preparation could have had an impact on the results. Furthermore, more critical verifications at higher thrusts absolutely had to be performed and these could only be done on the aircraft.

* * *

I don't intend to go into the detail of all the tests performed on the flight test bed. In addition to those mentioned previously, we might however mention a few others and first of all verifications specific to the engine. With low speed calibration, you can use pressure and flow measurements to compare the engine characteristics with those obtained on tests conducted on the manufacturer's bench on the ground thus providing a reference to evaluate performances, especially the thrust attained in flight.

Thanks to the engine acceleration and deceleration tests (full throttle, idle or intermediate thrust), the fuel flows can be adjusted so as to avoid rapid variations in thrust leading to extinction or stalling (blocking of the air flow). All these tests were repeated, in a lighter version, on the A380.

Some system tests had been performed on the ground in order to pave the way for those on the aircraft itself, including that for taking air from the engine for the cabin air conditioning system, and oil and fuel cooling.

Finally, there were also various measurements made relating to loads at work in the structure and vibrations, either on the engine strictly speaking, or on the engine pylon and its attachment on the wing.

* * *

A 'hot weather' test campaign finalised the first series of tests. This took place at Tozeur, in Tunisia, early in August 2004. The temperatures reached, at around 40°C, gave us a first idea as to how the Trent 900 engine would operate in these difficult conditions. The key issues to be checked out were ventilation of the nacelle and cooling of the oil and fuel systems. The results obtained led to various improvements on the aircraft being prepared, saving valuable time as we came up to certification.

* * *

As can be seen, the tests conducted on the flight test bed were many and varied, not all of them being mentioned above. They prepared effectively for testing of the Trent 900 once installed on the A380. Only some minor problems were encountered.

ENGINE TESTING

In addition to all the tests performed on flying test beds, we had to conduct many development and certification flights for the power plant on the A380. Let's take the example of the Rolls-Royce engine. Between tests on the flight test bed and those on the aircraft, modifications were made and new FADEC versions were developed, making it essential to have additional validations. In addition, on the A380, testing of the Trent 900 engine was performed up to a higher thrust of 80,000 lbs instead of 70,000 lbs on the bench. Some tests cannot be pursued on a flight test bed as the latter will never be sufficiently representative due to the many different interfaces involved (fuel, hydraulic and electrical circuits, air intakes, etc.). The engine is clearly a vital element of the aircraft and so its behaviour has to be validated in all operational situations, something that cannot always be done on a flight test bed. Hence the large number of flights on the A380. What's more, as far as the Engine Alliance engine is concerned, as tests on the flight test bed were not conducted by Airbus, we did not have such thorough knowledge of some of its characteristics, and no certification test had been performed, thus leading to a greater number of flights.

Not all the huge number of engine tests will be considered here. The reader will find descriptions of some of them in sections devoted to other issues including the flight test bed, icing, test campaigns in extreme temperatures, the fuel circuit or in the many development activities. Test techniques on the test bed and on the aircraft are identical and, with a few exceptions, we shall not return to topics already mentioned in the previous chapter, even if they led to flights on the A380.

Tests of the Trent 900 engine mounted on the aircraft, like those on the flight bench, were run by Jacky Joye and Emanuele Costanzo. Those on the Engine Alliance power plant were pursued by Manfred Birnfeld and Jean-Philippe Cottet.

* * *

Did we experience any unplanned engine shut-downs in flight during our testing?

The answer to that is obviously yes! My statistics over a little more than twenty years in transport aircraft testing show that our rate was about one engine stop for 1,000 flying hours. This is well above the usual values in commercial transport (of around one shut-down for 100,000 hours). Indeed, during development, a fairly high proportion of these failures are due to the test installation and many others are due to the 'torture' we inflict on the jet engines so as to define their operating limits. This time, we had three engine stops in flight. Personally, I had two, which flies in the face of statistical probability as I didn't total more than 1,000 flying hours on that aircraft!

The first took place on 24 October 2005 with the Rolls-Royce engine. We set out with Jacques Rosay for a flight devoted to measuring loads on the structure. On take-off, the EGT of engine # 1 was greater by 40 to 45°C to that of the other engines. It gradually went up and, as we climbed to over 10,000 ft to get to the test altitude, it reached the maximum value of 850°. We reduced power of the engine that stalled then shut it down and quickly returned to land. Analysis showed that one of the stages of a compressor had suffered damage and the decision was made to replace the engine.

On 22 March 2006, we had another failure, also on a Rolls-Royce engine. During a flight to test relights at different speeds, the engine ignited, then immediately switched to a degraded regulation mode. In that situation, you need to take precautions during increases or reductions in thrust as the engine is no longer availed of all the automatic protections. Shortly after relight, it cut off by itself and we had to return on three engines. The cause of this was the presence of water in a pressure measurement pipe providing data to the FADEC. The water had frozen as the engine had been held out of operation too long at high altitude and at pretty cold temperatures. Naturally, a modification was immediately implemented.

During a flight for Qantas aircrew training on one of our development aircraft equipped with the Trent 900, stalling of engine # 4 occurred on take-off leading to a return to Toulouse on three engines. This was due to a mechanical problem and the engine was replaced.

Although not as serious as an engine stop in flight, we did have a major problem on the Engine Alliance engine on flight # 11, after it having accumulated about 50 flying hours on aircraft # 9. This occurred on 18 September 2006 and I was again on board (maybe my presence put a jinx on things!). A big oil leakage occurred on an engine on the electrical generator drive shaft bearing (although this didn't require the engine to be shut down in flight). On the ground, inspection of the four engines showed that they were all damaged and the aircraft was kept on the ground for ten days to replace one engine and repair the others. The entire drive system had run through more than 4,000 hours of endurance cycles on a bench on the ground but, despite that, the bearing's design had to be reworked.

Apart from these failures, a number of inspections and detections of anomalies on the ground led to major repairs or engine replacements. Some incidents are recounted in other sections, but not all are mentioned.

* * *

Lapse rate take-off and nacelle ventilation tests, that had already been carried out on the test bed, were resumed on the aircraft and led to a large number of flights being conducted with the same test technique (see THE ROLLS-ROYCE FLYING TEST BED chapter). It is actually easier to run lapse rate tests on a four-engined aircraft than on a twin-engined one where things can turn out to be much more trying physically. Indeed, we generally carry them out at a relatively low weight so as to land again as soon as the test (lasting at most ten minutes) has been completed. On a light twin-engined airplane, with one of the engines on maximum thrust, the rate of climb is much greater, even with the other engine on idle. This vertical speed is thus reduced with full deflection of the flaps, the landing gear kept deployed, the airbrakes deflected, and possibly keeping the aircraft in a turn, to increase drag, whence handling difficulties. Such problems do not arise on a four-engined airplane where the thrust of the three engines not being tested can be modulated to obtain the desired rate of climb and keeping the usual flight configuration for initial climb.

Compared with what was done on the Rolls-Royce flying test bed, we ran additional tests at greater thrust values as we wanted to validate different settings up to 80,000 lbs. We therefore had to ensure that we didn't get too close to limits for EGT, temperatures in the nacelle and rotation speeds of the rotating elements. Unfortunately, during tests in Toulouse we discovered that the 80,000 lbs thrust couldn't be used due to too high an EGT. 'Hot weather' campaigns (see the corresponding chapter) confirmed this result. Modifications were needed for us to proceed with entry into service. This had no short-term impact as the first customers, with Rolls-Royce engines, had opted for 70,000 or 72,000 lbs.

As far as the Engine Alliance engine is concerned, given that the same level of knowledge had not been acquired on the flight test bed, a large number of tests were conducted on the aircraft, at different thrusts. They showed that the EGT margin and ventilation of the nacelle allowed the aircraft to be operated free from limitation with an engine thrust of 77,000 lbs and that is exactly how we proceeded throughout the development and certification tests. In addition, with a view to future developments, thrusts of 82,000 lbs were used during the hot weather qualification campaigns, as well as for noise measurements.

* * *

One of the engine operation checks that have to be conducted is to ensure that fuel gravity feed can replace failed pumps to draw kerosene from the tank. Validation of this double pump failure (triple on some aircraft) normally encompasses two different aspects, i.e. what happens for the tank itself and how the engine will operate. The first issue is examined in the section devoted to the fuel system. However, the situation is not always that simple as accompanying anomalies may disrupt analysis. An example of this was encountered on the A380. Thus, as we shall see in the FUEL SYSTEM chapter, an explosion on the ground led to some modifications being made to that circuit. Engine shut-down when supplied by gravity as experienced at the start of tests continued to occur, and it emerged that they were not due to the engine, but to the fact that the correction made to the fuel system was not totally satisfactory. The simple failure of a seal created a leak that prevented normal operation of the engine. This was revealed after a flight on 21 April 2006, an event that showed the difficulties involved in identifying the causes of anomalies. As it turned out, we had to find the guilty party – was it the fuel circuit or the engine?

Tests have to be conducted with both de-aerated and non de-aerated fuel. With non de-aerated fuel, the air dissolved in it tends to form bubbles due to the lower pressure at high altitudes and these can then disrupt operation of the engine. After about half an hour flying at high altitude, the evaporation process becomes almost complete and the fuel can be said to be de-aerated. You can then proceed to a higher altitude, especially when supplying the engines by gravity. The altitudes you can hold in these two situations depend a great deal on the characteristics of the engine and the fuel circuit and no general rule applies. For example, on an A320, having spent more than half an hour at a ceiling of 39,000 ft, you can stay there with gravity feed. On other aircraft, you have to descend to a lower altitude if you want to avoid instabilities in operation that could even lead to extinction.

When a failure occurs on engine fuel supply pumps shortly after take-off, the fuel will not be de-aerated and the maximum altitude flight will thus be lower. But whatever the authorised altitude, as the airplane will be less than one hour from its point of departure, there is still the option of heading back home. Later, during the flight, once the fuel has been de-aerated, the crew can push on to the destination, on condition that the aircraft's range so allows, as procedures may also require you to choose a lower cruise altitude, thus increasing consumption. The other solution is to divert, with no urgency attached, to an airfield that can be several flying hours away, but where the passengers can be looked after. Whether or not you can pursue your flight thus depends on the flight characteristics with de-aerated fuel and the moment when the second pump failure occurs. In both these situations, you have to provide the crew with procedures for descent to the maximum possible altitude, or to maintain cruise conditions without risking engine shut-down or even engine operating anomalies. Hence the need for tests with de-aerated and non de-aerated fuel to define these procedures.

These tests are impossible to conduct on an engine test bed as the design of the fuel circuit plays an essential role, so everything was done directly on the A380. These tests are easy to perform. All you have to do is cut off the fuel tank pumps at different altitudes when climbing and command engine speed variations between idle and full throttle. When the engine response is abnormal, meaning refusing to increase power for example, this means that the acceptable ceiling for that case of failure has been exceeded and you therefore have to set a slightly lower altitude where the crew can be sure of adequate behaviour.

On the A380, we had to conduct several test series due to modifications to the fuel circuit and the last of these took place in September 2007. With non de-aerated fuel, the results are different for failures on the pumps of the tanks supplying the inboard and outboard engines, with maximum altitude being 6,000 ft for the inboard and 14,000 ft for the outboard power plants. For de-aerated fuel, the ceiling in all cases is 30,000 ft for the Rolls-Royce engine and 29,000 ft for the Engine Alliance engine. Depending on the remaining flying distance and the fuel reserves, that fairly high altitude should allow you to reach your planned destination.

* * *

Nowadays, there is practically only one type of fuel used on commercial aircraft in the western world: JET A-1. This has not always been the case and other types of fuel were formerly available, especially for military aircraft. However, there is still one fuel, TS1, commonly used in the countries of the former USSR. As it is slightly more volatile than JET A-1, we had to check the behaviour of the engines with TS1 as this characteristic is likely to change operating procedures, possibly reducing ceilings for use of the aircraft when using gravity feed. Some tests were therefore needed.

The most critical case for the engine is where the fuel is very hot since this leads to increased volatility. High temperatures are obtained in operations when the aircraft remains out in the sun for a long time in hot countries with its fuel tank fully replenished for flight. Values of 50 to 60°C can be reached on the skin of the aircraft. That explains why ceilings for gravity feed are finalised with hot fuel so as to only have a single procedure rather than limit values according to the temperature that would be overly complex for the crew to manage.

In former years, these tests were implemented at Istres where there was a fuel heating station. However, given the reduced number of new certifications, it was rarely used and finally abandoned. For the A380, these tests with hot fuel thus took place at British Aerospace's Filton airfield near Bristol where they have an installation used for tests on the ground.

We made seven flights from Filton at the end of March 2007 with different levels of fuel. The goal was mainly to check the behaviour of the engine. However, we also seized the opportunity to check certain transfers of fuel from one tank to another by gravity corresponding to a case where the pumps that normally do the work have failed. Nothing new was discovered in this respect.

For engine flight tests, we take the precaution of only having one feeder tank containing hot fuel so as to preclude the risk of having a shut-down on more than one engine. On a four-engined aircraft, we therefore have to conduct tests on inboard and outboard engines on two different flights as the conditions for supply of each of them are different.

As TS1 fuel is not normally used in western countries, a first reference flight was made without heating it to check how the engine operated throughout the flight envelope, both with normal supply and gravity feeding. It transpired that there was no difference with the values established for JET A-1. We then resumed testing with fuel heated to between 50 and 55°C, first for an inboard engine, then for an outboard one. The manoeuvres were identical to those made with JET A-1 to check the engine's response with de-aerated and non de-aerated fuel. In addition, we climbed to the aircraft ceiling of 43,000 ft to ensure that with this type of hot fuel, engine operation still remains normal. The conclusion from all these flights was that there was no difference between cold JET A-1 and hot TS1 and so no modification in procedures proved necessary.

In July 2007, a similar test campaign was conducted with the Engine Alliance engine that also gave good results.

* * *

Drainage tests require an empirical approach, not that far from a DIY solution. Why is that the case?

Leaks of fuel, oil or hydraulic fluid can always occur in some zones of the engine nacelle or the pylon onto which the engine is secured. Fluids then have to be evacuated safely without being brought into contact with the hot parts and risking inflammation while also avoiding getting near to the various air inlets (ventilation, APU, etc.) or other parts of the aircraft. Their routing is often random, considering the pressures and ventilation in the various parts of the nacelle. The situation is not unlike rising damp in a house where the origin and paths taken are sometimes hard to determine.

We perform these drainage tests injecting water heavily coloured with eosin (a red disinfectant used for medical purposes) into the test zone for several seconds during different flight phases: during acceleration on take-off, in cruise, in approach and on landing with the thrust reversers in operation. Prior to this, the nacelle and fuselage are prepared with a colour revealing agent. On return from the flight, we check there are no excess traces of eosin under the nacelle and other parts of the aircraft, this being a proof of correct drainage. Many short flights are generally needed as there are so many zones to be checked. For some of them, trial and error is required before you find the right solution through improving sealing between the zones or simple changes to the routing of these fluids.

• See photo n°24 – page 241

The same drainage tests have to be repeated for the APU as the risk of fire has to be eliminated in the entire aft area of the aircraft.

One of the very last certification flights for the Rolls-Royce version was dedicated to APU drainage. A last zone remained to be validated and none of the solutions tested had worked. In desperation, following a new correction, Jacky Joye, with his usual humour, ended call-out of his test orders with the imprecation to "Overfly Lourdes!" I don't know if this was what worked the miracle, but in any case the last test was a success!

* * *

Oil consumption by the Rolls-Royce engines was extremely high during the initial flights. This was no surprise since we had already seen this during the ground runs before first flight where we had noted that it was roughly ten times greater than on other known engines. In flight, it was especially high during successions of phases with maximum thrust, then with idle. This was the case during flights # 9 to 12 that were devoted to measuring the climb rate with an engine shut down. Between two test points, lasting three to four minutes, we had to re-descend with three engines on idle to start the following stabilisation at the correct altitude, with take-off thrust on the same engines.

The quantity of oil diminished almost before our eyes, meaning we were led to shorten some of these flights. Thus, we interrupted flight # 11 when the low oil level warning appeared on one of the engines after only two and a half hours of flight despite the fact that we had topped up the oil before departure and had big capacity tanks.

In stabilised cruise conditions or with less significant and less frequent variations in thrust, the problem was not so critical. Nevertheless, if nothing had been done, we would have been obliged to install huge oil tanks, completely filled on take-off, with an increase in operational weight of several tens of kilograms. This would not have been taken kindly to by our airline customers. In weight terms it meant one passenger seat less and there was the cost of the oil itself. There had to be an improvement.

As from flight # 16, a first modification of some of the seals on engine # 2 gave spectacular results and the other engines were gradually equipped on the same lines. Shortly afterwards, smaller capacity, light-weight, oil tanks were installed on the aircraft. We weren't really masters of the situation, but finally Rolls-Royce got it right!

Globally, the oil consumption problem was settled. However, the engines we were using on our test aircraft were only modified in part in this respect and this led sometimes to some small disturbances when conducting certain flights. Thus, in November 2005, as aircraft # 4 was to leave for the Dubai Air Show, the consumption of engine # 1 shot up to almost 2 litres an hour. There were a number of possible causes for this, but given the uncertainty and pressure with lead-times, the engine was simply replaced.

* * *

The A380 has the special feature of being equipped with thrust reversers on the inboard engines only. The outboard engines are far from the central axis, more than 80 ft out and extension of reversers on these engines, on 200 ft wide runways, shoulders included, would have led to dust and pebbles on the edge being blown forward with the risk of the debris being sucked into the engine, especially at low speed. There was therefore a risk of damaging the fan blades. Transport aircraft cannot operate without thrust reversers as the brakes alone would not be sufficient to slow down the aircraft on landing on a flooded, snowbound or ice covered runway. To compensate for the fact that there were only two reversers rather than four, they had to be designed to be especially effective.

This reminds me of a story dating back to the Farnborough Airshow in 2006. We had invited Boeing President Jim McNerney and his commercial aircraft division Director, Alan Mullaly (a 'top engineer' who had taken part in the design of many aircraft) to visit the A380. Alan was seated on the left in the seat of the pilot-in-command and Jim on the right. Without me saying anything, Alan immediately saw there were only two thrust reversers and explained to Jim why such a design was really clever. There was a need for reversers, but those installed on the outboard engines would pose too many problems on an aircraft of that size. A 'blessing' indeed coming from a Boeing man!

Thrust reversers on all transport aircraft must have three safety levels to prevent them being deployed in flight as, according to the type of aircraft, this can have catastrophic consequences. Such an accident occurred many years ago on a Boeing 767 that only had two thrust reverser locking systems. Clearly, there is greater complexity when you multiply the number of safety systems and the risk of not being able to use the device when needed increases. Flexibility and variations in temperature sometimes lead to these precision mechanical devices becoming blocked when operating in a difficult environment, whence the need to conduct many tests in different conditions to ensure their satisfactory behaviour. You need to check there are no false alarms and that the reversers are available with a reduced likelihood of failure, in accordance with currently accepted values.

The thrust reverser was only rarely used on the flight test bed. First of all, we wanted to avoid damaging the engine during tests as it was installed very close to the ground and there was considerable risk of small objects being ingested as described above. Also, installation of the reverser on the flight test bed engine was not fully representative of real conditions.

During the first flight of the A380, we decided not to deploy the thrust reversers on landing. Firstly, we weren't sure that the two were going to be deployed as experience was limited. In the event of a failure, there would certainly have been some smart-arse to have picked up on the fact that only one came out and the news would have been on the Internet straight away! For the same reason, we had to be absolutely certain that both of them would retract after landing. So we weren't taking any risks!

Tests started as from the second flight and during development they were pretty varied in nature, including deployment and retraction at different speeds, determining the speed at which there may be re-ingestion of hot gases by the engine, the effect of the temperature on the manoeuvre, and so on.

Few difficulties were encountered. The main one related to the electric motors activating the system that, despite their considerable power, tended to slow down deployment when temperatures were very low. A few modifications settled the issue.

* * *

Tests also have to be run to check the engine's behaviour when there is a rapid variation in the angle between the flow of air and the engine. This is what happens on take-off. While it is still on the runway, the aircraft must pitch up to take-off. If this transition is extremely rapid, the airflow into the air inlet will become irregular and there will be a 'distortion', the tests themselves being known as air inlet distortion tests.

This test cannot be conducted in flight but has to be executed on take-off as the only solution to obtain null angle of attack at low speed is to have the landing gears on the ground. The manoeuvre simply involves pulling firmly on the stick at the moment of take-off so as to have a very high rate of rotation. Obviously, on small aircraft where this movement can be very rapid, you have to counter it immediately to avoid touching the runway with the tail. On a big airplane like the A380, there is less of a risk as, due to the sheer size and inertia, the rotation movement is relatively slow, even if the pilot sets the stick to the nose up stop. As a result, distortion is much reduced and so this test is not really critical for certification.

* * *

Some tests are not needed to obtain certification but necessitate to be scheduled shortly afterwards as they are considered useful by the airlines. This is true, for example, for engine balancing. The rotating elements behave much like the wheel of a car that has to be balanced after a change of tyre. So the rotating mass of the engine also has to be balanced as best as possible considering the huge rotation speed. However, with ageing, wear and impacts (bird strike, for example), conditions can change. What level of imbalance can be tolerated? The pilot has indicators on the level of vibration for each of the three sections. He therefore has to be provided with limits to indicate when mechanical intervention is needed. In practice, tests are limited to balancing the low pressure compressor, with the other rotating units being balanced in the factory.

We made the first tests at the end of January 2007, a little over one month after certification. Small weights were fitted to the fan shaft so as to create a slight imbalance. During the first flight, the pilots reckoned that the warning (at advisory level with indication of the value of vibrations for the section concerned switching from green to highlighted green, pulsating on the screen) came up when what was actually felt was already worrying enough. You really need to put yourself in the position of a flight crew member faced with the choice of deciding whether the level of vibration was too high to pursue the flight or not to settle the matter of the value to trigger the message. The matter remains subjective with no single solution to satisfy all. Our experience showed us that some pilots would cancel the flight when faced with the slightest anomaly while others would go ahead with it. We had to find a trade-off. Six flights were needed to finalise the choice of values. Pretty much at the same time, these tests were conducted on the other type of engine and, here too, six flights were needed to come to a decision.

* * *

Globally, all the engine tests went well. Both models are engines of the new generation, being much more fuel-efficient and less noisy. Such tests always lead to some surprises but you need to bear in mind that the treatment meted out to the engines is absolutely not representative of the way an airline pilot will use them. As already seen, we spend a great deal of time stopping them then relighting, cutting off the fuel supply pumps and pushing them to their very limits. Aircraft safety is our primary concern and no doubts can be allowed to subsist as to how things will work in extreme situations, and if the worst comes to the worst, the crews must be provided with special procedures.

PHOTO n°1: *Claude Lelaie and Jacques Rosay: departure for the first flight*

PHOTO n°2: *Toulouse-Blagnac, 27 April 2005: A380 # 1's maiden flight take-off.*

PHOTO n°3: Fernando Alonso with Chuck Yeager, the first pilot to break the sound barrier.

PHOTO n°4: Jacques Rosay receiving the 'Légion d'Honneur' from Louis Gallois.

PHOTO n°5: Jacky Joye

PHOTO n°6: Gérard Desbois on his jump seat

PHOTO n°7: Hugues Van der Stichel with André Turcat, pilot-in-command on the first flight of Concorde.

PHOTO n°8: Claude Lelaie with Max Fischl, pilot-in-command of the first flight of the A300B, the first Airbus.

PHOTO n°9: Flight test engineers' work station

PHOTO n°10: The cockpit

PHOTO n°11: The Iron Bird

PHOTO n°12: Tufts on the wing

PHOTO n°13: The engine nacelle equipped with strakes and tufts

PHOTO n°14: Take-off during a VMU test

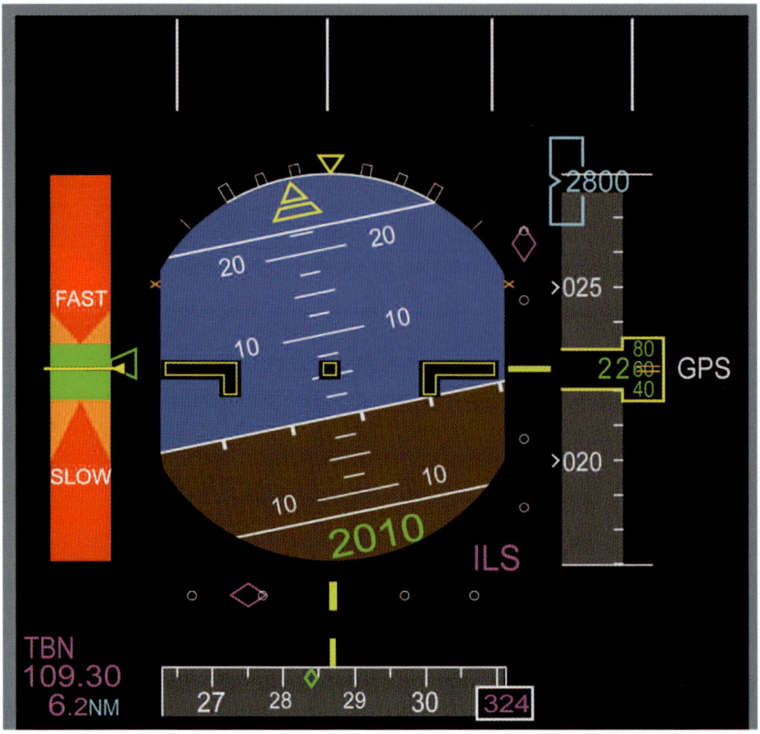

PHOTO n°18: Screen with the Back Up Speed Scale (BUSS) system.

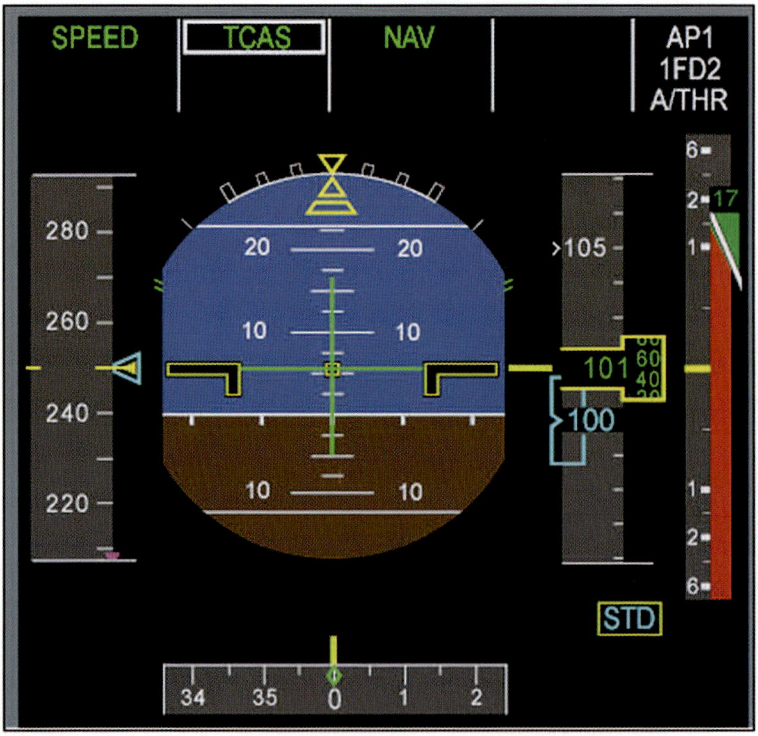

PHOTO n°19: Display with APTCAS active in avoidance mode.

PHOTO n°13: The engine nacelle equipped with strakes and tufts

PHOTO n°14: Take-off during a VMU test

PHOTO n°15: Rear skid during a VMU test

PHOTO n°16: Landing in crosswind conditions in Iceland

PHOTO n°17: The wing landing gear doors lost during a very high speed test

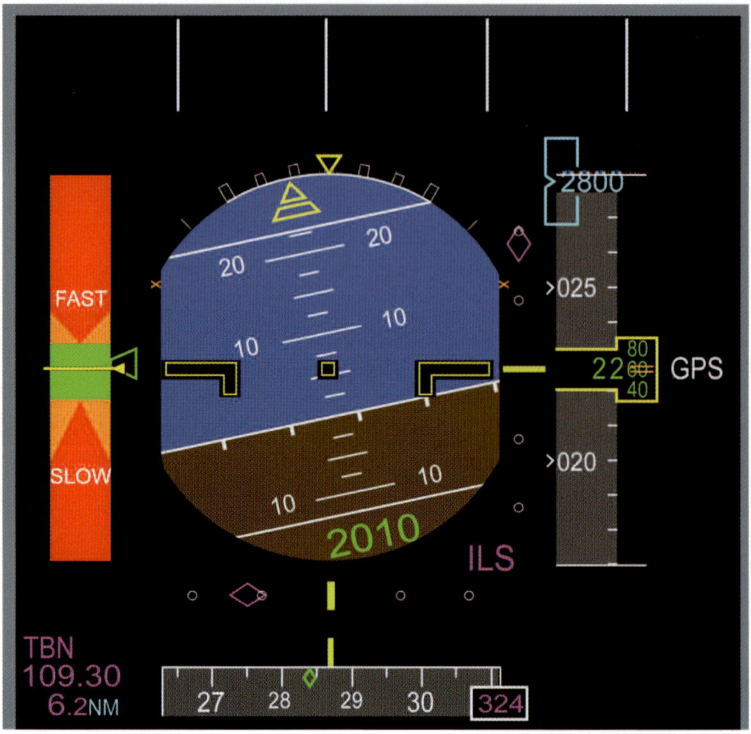

PHOTO n°18: Screen with the Back Up Speed Scale (BUSS) system.

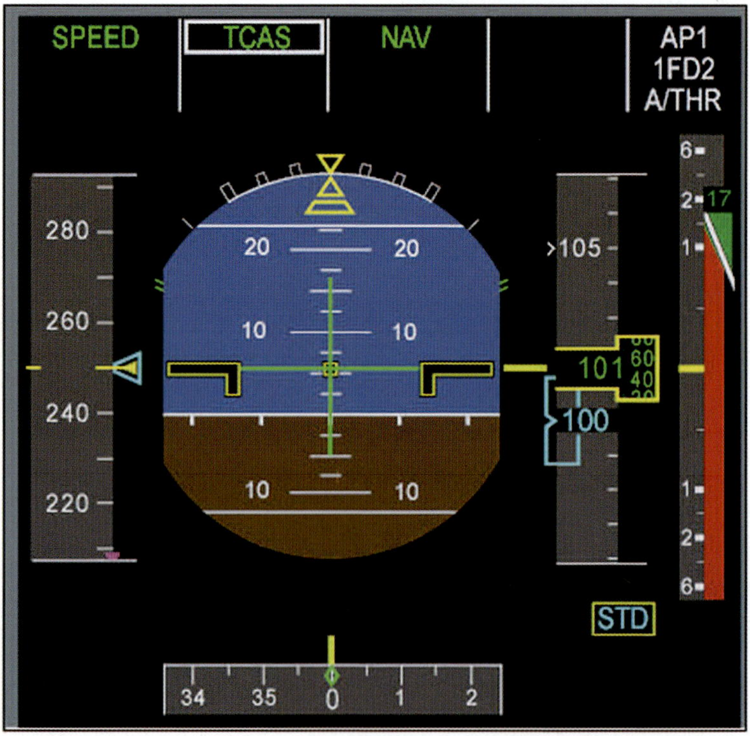

PHOTO n°19: Display with APTCAS active in avoidance mode.

PHOTO n°20: Firefighters cool the wheels after the maximum energy emergency stop test

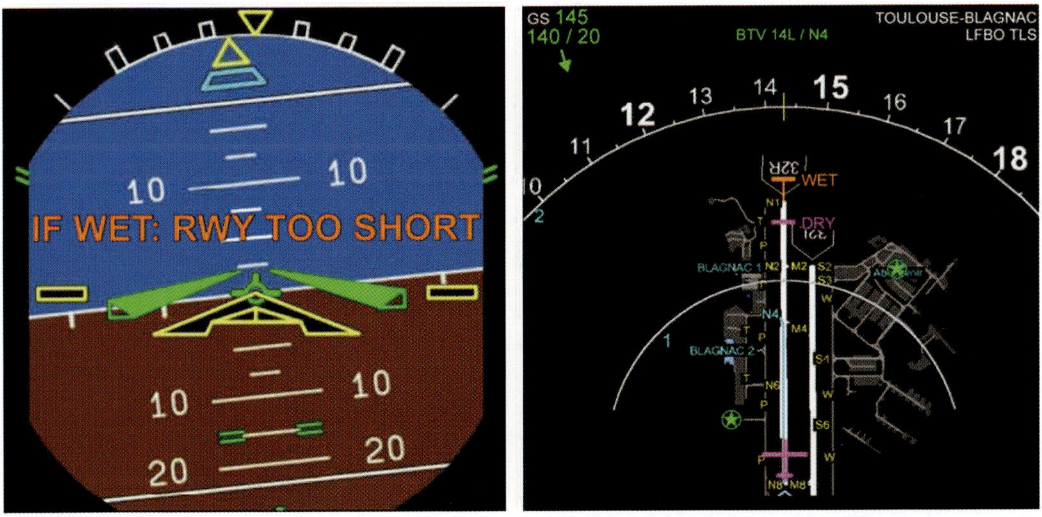

PHOTO n°21: Control screens with "if wet: runway too short" warning

PHOTO n°22: Rolls-Royce Trent 900 engine flying test bed

PHOTO n°23: Rolls-Royce Trent 900 flying test bed in flight

PHOTO n°24: Drainage test

PHOTO n°25: Artificial ice shape on the wing leading edge

PHOTO n°26: During the cold weather campaign at Iqualuit

PHOTO n°27: During the cold weather campaign at Iqualuit

PHOTO n°28: The trough

PHOTO n°29: Going through the trough with thrust reversers deployed

PHOTO n°30: Helicopter view of the trough entry test

PHOTO n°31: Vortices generated by a Boeing B747-400

PHOTO n°32: After the emergency evacuation.

PHOTO n°33: The crowd in the cockpit as we go over the South Pole

PHOTO n°34: The Patrouille de France escorts the A380 at the 2005 Dubai Airshow

PHOTO n°35: The A380 on static exhibition at the Oshkosh Airshow, Wisconsin (United States)

PHOTO n°36: The circuit to visit the aircraft at Oshkosh. To the right can be seen the water ballasts used to simulate different aircraft weights

PHOTO n°37: Flying over the bay and the famous Sydney Opera House

PHOTO n°38: Four A380s in formation flight

PHOTO n°39: Catherine Maunoury and Claude Lelaie back from their flight

PHOTO n°40: Singapore Airlines, first airline to operate the A380

AIRBUS A380 QANTAS – © Photo: Airbus

AIRBUS A380 LUFTHANSA – © Photo: Airbus

AIRBUS A380 EMIRATES – © Photo: Exm company / H. Goussé

AIRBUS A380 BRITISH AIRWAYS – © Photo: Airbus

Photographs

- Exm company / F. Lespinasse: photo n° 1
- Exm company / H. Berenger / H.Goussé: photos n° 22 and 23
- Exm company / H.Goussé: photos n° 8, 10, 38 and 40
- Airbus: photos n° 2, 5, 6, 7, 9, 11, 12, 13, 17, 18, 19, 20, 21, 24, 25, 28, 29, 30 and 33. Figures 1, 2 and 3.
- Airbus / P. Masclet: photos n° 3, 26, 27, 35, 36 and 39
- Airbus / P. Pigeyre: photo n° 4
- Airbus / S. Ramadier: photos n° 14, 15 and cover.
- Baldur Sveinsson: photo n° 16
- C. Brinkmann : photo n° 32
- Katsuhiko Tokunaga : photo n° 34
- G. Hall / Clay Lacy : photo n°37
- Vincent Rouquière: Illustration Hamburg Finkenwerder.

THE FUEL SYSTEM

The A380 has a complicated fuel circuit as 310,000 litres of kerosene weighing about 256 tonnes have to be accommodated. This complexity is mainly due to a safety analysis similar to that done for design of the flight controls and that has to cater for the possibility of a rotor disc of an engine shattering (see FLIGHT CONTROLS DEVELOPMENT section). To minimise the risk of losing too much leaking fuel, there are a large number of tanks and several lines for transfer to the engines or from one tank to another.

Since the A310, all Airbus airliners (with the exception of the A320 family), are equipped with a fuel tank in the empennage meaning that with intelligent transfers in flight you can achieve a centre of gravity position in cruise as far to the rear as possible. The aim is to slightly reduce drag and therefore consumption by being able to reduce the deflection of the control surfaces ensuring the aircraft's balance. Another advantage attached to this centre of gravity position is a reduction in certain forces acting on the structure whence less need for reinforcements and a lighter airplane. But let's be clear about the matter - the aircraft is no less 'solid', but the loads taken into account for structural fatigue can be reduced.

From one programme to another, this system has been improved on and now, on the A380, the computer managing fuel ensures the wisest distribution between tanks at any given time, including on the ground. When refuelling, all you need to do is inform the system of the weight and position of the centre of gravity once all the passengers and the freight have been taken on board and the fuel will be distributed around the tanks so as to have a centre of gravity as far aft as possible, while also reducing loads on the structure. Similarly, if the required quantity of fuel is already on board, the crew can press a pushbutton to order the computers to redistribute it to return to the optimum configuration.

The entire A380 circuit includes for each wing an inner, a mid and an outer tank. Each engine also has its own individual feed tank containing about 20 tonnes. Finally, the trim tank is in the tailplane.

Managing quantities of fuel in each tank is ensured by two identical computers. They take control not only of the supply of fuel to the engine feeding tanks, but also optimisation of performances and reduction in loads exerted on the structure as explained previously. For example, if the aircraft is not full of fuel on the ground, the outer tanks will only be partially filled. As from take-off, the system will command a transfer from another tank so as to fill them and thus reduce the in-flight loads on the wings. Similarly, the level of fuel in the trim tank is adjusted to maintain a generally rearward centre of gravity, within authorised limits. A whole series of pumps and valves are used to sequence these transfer sequences automatically.

The computers must also manage reconfigurations if an anomaly arises (normally with no need for the crew to intervene) as clearly a simple failure must not lead to the loss of fuel from a tank. Thus, if transfer between two tanks cannot be achieved through one duct, it will in principle be done through another one. There are over a hundred different failures that can be announced to the crew and many more that remain unannounced. In the event of such a failure, the computers have to manage transfers between tanks judiciously, taking each tank's capacity into account and determining the position of the centre of gravity. The circuit and its accompanying management systems are extremely complex.

* * *

At the time of the A380's maiden flight, tests of the fuel circuit on the ground were far from having been completed. In particular, many cases of failure remained to be examined. Furthermore, there was no complete test bed to precisely represent all the tanks. Such a bench would practically have required two real wings to realistically represent the geometry of the tanks. This investment was considered to be prohibitive, bearing in mind that, firstly, it would only have been available some months before the first flight and, secondly, it would only have allowed a small number of problems to be detected, with others likely to emerge only at altitude.

Some difficulties were encountered, not always being obstacles to certification but more often having an impact on reliability. Thus, on entry into service, after a period of three problem-free months, reliability became degraded for a number of reasons, with problems relating to the fuel circuit seriously contributing to the situation. Three years after the first commercial flight, practically all anomalies had been finally corrected and reliability came close to the set objective.

Flight tests for a fuel circuit involves first of all checking for failure-free operation, meaning that all transfers from one tank to another work normally in the different foreseen filling scenarios. You might imagine that all of this could be done during flights that are also devoted to other tests. However, in practice this is difficult as the test engineers often have good reasons to fly with a non-standard fuel load, for example to obtain the desired weight and centre of gravity position. Also, the flight profile, sometimes with a number of changes in altitude is not always compatible with such tests. During these flights with non-standard profiles and fuel loads, warnings are sometimes generated by the computers, apparently frustrated at not being able to work in normal conditions!

As for the engine, tests were run by the team led by Jacky Joye.

Let's take a look at just some of the more interesting tests and the problems encountered during them.

* * *

Development of the engine fuel supply transfer and feed pumps was a source of difficulties throughout development. Right at the start, and for a number of months, the pumps installed had to remain immersed in fuel for safety reasons. This constraint meant there was a need to have a minimum quantity in all the tanks. These pumps suffered from other defects, the first of which was that they were not always able to start up when commanded to do so. Finally, they frequently failed to correctly indicate their status as being ON or OFF.

Once this limitation on having a minimum content in the tanks had been removed through installing modified pumps, a first evaluation of the system's operation could be made. But the fact that the pumps were not always able to stop as soon as the tank was empty and end of transfer not be systematically detected disrupted the automatic sequencing of operations. In these conditions, it was difficult to effectively start fuel circuit tests and we had to wait again for a new standard for pumps, hoping this would resolve all the problems. In reality, the automatic transfer system was never fully reliable right up to entry into service.

Meanwhile, a mishap occurred on a pump ground test bed with an electric arc in the electronic control system. Theoretically, this is something that could have led to the outbreak of fire in the tank on a real aircraft. In reality, the risk is negligible as, firstly, combustion will only occur in the unlikely event of the air-fuel mix being in exactly the right proportions, and secondly, a certain number of protections were already installed to prevent an electric arc in the electronics part being able to propagate outside the pump. Nevertheless, even this tiny risk was not acceptable. EASA rightly reminded us of this early in October 2005, asking us to get it right. A new version of the pumps therefore had to be launched.

One modification was made after another and each time we were assured that all problems were going to be settled with the next pump. But although there was clear progress, minor defects subsisted to disrupt the operation of the fuel circuit. For certification, we had to establish dedicated procedures to use these pumps. These were acceptable, but we had to succeed rapidly in achieving fully automatic operation for normal operational use. On the aircraft's entry into service in October 2007, thanks to the latest pump models, operation was satisfactory and other modifications were made over the following months.

* * *

The search for what is known as unusable fuel is compulsory on any new airplane. The usable capacity of the tanks has to be determined accurately in order to establish performances, and more especially know the distance that can be flown. We also need to ensure that, when the engine is about to shut down, the quantity of fuel indicated by the gauge of the tank concerned is actually on zero so as not to mislead the pilot.

For the transfer tanks, meaning those that do not directly supply the engine, verifications may possibly be conducted in parallel with other tests, in so far as the latter do not lead to manoeuvres that are unrealistic for commercial operations, which could affect the results. Unexpected events sometimes emerge during such tests. Thus, for the A380, they allowed fuel pump anomalies to be discovered as, for the mid and trim tanks, the sensors did not always detect the drop in pressure on the pumps when the tank empty.

For the tanks directly supplying the engines, the aircraft is stabilised in the approach for landing conditions, and all you have to do is wait for the first signs of anomaly on the engine, for example a loss of engine power. As soon as they appear, the engine is set to idle, then shut down. Such tests obviously have to be performed on an inboard engine and on an outboard engine as the fuel routings between tank and engine are different in the two cases. Also, for obvious safety reasons, they cannot be done on the same flight.

On the A380, these tests started in March 2006.

* * *

One of the problems encountered very early in the programme was an anomaly in transfer of the outboard tanks by gravity. In the event of failure of transfer pumps, the idea of recovering that fuel in the inboard tanks had been entertained just leaving it to flow through as, in flight, the outboard tank is always higher than the other due to the wing dihedral. To reduce the loads on the wing, normal transfer towards the engine supply tanks only takes place relatively late, at the end of flight, and thus a failure will only be discovered shortly before landing. This gravity transfer worked much too slowly at a rate of 5 tonnes per hour, which did not cover the aircraft's consumption in cruise, leading to fears that the 16 tonnes in these two outboard tanks would be partly lost for the end of flight.

At the time this anomaly was discovered, early in September 2005, it had already been planned to install another transfer line to better protect the system in the event of an engine rotor disc shattering. But a test for this modification, on 8 March 2006, failed to give the expected results.

Another, theoretically simple, solution was recommended as from the start by our operations specialists: it sufficed to start this transfer earlier, so that even by gravity, and thus at slow speed, it can occur without the risk of losing out on fuel for the end of the flight. But the decision ran up against cumbersome 'administrative' procedures as we had to revalidate all the calculations for the wing in fatigue and modify the fuel computers.

Later, a new back-up system for transfer to the supply tanks of engines # 1 and # 4 was successfully installed.

* * *

On the A380, each feed tank includes a feeder, a small inboard tank, kept permanently full thanks to the feed tank pumps and that directly supplies the engine. If a pump failure occurs, then the feed tank will be at a high level, the feeder will remain full and indicate 1,300 kg. If, conversely, the feed tank is at a lower level and the pumps are no longer available to ensure the full complement of 1,300 kg, the crew see the quantity diminish until the levels of the feed tank and the feeder are identical.

This is what happened during an engine gravity supply test flight in October 2005. The crew had noticed that the feeder did not fill and that the quantity at the end of flight could very rapidly drop from 1,300 kg to 400 kg as the fuel was consumed by the engine, which worried them. A few days later, additional tests showed that the balance level was reached when the quantity was 300 kg in the feeder. As Jacky put it jokingly, that was hardly likely to "bring the pilot to prepare to meet his maker and repent for all his sins!"

There was no safety issue and, technically, there was no simple modification that could be made. So it was decided that the crew would be warned on failure of the two pumps so they would not be surprised by this rapid variation in the level of fuel in the feeder. In addition, a minor penalty on the unusable fuel quantity was added for the inboard engines.

* * *

On 27 February 2006, on aircraft # 1, an incident occurred on the ground that could have had serious consequences if it had happened in flight. This was during tests in preparation for validation of the engine relight procedures in emergency electrical configuration. Gérard

Desbois was performing a ground run and cut off the engine, as usual, with the fuel cock. Shortly after, at the beginning of a relight attempt on this engine, there was an explosion and an orange glow. The other engines were stopped, electricity cut off and a first examination of the aircraft showed that the rear part of the aft pylon fairing, a very hot zone, had exploded (looking "like a cauliflower" as Gérard put it!). There was other damage to the jet pipe and the fairing, but not to the engine as such.

What had happened? Shut down of the engine by this valve led to peaks in pressure in the fuel pipes and more especially in those that are to be found in the pylon bearing the engine. One of these ducts comes from the engine and the other from the tank and they are fitted into each other. Successive knocking had gradually led to them being pulled apart and the fuel had leaked out into the rear part of the pylon. It then caught fire and caused the explosion.

A first modification was made on the ducts but, after a special test, in April, it became clear that the change was not effective as a seal had broken and there was still a slight leak in the same zone.

Moreover, on 13 April 2006, during a flight to check gravity feed of the engines, we noticed that bringing the pumps back into operation led to a strong hammering effect in some engine supply pipes. These were similar to those that had caused the explosion of the rear part of the engine pylon.

Before finalising all the modifications, it was decided to conduct additional flight tests so as to gain a better understanding of these phenomena. On 29 July, during a dedicated flight, after having performed various manoeuvres, it appeared that pressure transients in the pipes could be extremely high when bringing the pumps back into operation at high altitude (42,000 ft). During one of the tests, the engine was about to shut down, but it finally restarted without the crew having to intervene. At that moment, the overpressure reached 32 bars, representing 32 times atmospheric pressure. The loads on the pipes were well above what was expected since the latter were sized to resist pressures of 10 bars. Obviously that flight was interrupted.

The pipe fittings in the engine pylon were again modified. What is more, the knocking caused when restarting the pumps had led to cracks appearing in some of the ducts and a number of other improvements had to be made to prevent this happening. These examples are typical of the difficulties that arise when developing a new aircraft and show that the discovery of unexpected problems on systems that would appear to be really simple is the near daily lot of test and design office teams.

* * *

On many long-haul aircraft where the weight on take-off is considerably greater than that authorised on landing, there is an in-flight jettison system that allows the weight to be fairly quickly brought down to a value close to that for normal landing to facilitate a rapid diversion. Clearly, in the event of extreme emergency, the crew can come back to land immediately and a few inspections then have to be conducted on the landing gear and the structure. Where the aircraft comes back to base although with no hurry to do so, it can jettison a quantity of fuel so that the wheel touchdown will occur in practically the same conditions as those that the crew is used to. Thus, on the A380, with maximum weights of 560 tonnes on take-off and 386 tonnes on landing, considering the expected fuel consumption for take-off and return to the airfield, it was decided to validate the jettisoning of 150 tonnes of fuel. This test took place over the Mediterranean on 11 October 2006 and lasted less than an hour. Ecologists may worry about such a waste, but we couldn't entertain the idea of letting a crew face the hassle of such an anomaly for the first time. Considering the altitude for the test, 25,000 ft, and the speed of the aircraft, the event can hardly be considered ecological disaster. Apart from those in the know at Airbus, the event went unnoticed and, due to the speed of the aircraft and the altitude, the fuel was dispersed in fine droplets before reaching sea level.

* * *

In the summer of 2006, certification was coming up and reliability of the fuel circuit remained a serious obstacle. Aircraft # 4 was worked on in late August to make various modifications mainly including all those intended to improve how this system operated. Unfortunately, as from the ground run at the end of the working party, it appeared that due to electrical problems, most of the corrections failed to give satisfaction. What is more, the fuel pumps did not always start up as planned, which disrupted the automatic transfer sequences.

In order to ensure certification, and temporarily, it was decided to develop a series of manual transfer procedures. Bringing the aircraft into service in this condition was clearly out of the question, but the overriding objective was for the A380 to be certified before the end of the year and to do so, it was perfectly acceptable to have special procedures, as long as safety was ensured. Thus, in October, a number of flights were devoted to these manual procedures.

On 23 November 2006, a fuel circuit certification meeting provided an opportunity to review many problems, many of which will be spared the reader here! EASA finally agreed to certification in the current condition with special manual procedures being applied. However, their engineers added one condition: that an aircraft completely up to standard with respect to the fuel system and electrical circuits be subject to review as our two test aircraft had only been partially modified. This was a perfectly logical decision.

So, early in January 2007, aircraft # 1 was taken out of the flight schedule for all the promised modifications to be installed. By the beginning of February, works had been completed and we were able to continue with the flights for the checks demanded on certification to be made with no particular difficulty being encountered.

However, airlines, quite rightly, demand an aircraft condition that exceeds the necessary minimum to ensure flight safety. Works were thus pursued to resolve other issues as they arose. In particular, several versions of the transfer pumps had to be tested before we managed to obtain the required level of reliability.

One of the lessons to be retained is the need for adequate preparation upstream, before the start of flight tests, so as not to be slowed down by additional developments of equipment that can have a major impact on the certification schedule, as also on reliability on entry into service, a key issue for the airlines.

ICING

Poorly controlled icing can incur risks, especially on fairly slow aircraft. Big transport aircraft are not so sensitive, but precautions have to be taken, especially for the engines. As the applicable regulations are the same for all, we had to conduct a considerable number of tests in this field on the A380.

In practice, two types of tests are performed. Firstly, we fly into icing clouds to accumulate ice on the various parts of the aircraft. We look at its effects on flight controls, on engine operation, what happens when sections of ice become detached and, of course, we validate the various de-icing systems.

But these flights in real conditions are not enough as we know that those situations that could be most critical from the aircraft handling point of view are extremely unlikely to be encountered. We therefore do additional tests modifying the profile of the leading edges of the wings and the empennages to simulate extremely severe icing. To do so we use artificial 'ice shapes'.

* * *

Certification regulations cater for two categories of natural icing: maximum continuous and maximum intermittent corresponding to very different ice accumulation rates. However, development also requires you to conduct tests where you need to accumulate ice outside any given time scale.

For maximum continuous icing, you have to accumulate about 25 millimetres of ice in 20 minutes. This quantity can be reduced if the accumulation is more rapid or, conversely, increased (within certain limits) if it is slower. In practice, you cannot go beyond 30 minutes without moving outside the certification criteria.

Maximum intermittent icing corresponds to a fairly significant and extremely rapid accretion, for example 6 millimetres in just one minute, with the same type of correction over time as in the case of maximum continuous, but without going beyond 80 seconds in duration.

The instruments the flight engineer has available mean he can determine not just the accumulated thickness but also the size of the droplets and the water cloud concentration. The latter must have a minimum value depending on the temperature and the size of the droplets for the accumulation to answer to the certification criteria. Perhaps it should be borne in mind that these rules were drawn up in the 1950s by NACA (the forerunner of NASA) in the United States that took into account observations from three weather stations and at that time 10% of extreme cases were eliminated. This explains why, although certification criteria are extremely strict, more severe icing conditions can be encountered around the world.

A variety of tests are conducted on the different systems. As far as the engine is concerned, you have to check de-icing of the air inlet leading edges and all the sensors contributing to regulation and see whether there is accretion or not on the fan blades and on the central cone (before the blades). You also need to ensure that, if pieces of ice become detached from the fan, the central cone or the air inlet leading edge, there will be no damage to the engine or, at least, no restriction on its capacity to provide thrust. The tests are first conducted with de-icing of the air inlet in operation, then with a delayed selection of that system, about two minutes later. At the end of each test, the engines are gradually accelerated up to maximum thrust to ensure that there are no abnormal vibrations for all rotation speeds.

For the aircraft itself, the icing effect is checked for each deflection of slats and flaps, with landing gear retracted or extended, and checking that its retraction and locking work correctly after a small accumulation of ice. We also have to conduct a qualitative evaluation of stalling after having accumulated 3 inches of ice. This no longer involves continuous maximum icing, so we can take the time needed to accumulate that quantity. In practice, these tests are not of any great interest since, as we shall see, they are repeated with ice shapes adhering to the wing leading edges and the empennages, which is a much more penalising situation.

Although it is highly unlikely that the aircraft be in electrical or hydraulic backup with extreme icing conditions, we still have to check for satisfactory operation of the Ram Air Turbine (RAT) in such conditions. Luckily, this test can be done without cutting off all the hydraulic or electrical power!

Operation of all the (numerous) de-iced systems must be checked: de-icing of certain parts of the wing leading edges, various sensors (speed, angle of attack, sideslip and temperature), etc. The fact of the matter is that these latter systems have already been validated on the ground by their manufacturers in accordance with the standards we provide them - much stricter, I should add, than those required by the regulations. Many other systems are also covered by verifications, the weather radar being an obvious instance.

* * *

Readers will certainly be interested in questions relating to the icing of the pitot probes on the Air France Rio-Paris aircraft that came down into the Atlantic and the certification tests preceding that accident. I think it is important here to clarify things technically. A pitot probe measures the flow stop pressure or total pressure. This can be illustrated by what you feel when you put your hand out of the window of a fast moving car as the pressure increases with speed. This total pressure is used to calculate the speed indications the pilots receive.

There are two types of icing. The commonest is that due to supercooled liquid water that freezes as it comes into contact with a cold wall. But there can also be icing related to the presence of ice crystals that can accumulate in cavities. This latter form is very rare and can only emerge at high altitude as it is associated with extremely low negative temperatures. At the time of the accident, there was no regulation requirement relating to these crystals and no aeronautical manufacturer took this condition into account in development.

The icing of certain pitot probes by ice crystals over a period of thirty seconds to one minute before the accident provides a likely explanation for the poor speed indications temporarily available for the crew. When we certified the A380, only a few cases of icing of this type were known on various types of aircraft and only for short periods. Neither the manufacturers nor the certification authorities had considered it worthwhile to modify the specifications and the regulation, or conduct special flight tests, bearing in mind that a handling procedure had been validated for these flying conditions. This had given complete satisfaction in response to the few known instances. In 2009 and 2010, after wind tunnel studies followed by two flight test campaigns on an A340 in tropical fronts, the possibility of icing on certain probes at high altitude by accumulation of ice crystals, for fairly short periods, was demonstrated. Airbus then launched the development of new pitot probes, with more severe specifications covering that specific case of ice crystal formation. A change in the regulation was also planned.

* * *

Three series of tests on the ground rounded off these flights, covering freezing fog, tests under snow and snow-bound runways.

In the case of freezing fog, there is a risk of ice accumulating at the base of the fan blades as also on the low pressure compressor stators when the engine is held on idle for a fairly long time. At the moment of application of power on take-off, pieces of ice can become detached and damage the nacelle or the compressor blades. On previous programmes, we have also observed internal engine damage. That explains just how important these tests were. The fog must be sufficiently dense for the test to prove significant, meaning maximum visibility of 500 ft and the temperature must be less than -2°C (regulation requirement). Such conditions are hard to find and not even the slightest opportunity must be missed during the winter prior to certification, going to the extent of going out the previous day to an airfield where such a condition has been forecast. Weather forecasts being what they are, the failure rate was far from zero! Luckily, on the A380 # 1 with the Rolls-Royce engines, it proved possible to perform the tests in Toulouse on 23 December 2005. The conditions were ideal, with freezing fog and a temperature of -5°C, with no anomaly being encountered.

Tests in freezing fog on the Engine Alliance engines were in two stages, the first being in Toulouse in late January 2007 and in satisfactory conditions. Additional work remained to be done that we managed to complete on the eve of certification (sighs all round!) on 13 December 2007, in Oslo. Aircraft # 9 had been prepared the day before and was back for certification on the 14th.

Tests under snow and on snow-covered runways are mainly used to validate operation of the engine and APU but we also take advantage of them to define the procedures for use of the thrust reversers as there is a major risk of a great quantity of snow lying on the ground being sucked back into the engine when used at too low a speed. In this case, the engines may become extinguished – something I once experienced while conducting tests on an A320 as we were clearing the runway (we were in almost total darkness as only the batteries were connected, having made the mistake of not starting up the APU early enough!). For that same reason, you are frequently asked not to use the reversers (even on idle) below a certain speed, when the ground is covered in snow.

* * *

Tests in natural icing conditions require a dedicated installation on the aircraft. First of all, you need sensors to measure the size of the droplets in suspension and their quantity. We also install what we call the CEV sensor on the left side of the cockpit in view of the pilot in command when he turns round to look at it. This tool was used in the past when there were no reliable instruments to measure the size of the drops and the quantity of water. All you then had to do was to check on this sensor (a simple stick on the outside of the airplane) that the rate of accretion of ice was sufficient. Thus, the pilots knew that they needed to accumulate roughly 1 inch of ice in 20 to 25 minutes for them to then follow the test's progression. Nowadays, it is no longer normally used for this validation, but it is still available in the event of failure of other instruments. It also helps make decisions rapidly when running tests, for example if the pilot realises that accretion is clearly too slow and there is no chance of making a success of the test. He can then seek agreement from the flight engineer to call it off. Cameras are also installed at different locations in order to monitor deposits of ice on a number of elements that are otherwise hard to see.

Now a few words to better understand the test technique. The impact temperature is that measured at the point where the aerodynamic flow is suddenly halted leading to an increase at this point in the temperature as compared with the ambient temperature. In icing tests, this impact temperature must be negative, otherwise water will not turn into ice as it touches the aircraft. In practice, we aim at a value of about -2°C to cover the small fluctuations you usually encounter in clouds.

The difference between the ambient temperature and the impact temperature increases with speed. Thus, on a wide-bodied aircraft that cannot be flown too slowly, the temperature of the cloud must be about -10°C. The lower the temperature of a cloud, the less water there will be in suspension, which explains the difficulties encountered in finding the ideal conditions to pursue the tests on our airplanes as compared with slower ones.

Weather forecasters can reliably predict whether clouds will be icing or not. Unfortunately, they are unable to give us the water concentration of such clouds, this being the fundamental criterion for making a success of this type of test flight. We are thus left to our own devices to decide whether to commence an icing flight. Finally, it comes down to experience and a bit of crystal ball gazing!

At Airbus, the most commonly used method (but not the only one) to accumulate ice involves quite simply flying around inside big cumulus or small cumulonimbus (storm) clouds. You can never be sure things are going to work out. Merely looking at the clouds, whether with the naked eye or using radar, will never tell you the degree of saturation in water on which the success of the test will depend. However, a little experience of this type of flight means there is a good chance of choosing the best options.

The flying technique used in these clouds is similar to that adopted by gliders to take advantage of an ascending current. We aim to turn around in the yellow zone indicated by the radar. We also attempt to avoid the central red area (with the greatest risk of lightning strike) and the outer edges shown green (where the cloud is less dense). But, according to the information given in real time by the engineer on the rate of accumulation, you adjust your bank angle, tightening the turn when you appear to be leaving the favourable zone and straightening out when this is not the case, meanwhile slightly adjusting the altitude to maintain the ideal temperature. You have to be pretty affirmative in the way you fly the aircraft and take advantage of all the airplane's manoeuvring capabilities. However, as the aircraft enters the cloud it actually heats it up. It is therefore practically impossible to do a test point with an aircraft of the size of the A380 turning around inside a small cloud.

Obviously, lightning strikes are a frequent occurrence. For example, during the 12 May 2006 flight on the A380, there were 20 to 25 of them. They are not particularly dangerous and when you look at all the icing campaigns conducted by Airbus, the most serious risk relates to the radome (the tip of the aircraft's nose) and the weather radar. However, every strike very slightly damages a few rivets, something that can be seen from a very small dot on their heads. When you get to the stage where you reckon too many have been affected within a given zone, they are replaced at the end of the campaign. During the flight mentioned above, we noted 116 impacts, this still being far from the record established on another programme!

A lightning strike can sometimes, although rarely, lead to an engine stall. This happens when the electric arc of the strike passes in front of the compressor, creating a huge, transient increase in temperature and distortions in air circulation, thus temporarily blocking the flow in the engine.

* * *

As always in flight tests, you have to try, whenever possible, to get another step ahead to proceed rapidly with the first verifications, identify potential problems and introduce the necessary modifications. When attempting to address the icing problem, we had already had our fingers burnt in the A340-600 programme (equipped with the Trent 500 engine). Indeed, for this airplane, the first engine flight tests had also been conducted on our first A340-300 used as a flight bench and we had decided to round off the campaign with a search for icing conditions. We encountered a number of problems due to icing and the engine was damaged during these tests, leading to a number of modifications being made.

The engines are certified in icing conditions in accordance with a specific set of regulations, once technical justifications have been provided and a certain number of checks have been conducted on the test bench on the ground. In the case of the Trent 500, a single test on the bench had been conducted, no doubt due to the similarity with other engines (Rolls-Royce RB211 and Trent families) and the experience that had been accumulated in flight. At that time, flights on the flight bench and on test aircraft had clearly shown that there was no direct correlation between certification of the engine and that of the aircraft. The engine was certified, but unusable in flight, as operation was not acceptable in icing conditions. You may well imagine the discussions around the contract this can lead to, with the engine manufacturers arguing that they had nothing further to do in the matter (and above all nothing more to pay out) as the engine had been approved and the aircraft manufacturer demanding changes (free of charge) to the engine for the aircraft to be certifiable. In parallel with our quest to find a technical solution to the problem, we had asked EASA to take a fresh look at its engine certification regulations to avoid again being faced with this totally absurd situation. A few changes are in the pipeline and the future will see whether they are adequate or not.

Obviously, we decided to conduct a dedicated icing campaign on the flight bench on which the Trent 900 engine was installed. The initial aim was to ensure that there were no anomalies on the engine in the certification conditions required for the aircraft. But to gain time over the development cycle, we also intended to proceed, where possible, with certification points relating to the engine and the air inlet de-icing system so as not to have to again address the same issues on the A380. In practice, as there were tests specific to the A380 that had to be scheduled, repeating a small proportion of the tests was not too bothersome. However, doing them all again could have led to having to conduct a number of additional flights and above all further immobilisation of the aircraft, changing weather conditions also taking their toll.

This dedicated icing campaign took place from 25 April through to 14 June 2005, simultaneously with the first flights of the A380. At that stage, we could still entertain hopes that the engine would not have to undergo major modifications, so that the results obtained on the engine bench could be used directly for certification. Finally, everything went as planned. Eight flights and thirty hours were needed for these tests, remembering that a substantial share of flying hours was not devoted to icing as such, but to actually getting to the zone where the favourable meteorological conditions were predicted.

The main problem encountered during the flights and that had not been seen during tests at Rolls-Royce was the difficulty of getting rid of the ice that accumulated on the fan blades when the engine was run for a fairly long time in icing conditions at low rpm. Usually, an acceleration towards high level thrust (for example to climb) was sufficient to use the centrifuge effect to expel the ice. On the Trent 900, this just couldn't be done at the time. As the ice didn't stick to the blades in a perfectly uniform manner, this gave rise at high engine rpm to strong vibrations due to the imbalance. These strong vibrations of the fan were announced to the pilots, but two questions arose: was there a risk of damage to the engine and were the vibrations acceptable for the crew? In answer to the first question, tests were conducted on a bench on the ground, but unfortunately with an imbalance four times greater, leading to an engine failure. A modification was introduced on the engine to avoid such a situation arising. Acceptability in terms of the vibrations felt physically by the crew was also checked by a pilot from the authorities on a dedicated system, with a favourable response being received.

The intermediate compressor blades were damaged in extreme cases, but within acceptable limits. This anomaly had already been found during tests at Rolls-Royce and modifications were in process.

* * *

Icing flights commenced on the A380 # 1 in May 2006 after preparation works lasting more than a month. In fact, not all the modifications were intended for this campaign as, normally, installation works for new systems are grouped together. As is sometimes the case, it was no easy ride once works had been completed. The first icing flight had been scheduled for 4 May but was interrupted quickly due to electrical problems. However, there was some good news when the installation dedicated to icing tests proved to work correctly. But a whole week was needed to sort out all the electrical anomalies. The next icing flight on 11 May was again interrupted, just after take-off, as one of the landing gear compartment doors failed to lock.

Serious matters started on 12 May, with a few successful test points during a flight over Brittany, although once back to base, an examination of the engines showed that the results weren't as brilliant as expected as three engines, and especially # 2, were damaged. This was very probably due to centrifuging of ice on what are known as the 'abradable' parts of the engine, that is the zones on the casing, close to the fan, on which friction from the blades is possible - leading to wear. Authorisation to continue with flights for a limited period was granted to us by Rolls-Royce.

The campaign was completed at the end of the following week after four additional flights. Overall, we could say that it was performed in record time since, on other programmes, it had sometimes taken several weeks or even months with having to wait for favourable weather conditions. The succession of a series of weather fronts across Europe meant we were able to make rapid progress. Added to this, all the tests conducted on the engine test bench meant the number of certification points on the A380 could be reduced.

Among the other anomalies encountered was the ice detection system failing to operate. This system asks the pilot to activate de-icing of the engine air inlet and that for the wing. Additional development work settled the issue.

Shortly before certification, a number of flights in icing conditions had to be made as we had discovered that what are known as the packs (basically air conditioning systems) could draw in ice. Their air inlets were therefore re-designed to reduce this risk and we then had to validate the resulting modification. The aircraft came out of the working party on 27 November 2006 to commence flights. Unfortunately, weather conditions were far from being as favourable as for the previous campaign as many flights had to be made, with pretty limited accumulations of ice.

Tests on the version equipped with the Engine Alliance engine started on 21 March 2007. They were obviously devoted solely to that engine. Here too, conditions were not as favourable as for the first campaign with the Rolls-Royce engine. The only noteworthy event was the loss of three flight controls PRIM computers after a lightning strike. The flight was interrupted and, obviously, a modification of these computers was introduced immediately to prevent a re-occurrence.

In June 2007, we had to make a few additional tests not directly related to icing. Indeed, for the first time, we had a new type of weather radar capable of recording the map of all active clouds in relation to the ground in three-dimensional space. In normal flight, the pilot only had the clouds that were on his flight path displayed on his screen, but he could, by changing an altitude setting, take a look at what other flight levels had in store. Unfortunately, while in dense clouds (like those we were seeking out for icing) the data were reasonably useful, they were sometimes misleading for less active clouds. Thus, we detected high altitude clouds at great distance that were of no real interest for the pilot. Meanwhile, for other clouds, when the distance to the cloud came to less than 40 NM, the visual data were abnormally impaired. We thus performed a number of dedicated flights with the A380 to attempt to identify the anomaly, but above all we asked its designer, Honeywell, to intervene by analysing our recordings and conduct their own development flights. In my view, flying an A380 to develop a new radar was not a reasonable idea considering the flying hour cost and the fact that, with a small twin-jet, you could do exactly the same work. Furthermore, it was up to the company selling the radar rather than Airbus to ensure such development work. Additional flights were thus performed by Honeywell and an improved version of the weather radar was delivered to us a few weeks later. In addition, this new type of radar was then offered across the whole Airbus range.

* * *

Not all handling tests can be conducted during flights in natural icing conditions. Furthermore, the profiles of ice formation on the wings and the tail section are not the most severe. As a result, the aerodynamics specialists calculate, for each flight phase, the most critical 'ice shapes' for the leading edges for the certification conditions (duration and severity of icing).

These artificial ice shapes are built of polystyrene covered with small bonded grains that simulate the granularity of ice and are secured to the leading edges of the wings and the empennages. We then perform a few identification flights, especially with stalling to check whether there is a need to increase landing speed in order to retain the usual safety margin. Thus, we may be led to recommend a slight increase in the case of extremely severe icing.

• See photo n°25 – page 241

In addition, we check the setting of the flight controls protections for low speeds. If it appears that the aircraft is insufficiently protected because, with the stick on the back stop, at maximum angle of attack, there is too much buffeting, or if the aircraft becomes laterally unstable, the setting of the protections can be adjusted by a slight reduction of 0.5 to 1 degrees of the maximum authorised incidence. This generally has no effect on take-off and landing performances.

* * *

On the leading edge of each wing of the A380, the slats are divided into seven sections and only one of which (known as slat # 4) is de-iced. This slat is located next to the outboard engine, on the fuselage side. De-icing is implemented by circulating hot air taken from the engines. In the flight tests team, we were persuaded that this was completely superfluous. Indeed, what could an inch or so of ice do on such a huge leading edge? By dispensing with de-icing, we could reduce the weight by about 130 lbs (half a passenger!). Although we had not received a formal request from the Design Office, we decided to take a closer look at this configuration.

The first flight with these ice shapes took place on 26 June 2006 on aircraft # 1. We therefore placed all our bets, setting out with artificial ice installed on all the slats, including # 4. The first tests were for stalling. With the flaps retracted, the results were good but gradually and distinctly worsened with the deflection. For the landing configuration, there was a notable drop in lift and the aircraft tended to pitch up by itself as stalling was approached. This was not good, but Jacques and Fernando suggested we all three make a flight to come to a decision as, although stalling was not perfectly satisfactory, if the protections at low speed could remain unchanged, we could perhaps accept this degradation to make some weight saving. There was no point seeking perfection in a flight condition that would in all likelihood never be encountered in the lifetime of any of the aircraft. We simply had to ensure safety. But we had to face up to facts: due to this pitch up phenomenon, the loss of angle of attack on the protections at low speed, with the flaps deflected, was of the order of 2 degrees, which was much too great. The flight tests team had been mistaken and the aerodynamics specialists had got it right!

Tests were continued once the ice shape had been removed from the de-iced slat. We made a number of flights to validate the entire operational domain. The protections at low speed were adjusted for deflections of flaps 3 and landing, reducing the maximum angle of attack by 0.5 degrees. This was a good result as it had no consequences on aircraft operations.

Tests were rounded off with what we call the failure configuration. For this, slat # 4, usually de-iced, was equipped with an ice shape smaller than that on the other slats. It is accepted that the de-icing system operates normally as you enter icing conditions and, as a matter of procedure, the crew is then asked to quit them when failure occurs (change in altitude for example), whence a thinner layer of ice. Protection at low speed was good with the flaps in position 3. With normal deflection of the flaps for landing, the aircraft was a little unstable in roll. For us, this was acceptable as we knew that such an accumulation would never occur in operation and if, by chance, it did arise, no pilot would hold the stick against the stop for a number of seconds. Even in such an eventuality, he would not lose control of the aircraft. But the certification rules had to be respected. We thus left it up to the judgement of the pilot in charge of validation, knowing that if it was not accepted, we still had the option of asking for landing to be made with deflection of flaps 3, which was not penalising. But common sense prevailed as the landing configuration was approved.

HOT WEATHER AND COLD WEATHER TEST CAMPAIGNS

Development and certification of an aircraft require all operating conditions that can be encountered at different airfields (especially at temperatures and very high altitudes) to be validated. The experience we had accumulated through Airbus programmes showed that it was possible to extrapolate certain results, in particular those relating to performances on take-off. But we still had to check the aircraft's operation in all these situations.

A number of different types of campaigns are performed, including in 'cold weather', possibly with various temperatures, 'hot weather', at low altitude, and tests at high altitude airfields, generally associated with a hot climate.

* * *

The aim of cold weather tests is mainly to validate procedures to restart the aircraft after an extended period of stand-by on the apron at very low temperature. Sometimes, problems are discovered on systems or on the engines and the A380 was no exception to the rule.

The cold weather campaign often proves problematic when it comes to drawing up the flight tests schedule. Firstly, when a development programme is expected to take about fourteen months, you need theoretically to be able to place this campaign at a time when development of the aircraft is sufficiently advanced for it to be profitable. But this must not be too late to prevent any necessary modifications being made before certification. Unfortunately, the cold weather 'window' only lasts a few days and you are left with little choice in the matter! For the A380, as about twenty months had elapsed between the first flight and certification, we were able to take advantage of the winter period, early in 2006, to conduct a series of useful tests, with an already well prepared aircraft.

We conducted the first campaign at the Iqualuit airfield (formerly Frobisher Bay), in Canada, in the land of the Inuits. This field was used at a time when flights between Europe and North America required a stopover and it had a sufficiently large apron to provide parking for large aircraft.

Tests are pursued gradually, increasing the times of exposure to the cold, up to a complete night. You then need to check how the aircraft 'wakes up' following all the planned procedures. As a general rule, when the temperature is less than -15°C and the aircraft immobilisation time is greater than about eight hours, the aircraft batteries are removed after shutdown and stored in the warmth of a hangar. They are replaced on the aircraft before start-up, while all the other electronic equipment is heated by air sent by benches on the ground. Indeed, some computers do not operate in intensely cold conditions and have to be started up at a reasonable temperature so as not to fail definitively. The next action is to start up the APU so as to ensure the aircraft's autonomy in electricity and hot air. Finally, once the entire aircraft has got back to a fairly normal temperature, the last potential difficulty is to start up the engines.

The minimum temperature aimed at for this first campaign was -30°C. Considering the forecast temperatures, we sent out with aircraft # 4 to Iqualuit on 6 February 2006. On arrival, the first complete shutdown of the aircraft lasted seven hours, with no particular difficulty being encountered. The following test was longer, lasting one full night. Unfortunately, the next morning, the thermometer showed just -20°C. On Wednesday 8 February, testing was called off due to an extremely powerful blizzard.

• See photo n°26 – page 242

Tests were resumed on 9 February at -28°C. On start-up of the first engine, its oil pump broke, even though this was a modified version to improve on the previous ones that broke towards -5°C. Hearing this, we gave out precise instructions to the entire team: in the event of a new failure, the pumps were to be pre-heated before start-up and we would immediately return to Toulouse as the stock of spare parts on board was likely to prove inadequate.

On 10 February, with an outside temperature of -30°C, none of the pumps broke and a short local check flight was performed. The only operational difficulty encountered was on use of the thrust reversers due to the undersizing of the electric motor activating them.

The following day, it snowed and, as almost all the tests had been completed, the crew decided to head for home. That return flight to Toulouse included flying over the North Pole to validate the navigation system in that region.

* * *

A similar campaign obviously had to be pursued with the second type of engines from Engine Alliance. Aircraft # 9 thus left for Iqualuit on 10 January 2007. This airplane had already been flown for a number of months now and it benefited, from a systems point of view, from all the experience garnered on the other version that had just been certified.

Problems started the very next day with the APU. A modification made after certification and insufficiently validated led to difficulties starting up cold. But that was not the most serious issue. While the temperature stood at -31°C, two oil pumps failed on the engines. We decided to interrupt tests and sent ten pumps to Iqualuit on a special flight. The damaged pumps were replaced in tough conditions, with the teams having to work in relay due to the intense cold. The engines were started every three hours so as to keep the pumps hot and thus minimise risks of failure.

A short inspection flight was made on 13 January and the aircraft then returned to Toulouse.

* * *

On all the other programmes, we had got used to conducting a second cold weather campaign seeking out a temperature of around -50°C. Practically the only chance of finding such conditions at an airport is to go to Siberia, to Iakoutsk, for a very short period of just a few days, generally at the end of January or early in February, knowing that temperatures of less than -45°C are not systematically reached every year. In addition, we try to conduct this campaign on an aircraft equipped with the passenger cabin, so as to check operation of all the commercial systems. On the A380, we finally decided not to do tests at such temperatures. The main reason for this was that we were concerned as to the possibilities of repairing the aircraft in the event of a serious problem, as with the need for engine replacement. Next, for a four-engined airplane, the urgency of being diverted to such an airfield is reduced as compared with a twin-engined one, even if the causes are not all related to failure of an engine. Finally, adding up the very slight probability of diversion to that airfield of the extreme with that of encountering a temperature of -50°C on that day, we concluded that the whole thing wasn't worth the bother.

Despite that, we did conduct another campaign, once more at Iqualuit, in January 2007, with aircraft # 2, equipped with its cabin. The main aim was to validate how the cabin worked. Some technical improvements had been made since the previous tests, especially those relating to the engine oil pumps and start-up of the APU towards -30°C.

The aircraft was ferried on 29 January 2007. Globally the same types of tests as during the first campaign were performed with a number of different periods of exposure to the cold. No major difficulties emerged, even with the cabin. The minimum temperatures reached each day were around -28°C. The crew made two local flights and flew the aircraft back to Toulouse on 1 February.

• See photo n°27 – page 242

* * *

Hot weather campaigns at low altitude mainly allow engine operation to be checked, especially its temperatures and cooling (nacelle ventilation, fluids, etc.) and ensure that the air conditioning behaves satisfactorily in that extreme situation. The temperature aimed at is +50°C at sea level and slightly lower if the airfield is at high altitude.

A significant number of flights are devoted to lapse rate take-off tests whose principle is explained in THE ROLLS-ROYCE FLYING TEST BED chapter. In very hot weather, EGTs are higher as the hotter air penetrates the engine and, at maximum thrust, values get close to the authorised limit. A comparison between those obtained in standard temperatures and in hot weather allows the extrapolations made for outside temperature to be validated and to ensure that, for a given thrust, the margin on EGT in relation to its limit will allow the engine to be operated for a considerable number of flights.

We take advantage of these lapse rate tests to measure the temperatures at many locations on the nacelle so as to ensure that its ventilation is adequate. Similarly, we check the oil temperatures and, if necessary, an extrapolation allows the maximum value to be estimated to ensure it remains within acceptable limits with an outside temperature of 50°C or more.

All the systems are also checked as the aircraft are equipped with a large number of computers and poor ventilation could disturb their operation when temperatures get too high. The 'electronic bays' are thus carefully monitored.

We focus in particular on the air conditioning. Indeed, being able to know changes in temperatures and air flows in extreme conditions is essential for qualification. Initially, this mainly means validating operation as the test aircraft used for this campaign is not generally equipped with a cabin. The check must relate to normal operation, but also failure situations, and you need, for example, to answer the following questions: can just one of the two air conditioning systems alone satisfy demand when it is very hot? Does the APU allow for sufficient cooling on the ground and in flight? Additional tests relating to temperatures and air circulation with a cabin installed are conducted during other campaigns or commercial flight rounds. Air conditioning tests are discussed in the CABIN COMPLETION chapter.

As far as airfields at altitude are concerned, verifications cover the start-up of the engines, their behaviour and cooling, bearing in mind that practically all these airfields are located in hot climatic zones. We also check performances on take-off to validate the calculation models in relation to altitude.

* * *

One of the difficulties encountered is that, due to local conflicts, there are few airfields where these tests can be conducted in complete safety. There are countries that were 'accessible' at the time the A340 and A330 were tested in their different versions, but where we couldn't contemplate sending out teams for the A380, including the Yemen

(Sanaa) and Sudan (Khartoum). Finally, we managed to identify two airfields as being satisfactory for our tests at low altitude, with Al-Ain in the United Arab Emirates, near Abu Dhabi, and Tozeur in Tunisia. Al-Ain has the advantage of having a very long 200 ft wide runway with very little traffic, while Tozeur lies close to Toulouse.

For the version with the Rolls-Royce engine, we conducted tests at high altitude from Medellin in Columbia (7,000 ft). This choice seemed a strange one to some considering the local guerrilla activity, but the organisation allowed everyone's safety to be ensured. For aircraft # 9, equipped with the Engine Alliance engine, our choice went in favour of Addis-Ababa where the airfield lies at some 7,300 ft.

The question arose as to whether to conduct some tests at La Paz whose airfield is at 14,000 ft in altitude. However, we gave up this idea as, considering the altitudes already authorised by the demonstrations conducted at Medellin and Addis-Ababa, this would make no contribution for operation of the aircraft over the first years as none of the routes planned by our first customers went over airfields at such an altitude. In addition, the parking apron was too small and site work would apparently have been required. We thus avoided devoting precious time to these non-urgent tests.

* * *

So we left for Medellin to pursue the first test campaign at an airfield at high altitude on 10 January 2006. The tests were mainly devoted to lapse rates, but also to a first evaluation of automatic landings at higher altitude. Six flights in three days. On the way back, we stopped over at Pointe-à-Pitre and Fort-de-France, between 14 and 16 January.

For the Singapore Airshow, late in February 2006, a number of hot weather tests were conducted, mainly on the ground, with aircraft # 4. These were followed by a flight.

The following campaign, covering hot weather tests, was again performed on aircraft # 4, in the Emirates, with Al-Ain as main base. The ferry flight was on 23 July 2006 and ten test flights were pursued.

We checked operation of the Rolls-Royce Trent 900 engine up to maximum thrust of 80,000 lb, especially during lapse rate tests. At that thrust level, the EGT reached its limit (meaning the engine had to be reduced slightly) and, what is more, ventilation of the nacelle was marginal. Modifications had to be introduced to be able to use the engine at that thrust. This had no consequences for initial certification as, for the thrust values chosen by our first customers (from 70,000 to 72,000 lb), the results concerning EGT and ventilation were satisfactory.

Anomalies were also highlighted on the air conditioning system (see the CABIN COMPLETION chapter).

Having looked closely at certain test results, we realised that there may be a new potential problem that had not been encountered on previous aircraft types. The A380 is equipped with electro-hydraulic servo-controls of two different types: EHA and EBHA (see the explanations in the FLIGHT CONTROLS DEVELOPMENT chapter). We wanted to check that, when taking off when these servo-controls are extremely hot, in-flight cooling was adequate to ensure normal operation. To do so, we decided to take-off for Tozeur in Tunisia, park the aircraft out in the sun for a number of hours to obtain as high a temperature of the wings and tail sections (and thus of the servo-controls) as possible on the ground. On return, we could conduct our checks.

Once we had received all the necessary authorisations, aircraft # 4 took off for Tozeur on 22 August 2006. But our operations department received a telephone call from the Tunisian authorities informing them that the Civil Aviation Authority no longer wanted the A380 to land there. Discussions led nowhere. We therefore had to call back the aircraft that was already in contact with the Algiers Control Centre. Negotiations were pursued over the following days to obtain authorisation to land at Tozeur. We ended up resorting to blackmail, telling them it would be the next day or never! And then permission came in on Monday 28 August towards 4.00 p.m. The aircraft left the next day. Unfortunately, the crew were unable to retract the landing gear, leading to the mission being definitively cancelled. French national radio (France Info) announced this world-shaking news! We then decided to proceed with these checks on aircraft # 9 that was to leave for the hot weather campaign on the Engine Alliance engine a few days later.

Aircraft # 9 started its tests on the high altitude airfield at Addis-Ababa in October 2006. We then ferried it to Al-Ain for other checks at high temperatures, at an altitude close to sea level. Throughout this campaign, ten flights were made, mainly devoted to engine tests, including lapse rate and nacelle ventilation. Some autolands at Addis-Ababa were added for development purposes. During these tests, the Engine Alliance engines, used up to a thrust of 82,000 lb, showed extremely good operational characteristics, especially in terms of temperatures and cooling. Anecdotally, during tests at Addis-Ababa, a ground cart (probably something of an antique) providing air pressure for start up caught fire and had to be hauled away from the aircraft.

We conducted a new hot weather campaign at Al-Ain late in July 2007 for certification of the equipped version of the Engine Alliance aircraft, with a further ten flights, but aiming at higher temperatures. We also took advantage of the circumstances to conduct a few training flights for Emirates, the first customer to receive the aircraft equipped with these engines.

WATER TROUGH TESTS

What are the strangely named 'water trough' tests meant to do?

In operation, an aircraft may have to take-off or land on a flooded runway after a violent storm. The ground may also be covered in slush. You therefore have to check that everything will work perfectly in such difficult conditions. Among the potential problems there is the worry that there will be a variation in the thrust of the engines after ingestion of water, that mechanical parts or hydraulic pipes be broken by the force of the water projected, or electrical wiring ripped out. A sufficient number of tests representative of different situations has to be conducted so as to ensure there will be no damage or malfunction. Indeed, such tests are required under certification.

These tests are spectacular to watch and videos taken of them have been broadcast widely. However, there is an element of danger in them. Everything finally went well during the campaigns conducted on Airbus planes but a partial or total loss of braking always remained a possibility, with potentially serious consequences.

* * *

The regulations require these tests be performed with a clearly defined level of water, i.e. ¾ inches. Just waiting for a suitable storm to break out to obtain the precise quantity of water for testing was clearly not an option. Also, for development, as long as you cannot be absolutely sure everything will work correctly, it remains preferable to have a dry stretch of runway before and after the flooded section to ensure safety through more effective braking.

For these reasons, we were led to build a water trough for the tests on each aircraft. For the A380, it ran to a length of 330 ft and was 132 ft wide. Its shape had to be adapted to the position of the landing gears and the engines.

One of the construction difficulties lies in the fact that runways have a crowned (or sometimes sloped) surface to ensure good drainage. So it was impossible to build a pool in one go perpendicular to the runway as there would have been very little water on the central axis, where the nose gear makes contact, and above all much too much of it under the outboard wheels of the wing landing gears and under the engines. For the A380, it was therefore decided to divide the pool into nine parts perpendicular to the runway, with one for each of the landing gears (nose, fuselage and wing) and widths varying between 10 and 15 ft and two under each inboard engine 18 ft wide. The outboard engines were not catered for as they are placed higher than the inboard engines, meaning that if the latter proved problem-free, they too could be considered unaffected. A trade-off had to be found on the different widths as, for example, while the mean height in each basin respected the regulation, due to the crowned shape of the runway, the water level on the outer flanks of the outboard wheels was just over 1½ inches. This was unrealistic in known meteorological conditions and technically penalising.

• See photo n°28 – page 243

The side edges had to be placed so there is no risk of running over them during tests, even with a slight deviation from the central axis. We also needed to make sure no other aircraft entered this water trough. Having a different geometry to that of the A380, they would certainly damage the trough on entering it.

Lengthwise, each basin was divided into four parts measuring 80 ft so as to have water levels as uniform as possible and facilitate filling.

The edges of each small basin were made of rubber and their lower part was inserted into a shallow groove cut into the runway. The whole set-up was then sealed to minimise leaks. Between two tests, about thirty minutes were needed to replenish the water.

* * *

The tests were scheduled to take place at Istres, over a week-end. Indeed, the runway had to be taken over for quite some time, firstly to build the basins, then to conduct all the tests. Although there were only ten or so to do, each one of them demanded quite substantial preparation time, replenishing the basins, checking the aircraft and then setting everything up. High pressure weather conditions also had to be chosen so as to have the lightest possible winds.

Where the trough was to be positioned led to much discussion, despite the fact that the Istres runway is about 16,000 ft long. There had to be enough room for a normal take-off before coming in for the landing tests. The distance beyond the basins also had to be sufficient to be able to come to a halt safely after an emergency stop at high speed or a landing, whatever the direction of the manoeuvre. Our activity normally involved closing down the Istres runway. However, we arranged matters for there to remain a sufficient take-off distance for other aircraft on the air base (Mirage 2000s and KC135 tankers).

One minor vexation involved a colonel who asked us to empty and even remove the water trough as he was to take off the next morning with a KC135 although he had 12,000 ft of free runway available to him. We didn't know what his weight on take-off would be but, in the worst case scenario, even in the unlikely event of having to perform an emergency stop at very high speed, he would have entered the trough slowly, with no risk whatsoever. But he stuck to his guns, even trying to lecture us on what an emergency stop and a minimum control speed involved in the event of engine failure! We didn't take the trough down, but we lost a lot of time in fruitless argument. What's more, his flight was finally cancelled!

* * *

Tests were intended to validate that there were no problems due to water being sprayed or drawn into the engine, whatever the flight phases, whether taxiing, braking, taking off, landing and for all speeds. The thrusts considered were to vary between that adopted on take-off and the value authorised with reversers according to the speed.

The point for the start of acceleration that determined the speed of arrival in the water trough had to be determined accurately. Indeed, when making an acceleration with all engines at maximum thrust, a minor deviation on the initial position could lead to a speed very far from the one desired.

A large number of cameras were placed on the aircraft and around it to determine the shape taken by the spray of water from the different landing gears, both horizontally and vertically. Thus, we were able to create models so as to be able to validate other situations, if necessary.

A helicopter in stationary flight was used for each of the tests with the aim of obtaining the best video visualisation of the lateral projections, especially those from the nose gear.

For all these tests, a near null crosswind was required, first and foremost for safety reasons as otherwise, above the aquaplaning speed, lateral control of the aircraft could become marginal. Furthermore, the water spray models established hadn't to be skewed by a wind effect.

* * *

This type of testing requires tremendous organisation. The man orchestrating all this was Robert Lignée. The two pilots were Hugues Van der Stichel and François Barre. Alongside the technical team in charge of the aircraft and possible repairs, major additional resources had to be set up to prepare the water trough and clean the runway with sweepers to deal with projections of pebbles or debris. Teams to implement the measurement installations and video systems were also needed, plus specialists from the Design Office in various fields (engines, landing gears, brakes, performances, etc.), safety services, teams from the air base to coordinate with other activities, the certification authorities, etc. We also planned for spare parts to be routed in the event of serious problems arising.

The first tests took place on Saturday 8 October 2005. The day before, aircraft # 1 came in to Istres for a LIDAR test flight intended to take measurements of the wake vortex and it stayed on.

The first stage, on Friday evening, involved building the water trough. This commenced late in the afternoon as flights from the air base came to an end for the day.

The tests started on Saturday morning. The first was an acceleration at maximum thrust so as to enter the trough at a speed of 60 kt. The second, an acceleration followed by a deceleration such that, in the trough, the speed would be 70 kt with both thrust reversers deployed and at their maximum power. This case corresponds to the lowest speed at which the reversers can be used with that thrust. Between each test and during filling of the trough, the aircraft was thoroughly inspected, especially its landing gears. After this last stop, it transpired that on each wing gear bogie, a part supporting hydraulic ducts related to the braking system had bent by 90 degrees, leading to a risk of them failing. This already led to a very minor leak. There was also damage to other supports and pipes. The aircraft was no longer in flying condition. Such damage could have been critical during tests at greater speed or with projections of water with a different shape. It also became clear that the routing of some of the wiring was far from ideal.

• See photo n°29 – page 243

There was no other solution but to stop tests and then resume them after analysis of all the anomalies and implementation of the necessary modifications. However, there was one we could proceed with without any risk. This was a manoeuvre of the thrust reversers with the aircraft stationary in the water trough. The reversers were held on idle as for usual use where they are brought back to minimum thrust at a speed of about 70 kt in deceleration.

Meanwhile, we immediately had a small Corvette twinjet business aircraft flown over from Toulouse to bring us the parts needed for repair. That very evening, the A380 was ready to fly again although it stayed on at Istres to continue the LIDAR tests campaign.

* * *

Modifications on the landing gear were relatively simple and additional tests could have been conducted very quickly in the following weeks. However, we were reassured as the problems had been clearly identified and, early in 2006, there were other priorities that sometimes had repercussions on programme lead-times where other anomalies emerged. This explains the decision to postpone the new series of 'water trough' tests for the following autumn.

The second campaign was conducted on Saturday 30 September 2006. At that time, the series main landing gears of the aircraft were not yet available and were only to be installed on aircraft # 7 that was supposed to fly in the weeks to come. It had therefore been agreed that their qualification would be done later. Thus, tests were devoted to validation of the nose gear, analysis of water projections and checking the risks of ingestion by the engines at different thrusts. For that reason, only the basins where the nose wheel went through and those under the inboard engines had been filled.

Four tests were performed. Manoeuvres were limited to acceleration stops at different speeds, up to 140 kt, with the engines on take-off thrust, as well as an entry into the water trough with the thrust reversers on maximum power.

No anomaly was encountered during these tests

• See photo n°30 – page 244

* * *

The third series of tests was run on # 7 on 16 February 2007 to validate all the modifications introduced on the landing gear since the concerns encountered during the first campaign. For this reason, only the strips in the trough where the four main landing gears drove along the runway were filled with water.

Three acceleration stops were conducted including the fastest at almost 130 kt. Considering all the data collected since the first campaign, it was not considered necessary to perform take-offs and landings in the water trough as initially planned.

The results were good. Only a few parts not completely up to series standard were very slightly damaged, without safety being brought into question. The necessary modifications were made before the first delivery.

NOISE MEASUREMENTS

Aircraft noise is a critical issue nowadays. Many airports base authorisation for take-off and landing of an airplane, at a given time in the day, on the noise it makes. Some airlines thus make this feature one of their criteria for the planning of their fleet so as to limit any operating restrictions, minimise difficulties with the people living alongside their main airports and, in some cases, reduce taxes.

It is interesting in this respect to take a look at the example of London's major airports considering the scale of traffic to be catered for with a limited number of runways (just two at Heathrow and one at Gatwick). For given time intervals and depending on the season, Quota Counts (QCs) are assigned. According to the noise emitted on take-off and landing, each aircraft is placed, for each phase, within one of seven categories with a QC of 0.25, 0.5, 1, 2, 4, 8 or 16. In practice, the last two correspond to older airplanes of which only small numbers remain in service. Each airport can allocate flight slots to aircraft within these intervals but such that the total obtained when the QCs for all aircraft are added together do not exceed the assigned quota. Let's take an example: if an airline replaces QC4 aircraft on take-off by QC2 aircraft, it can double the number of movements within a time interval in so far as this remains compatible with other constraints. The main advantage is not necessarily in being able to add more flights, but to gain greater flexibility in being able to schedule during more critical periods noise-wise.

* * *

Noise levels for certification are measured at three points: one under the approach path, the second under that for take-off and, finally, a third on the side, also for departure. For landing, the approach point is located nearly 7,000 ft from the start of the runway and along the axis. On take-off, that known as fly over is to be found just over 24,000 ft from the point where the aircraft starts to accelerate and on the departure axis. The last, side line, is positioned laterally nearly 1,500 ft from the take-off path and at the point where the aircraft is at a height of nearly 1,000 ft.

Noise levels are normally expressed in decibels (dB). This is a logarithmic unit, meaning that a reduction by three decibels corresponds to the power emitted being divided by two. The Effective Perceived Noise decibel (EPNdB) used to measure the aircraft noise nuisance is a variant of this that takes into account various other noise-related parameters, including disturbance for the human ear, emergence of pure frequency and exposure time.

The global level as used in the OACI international classification is the cumulative noise, which is the sum of noises measured at the measurement points. It has no physical meaning.

* * *

The sound level is taken into consideration in the design of any new aircraft and specifications going well beyond the certification regulations are established as soon as the project commences. The aim is to optimise the entire airplane and not just the noise emitted by the jet engines. Indeed, in approach, noise is pretty much divided half and half between the engines and the aerodynamics, with the wings, landing gear, etc. Indeed, you will hear a distinct increase in noise as the landing gear is extended. Although, on take-off, noise mainly comes from the engines, the aircraft's specific characteristics play a major role. Thus, a large surface area wing with low drag allows for a shorter take-off distance and a greater rate of climb so you go over the measurement point at a greater height, whence reduced noise. This shows how important it is to work not just on the engines, but also on the general architecture of the aircraft. A good trade-off has to be found between performances and reducing the sound level.

A few weeks before the A380 programme was launched, in mid-2000, Airbus Management decided to review the cumulative noise specifications, reducing the acceptable level by 6 EPNdB, and above all change QC for London (QC2 on take-off instead of QC4). The Airbus Design Offices and the manufacturers of the two engines thus had a rush job on their hands to rework the design of the aircraft and the engines and initiated major modifications. Without going into the details, note for example that for the Rolls-Royce engine, the diameter of the fan was increased by 7" and the nacelle was lengthened. Acoustic treatment of the engine air inlet was also improved with a new system that made for a saving of 0.4 EPNdB.

In addition, on the aircraft, the operating principle for the wing leading edge slats close to the fuselage was changed (drop nose). Deflection downwards of the ailerons when the flaps are deflected on take-off was optimised. Finally, the small surface perpendicular to the wing at its wingtip was modified, again to improve performances on take-off.

A number of details were reviewed in design of the landing gear: for example, some small cavities that could have caused a whistling noise were plugged. Noise reduction sometimes comes down to attending to the minor details!

In parallel with this, a research campaign was launched on the landing gears. Firstly, a landing gear from one of our A340s was placed in the wind tunnel to analyse where the noise came from. Then, our A340-300 # 1 was equipped with different small fairings on the landing gear. The idea was to prepare modifications for the A380 where the specifications proved to be unsatisfactory. After these tests, we found solutions to make a saving of about 0.3 EPNdB, but we didn't take advantage of this for the A380 as, firstly, the sound level was much better than expected and, secondly, such an installation would have led to very slightly increasing the aircraft's weight and causing a few maintenance constraints.

* * *

Organising a noise measurement campaign is a complex issue as, first of all, you need to find a quiet airport, which is difficult enough in itself. Thus, tests in Toulouse were abandoned early in the 1990s due to the air traffic that reduced the number of flights that could be fitted in. But, more generally, all the background noise from cars and lorries can make measurement impossible. For Airbus testing, two airfields are now used, Tarbes-Lourdes and Morón (a military base near Seville).

Acoustic acquisition systems (microphones, analysers, weather sensors, etc.) are installed on each check point in accordance with precise rules.

A noise control station is entrusted with overall coordination. Each test is validated by this control position taking into account not only data relating to the aircraft in terms of acceptability in relation to the respect for tolerances when passing over the check points, but also all other constraints (including noise not related to the aircraft, weather conditions, ground recordings, etc.).

A Differential GPS system (DGPS) is installed. This involves the equivalent to a satellite on the ground. As its position is determined exactly, it can be used to obtain the path of the aircraft to within accuracy of a few inches.

We have to send out weather balloons regularly, at intervals of less than one hour, in order to validate the weather conditions at the height of flyover above the check points, as well as between the ground and the aircraft, with the aim of making attenuation corrections at a later stage.

* * *

The certification regulation specifies the permissible values for temperature and humidity to perform a noise measurement test campaign. However, we do not simply make do with remaining within that acceptable zone, but rather wait to have those conditions that will give us the most favourable results, considering the correction methods imposed and bearing in mind that all measurements must be corrected in line with a temperature of 25°C and humidity of 70%.

There are also turbulence and wind conditions to be respected. The wind must not be stronger than 12 kt on the axis and 7 kt cross-wise.

The advantages accrued from all this measurement optimisation and calculation of correction is extremely significant, of the order of 0.5 at 1 EPNdB for each check point. To provide an order of magnitude, 0.5 EPNdB on calculation of the QC in London corresponds to a greater weight margin on take-off of 18 tonnes for an A380.

* * *

For tests on landing, the aircraft must pass precisely over the check point with extremely accurate setting of the engines and tolerances are about 100 ft laterally and vertically. Using the data provided by the DGPS, the flight engineer ensures guidance to hold the ideal flight path. Thus, the pilots come in on the axis and go into descent when the engineer gives the signal, knowing that the lower the thrust, the later descent has to start to go over the same point at a given height. But the speed must also be held to within an accuracy of 3 kt. If, for example, the speed is too low and fails to meet the acceptability criterion, there will be no other solution than to dive slightly to increase it, hoping that the passover altitude remains within tolerances. As the criteria have to be respected without any possibility of making adjustments using the thrust, the start-up conditions for each measurement point must be near perfect.

* * *

Three series of measurements are made for take-off, all from go-arounds so as to facilitate execution and save time. We start each one at maximum take-off weight and aim to fly over the check point at different altitudes and with varying levels of thrust.

Guidance has to be just as strictly controlled as in approach and the test engineer also gives all the necessary data. Setting the targeted level of thrust and starting climb takes place at an extremely precise position and from a height that allows you to fly over the check point in the required conditions. Here too, there is practically no margin for adjustment available to the pilot.

In addition to establishing levels of noise for certification, all these tests allow the noise reduction procedures on take-off to be refined. Thus, generally, there is first an altitude where you can reduce thrust, then another higher one where acceleration towards the climb speed can be commenced. There are also more complex cases where the zones over which you have to reduce nuisances are further away from the airfield. The whole set of measurements made allows a solution appropriate to each situation to be determined.

* * *

Independently of the technical difficulty involved in performing these tests, there are safety issues as we have to fly at very low altitude, sometimes quite some distance away from the runway. Furthermore, to establish the effect of certain parameters, we go into some quite unusual configurations. Let's take an example: to clearly identify the noise related to the aircraft's aerodynamics, we have to perform a test in the landing configuration, but with minimum engine idle, that usually obtained when the flaps are retracted. The rpm is then lower and there is less noise coming from the engines. However, the time taken to accelerate towards full thrust increases significantly, coming to fifteen to twenty seconds, rather than eight, with quite some time spent at low engine thrust. This idling is forced thanks to a modification in the logic of the engine computers. Descent is pursued with the engines on idle (as explained above) and, as soon as we come out of the measurement window, at less than 250 ft, we can increase thrust. But the engines do not always accelerate in completely identical fashion and, for example, on an A380, while both engines on the same side may be close to maximum thrust, the other two could still be close to idle. That sort of situation could be extremely awkward as, at that instant, we will find ourselves below the minimum control speed on two engines (VMCL-2). We therefore take the precaution of accelerating the engines in two stages. Despite this, there is always a tense moment in the cockpit on go-around as full thrust on all four engines is only established once the aircraft is really low, just above the runway.

For many years, considering the risks involved, all noise measurement tests were conducted only in daytime. After analysis and taking special precautions, we finally decided to conduct them at night-time if necessary, this being the best time to find optimum test conditions.

* * *

Usually, two measurement campaigns are conducted for each aircraft, one for development and the second for certification. The A380 was no exception. Certification tests have to take place fairly early in the programme as analysis takes a long time and results are only obtained some months after the measurements.

Immediately after the first campaign, we get an idea of the sound level. Above all, these tests give a frequency analysis of all the noise emitted to detect its various origins and, if necessary, make changes to reduce it. Thus, we made the first tests on the A380, at Morón, between 19 and 21 October 2005. Four flights were enough for us to cover the entire programme. Right from the start we saw that we were easily going to fulfil our commitments towards our customers. However, various parasite noises were identified. The biggest source proved to be the engine control valves. A change to the opening system of these valves cut noise on approach down by 3 EPNdB.

The second campaign was run in Tarbes from 27 June to 1 July 2006 to conduct all the tests described above for certification. We made five flights as the two first had to be interrupted due to worsening weather conditions. For the first time, on 30 June, two flights were made at night. All the results were finally better than expected. Thus, cumulative noise showed a margin of 27 EPNdB as compared with the regulation limits. For London, progress was spectacular since QC1 was predicted for landing, but finally the A380 achieved QC0.5. The figures here tied in with the impressions of observers on the ground: the A380 is indeed a quiet aircraft.

* * *

Noise tests are not limited to measurements with the aircraft in flight. Other data are needed to ensure operations, especially for airports. Thus, the noise map on the ground with just the APU running or with engines on idle has to be prepared.

Other, completely different tests are performed in flight. This involves what is known as acoustic fatigue. Indeed, the noise generated by vibrations is capable of creating alternating forces on different parts of the aircraft and possibly damaging them in the long term. Generally, the elements most concerned are the flaps and their supports, the horizontal tail section, fairings, etc. Measurements are made using microphones secured to the structure and the results are obtained by tests on the ground and take-offs at maximum thrust. The level of vibrations on the various sensors gives a picture of the fatigue life for the various elements so you can then define locations where modifications have to be applied. For the A380, the level of forces at work proved to be in line with the specifications.

WAKE VORTICES

The matter of wake vortex (the turbulence that forms behind an aircraft as it passes through the air) aroused much comment in the press and speculation from broadcasters during development of the A380. All this brought a price tag with it for us, mainly due to the increased number of flying hours (about 390, while certification itself was obtained after some 2,500 hours) as we responded to allay concerns. The incredible number of meetings and the travel involved also added to the cost.

First of all, let's take a look at what causes wake vortices.

To ensure the aircraft's lift, you have high pressure on the lower surface of each wing, and low pressure on the upper surface. At the end of the wing, the overpressure under the wing will deviate air outwards to respond to the low pressure of the upper surface. This pressure differential leads to the passage of air from beneath the wing upwards, a rollup of the airflow that creates swirling air masses (vortices) behind the wing. Due to the high speeds of these swirling air masses, any aircraft crossing the vortex is subjected to a sudden and brief bout of what can sometimes be very strong turbulence. There is a vortex for each wing, turning in the opposite direction to each other. Past studies and experience have shown that, generally, the bigger the aircraft, the stronger the vortices will be, thus presenting a potential risk for the following aircraft if it enters this wake.

As from 1998, Europe and the United States initiated studies in various fields within the scope of the VLTA (Very Large Transport Aircraft) project. The issue of vortices was addressed and conclusions were to be submitted in 2003. However, meanwhile, Boeing withdrew from the project and it all failed to come to fruition.

When the A380 programme was launched, the FAA decided to impose certification of separation distances between aircraft. Such distances should have been attested for the airworthiness type certificate to be issued. But, finally, this was excluded from certification and became an operational requirement to be established for entry into service of the aircraft.

In 2003, the ICAO set up an international working group to study the vortices of the A380 including members of Airbus, the FAA, Eurocontrol (in charge of the European air space) and the JAAs (the European certification authorities present at the time the A380 programme was launched). A project steering group supervised the working group's activities, with at its head two co-chairmen, one from Eurocontrol and the other from the JAA. This organisation's goal was to submit a report to the ICAO, who would in turn publish recommendations for the whole world. Later, as from 2009, EASA, the European Air Safety Agency, played an active role in these studies, replacing the JAAs.

This working group's remit was not to decide on separations behind the A380 in absolute terms, but to propose solutions that would provide a level of safety equivalent to that already existing. As a result, the basic aircraft used for comparison was the biggest airliner existing before the A380, the Boeing B747-400, that has a maximum take-off weight of 396 tonnes in its passenger version.

The most active participants in this working group were Airbus and various departments of the FAA. It should be noted that, until recently, according to its articles of association, the FAA was called on not only to define the regulations and check their application, but also to support the American industry. This sometimes raised doubts as to the reasons behind certain of the positions it adopted. The co-chairmen took part only in the steering group, and there was nobody from the JAA or EASA on the working group for the first five years, although there was one Eurocontrol representative. Most of the time, over the first years, working group meetings involved a discussion between five or six FAA engineers and Airbus participants, sometimes less in number. The European authorities gave the impression that they were only remotely concerned by the success of the A380 and they only provided us with minimal support.

This was not the first time such a topic had been addressed. Indeed, major experimentation had been conducted by Boeing in 1970, with assistance from NASA as the B747 came into service. This aircraft was compared with the B707, the first long-haul jet. At the time, data were obtained empirically as the measurement systems had extremely limited capabilities compared with what we now have available to us. But interesting results had been provided, for example that on approach, there can be seen no significant difference between the vortex of a B707 and that of a B747. For information, the Boeing report was drawn up by Phil Condit, later to become President of the company.

* * *

Several decades ago, the ICAO defined the rules relating to wake vortices. The minimum separation distances in cruise were identical for all aircraft, whatever their size: 5 NM in the horizontal plane and 1,000 ft in altitude.

For take-off, approach and landing, there are three categories of aircraft: Heavy with maximum take-off weight of more than 136 tonnes, Medium between 7 and 136 tonnes and Light, less than 7 tonnes. The minimum separations in these flight phases are determined from the categories: that of the preceding aircraft and that of the following one. Thus, behind a Heavy, another Heavy can follow at 4 NM, a Medium has to be at more than 5 NM and a Light at at least 6 NM. If a Heavy follows a Medium, there is no limitation other than that which the controller may ensure thanks to his radar screen. This is what is known as minimum radar separation, of around 2.5 NM.

Some countries have set themselves apart with separations different to those of the ICAO, among them the United States. This is also the case for the United Kingdom that, over time, has increased the number of categories to lead to separations that no doubt better correspond to the risks involved. Indeed, the ICAO regulations are impaired by the fact that they assimilate an 8 tonne business twin-jet with a 130 tonne transport aircraft.

• See photo n°31 – page 244

* * *

When the working group commenced its activities, the idea was to determine two values: the likelihood of entering a wake and that of exceeding a severity criterion. Multiplying these two figures gave the global probability of running into a risk situation in a vortex. The first mathematical model was provided by the FAA. The second derived from European studies and was based on a simulation of the influence of these vortices on various types of airplanes, whether with or without piloting. But serious doubts quickly emerged as to the validity of the entire modelling system and it was finally abandoned.

The group then decided to base its comparisons between the A380 and B747-400 on measurements obtained with a LIDAR (Light Detection And Ranging) system that records the movement of particles drawn by the air and tangential speeds in the vortex over time. Using a series of calculations, this means you can deduce the vortex's force. One of the LIDARs was based in Frankfurt to do recordings on the B747-400 while the other, for the A380, was either in Toulouse or at Istres.

It emerged that these measurements, made in different environments (positioning of the LIDAR, weather conditions, etc.), failed to provide anything formally conclusive.

So we were left wondering what to do.

* * *

That's when, on the eve of departure for our first commercial tours to Singapore, Australia and Dubai, in November 2005, we had the impression the sky had fallen in on us. Without the flight tests team being previously informed, the ICAO published a recommendation: at high altitude, all aircraft were to keep to a distance of more than 20 NM (about 37 km) behind the A380 and vertical separation was to be more than 1,000 ft (and thus, in practice, 2,000 ft!). The absurdity of this situation is patent: in cruise, a small twin-jet business aircraft could fly at 5 NM behind a B747, while a B747 had to keep to 20 NM behind us. In approach, all aircraft, whatever their type, had to be at least 10 NM behind us. Who could possibly have suggested something so unbelievable? Apparently, the person behind it was an expert from the FAA consulted by the ICAO. As I learnt some years later, some Airbus Top Managers gave their agreement. There was no way of overriding such a safety instruction as it had immediately been adopted by all Countries.

We were furious with this decision. It had no real consequence on the sales promotion round, but we were extremely worried by the negative message our customers were going to get. There were two solutions: either we vehemently denounce this technically indefensible position, or temporarily accept it and act to have the limitation removed. On the flight tests side, we were in favour of the first solution, but we received the order not to make comment. With a few years hindsight, and seeing the time it took to deal with the issue, I wonder if we weren't right at the time.

* * *

Now, it became a matter of urgency as it was out of the question for the aircraft to enter into service with such penalties. At the end of 2005, we therefore decided to make comparative LIDAR measurements in the same conditions, firstly on the A380 and the B747-400, as also on the A380 and B777. The principle behind these measurements was as follows: the LIDAR was installed on the approach axis of the Istres runway and we had each of the two aircraft pass at an interval of five minutes – with the A380 and B747 for example. During the first test campaign, for each group, about 75 passes were made per aircraft in conditions simulating landing or take-off.

Thus, late in December 2005, we had on our hands a considerable number of measurements on the three aircraft, the A380, B747 and B777. There was some scatter in the results. However, the tests had been conducted in conditions that all pilots would consider to be perfectly calm. This observation was not enough as the LIDAR saw some particles moving outside the passage of the airplanes and some people in the working group judged it necessary (and later imposed) that the tests be performed in conditions of absolute calm so as to be absolutely certain of encountering the most critical scenario. Only a few passes made during flights with the B777 were acceptable, but unfortunately, this was not the aircraft we were so much interested in as it was lighter than the B747. But initially, we just had to make do with those results.

Comparative analysis of the A380 and B777 tests gave rise to endless discussions, especially with certain experts from the FAA. At each new meeting, they proposed an interpretation even less favourable to the A380.

Without going into the detail, it is interesting to see how we were treated. First of all, on 29 May 2006, the FAA clearly explained that it was seeking near total safety, demanding absolutely perfect results. Frank Ogilvie, my colleague on the steering group and eminent aerodynamics specialist (one of those to whom we owe the existence of the marvel that is the A380) and someone more inclined to diplomacy, this time openly took issue with the FAA's negative attitude. At that time, there was the vague idea of penalties in relation to the ICAO Heavy category of +2 NM for Heavies (6 NM instead of 4 NM for a jumbo behind an A380) and +3 NM for Mediums (that is 8 NM for a A320 behind an A380 instead of 5 NM behind another Heavy). On 9 June, during a meeting in Montréal, the atmosphere cleared a little, with the FAA recognising certain errors and separation reductions seemed a possibility. But at the following meeting towards the end of July in Bremen, the message had changed. Penalties were back to +2 NM for Heavies and +3 NM for Mediums. At Airbus, the American employees were ready to follow the FAA. The same applied to certain Directors. The members of the vortex group and the flight tests people were totally opposed to these decisions. Late August, in Toulouse, we were in a stalemate situation. A new meeting in Boston on 7 September unfortunately confirmed this situation, as endorsed on 22 September.

On 9 October 2006, recommendations were published by the ICAO in a 'State Letter' put out to all States. The A380 was to be qualified as 'Super' with specific separations. While the name assigned sounded pretty positive, the consequences were less so as all these separations represented a serious penalty for operation of the A380. The airlines and airports would pay the cost, but we were unable to do anything against the hostile fire coming from the FAA.

* * *

Work continued to use all the LIDAR passages in order to reduce these penalties. Many meetings and teleconferences took place early in 2007. To force the working group to conclude, the steering group had set a deadline for 28 August to draw up the final report or at least to obtain a formal agreement as to its content. Come August, there was a sustained rate of teleconferences, these taking place every Friday. We were alone as we faced up to the FAA: six against six! On 3 August, whereas we were already using four different methods of statistical analysis, the FAA demanded a new one be added called 'Student T'.

This derived from the name of an Irish student who had developed this process in a pub to validate the quality of the beer. The FAA was to send us the entire analysis in accordance with this process for 9 August, before the following meeting. We only received it on the 17th, just before the weekly meeting. Without going into a detailed analysis, it was clear that it was unacceptable as it compared the 'best' of the B777 with the 'worst' of the A380. This gave penalties close to 3 NM for all categories. The principle involved was absurd. Let's suppose we were to compare an aircraft with itself: a penalty of at least 1 NM would have to have been added as a difference had to be made between the 'worst' measurement and the 'best'! We argued that there was no question of using an evaluation method that did not satisfy the plausibility test of null deviation for an aircraft being compared with itself.

On 24 August, the FAA's threats became more specific through a series of e-mail exchanges. In substance, they were saying "Accept what we have to offer or there will be no positive outcome". To which was added: "No negative comment, please". The tone of the debate became more heated during the teleconference as an agreement had been reached on the methodology and 'Student T' was not included. On 27 August, relations became even more embittered: one of the FAA representatives sent an e-mail to the two co-chairmen of the steering group that failed to explain the reasons for the disagreement, but that tried to set the cat among the pigeons in the opposite camp by explaining that, at Airbus, not everyone was on the same wavelength (untrue). He even added: "The new Airbus management seems to be on a learning curve, or perhaps does not agree with what was agreed in the working group last year". Clearly, I was in his sights as, without having changed position, I had only started to take part in certain meetings for about a year! We resolved internally to submit to the authority of the co-chairmen.

That meeting with the two heads of the steering group, plus the FAA and Airbus, took place on 17 September 2007 at Hoofddorp, near Amsterdam. Again, there were four 'Airbus people' facing four from the FAA! The decision was made to keep to the methods that had been accepted a few months ago. The FAA was to withdraw its requirement to compare the worst of one aircraft with the best of the other. It was planned to sign the document on 1 October. But this didn't happen as we reckoned we had hit on a better strategy.

* * *

The first reduction in separations on approach and take-off had been obtained by comparing the B777 and the A380, which penalised us as we had seen that the LIDAR signature of the B777 set it below the B747, and thus further than the A380. Considering the FAA's determination, we decided to repeat alternating comparative passages with the A380 and the B747, but in absolutely perfect weather conditions. Two flights were organised over two days, on 5 and 18 April 2007, bringing in on each occasion a B747 from Lufthansa based in Frankfurt. For these tests, we chose Tarbes-Lourdes Airport as there was less likelihood of turbulence there than at Istres.

Both flights gave results we considered extremely satisfactory. Reducing the penalty to 1 NM for all Heavy and Medium aircraft should be possible. The results were analysed for several months and the working group met for two days in Madrid in January 2008 to discuss them. Right from the start, cold water was poured over all our hopes: "Too many uncertainties" an FAA representative told us. Then he provided a few explanations: firstly lack of accuracy as to the height of passage of the aircraft by some 15 ft plus too great a stratification of the atmosphere, meaning there was no turbulence and too rapid a dissipation of the vortices. We were bowled over by this. In flight tests, a deviation by 15 ft over the height of passage of an aircraft some 65 ft high like the B747 and more than 80 ft like the A380 is negligible. But as these experts from the FAA were not flight test specialists, our recriminations served no purpose. What's more, as far as turbulence was concerned, this was even more unexpected since it was precisely a lack of turbulence that we wanted for these tests. Unsurprisingly, the measurement points the FAA asked us to eliminate were those favourable to the A380!

How many meetings, telephone calls and trips were to be needed to reconcile the opposing positions? To summarise, for travel alone there had been two days in Madrid, two days in Boston, two days in Seattle and several days in Montreal to finalise the dossier and present it to the ICAO.

Meanwhile, at each meeting or teleconference, the FAA came in with new demands. Thus, to give a picture of the Alice in Wonderland side to things, in addition to the four statistical methods already used, they added two others with their variants with the 'Wilcoxon rank-sum test' and the 'Kolmogorov-Smirnov test'. I'm not at all au fait with these methods, but statisticians reading this book might be interested!

By how much did we manage to reduce the penalties? By nothing at all on Heavy aircraft and 1 NM on Medium ones! We had to make many concessions. But we remained unconcerned as we were counting on a new testing technique (as explained below) and didn't want to hold up the results by prolonging discussions. This reduction was finally confirmed by a new 'State Letter' from the ICAO on 8 July 2008.

* * *

In parallel with this, the problem of separations at high altitude had to be resolved, this being, from the air traffic perspective, at least as critical as distances on approach and take-off. With Hugues Van der Stichel, we decided to conduct a test mobilising an A380 for twenty minutes during one of its flights. At high altitude, on some days, vortices can clearly be seen in the trails of condensation from the engines just behind the aircraft. That's why even four-engined aircraft still only have two contrails at a short distance behind, corresponding to the vortices of the two wings.

We thus boarded an A318 with Hugues. During this first attempt to tackle the issue, on 11 January 2006, we first of all sought to determine by how far these vortices could descend. The contrails were no longer to be seen beyond 8 NM, but it was clear that they could go down as far as 1,000 ft below the level of departure. This was rather unsettling as the FAA specialists had claimed that behind the B747, they didn't descend below 800 ft, bearing in mind that the vertical separation between aircraft in cruise at high altitude is 1,000 ft.

We then entered the vortex directly at 5 NM for the first test. This was the minimum authorised distance for other aircraft and the effect was pretty violent! We crossed both vortices. On the second, Hugues had to let go of his video camera despite the fact that it was resting firmly in the cockpit coaming! Other encounters with vortices were pursued at slightly greater distances in order to try and get a clearer picture of the reduction in severity with separation. The result of the flight was clear: this method with encounters in contrails could be used to characterise vortices.

However, major preparatory work was required as such tests had never been conducted on a major scale since those done by Boeing in 1970 and we needed to fine tune flight techniques that would allow for characterisations and comparisons. Thus, we had to answer a number of questions. What flight controls law should be used? What should the sideways penetration angles be? Should we penetrate in the vertical plane? Should we release the controls on entering the vortex or, on the contrary, counter to observe maximum deviations in relation to the planned trajectory? It should be remembered that the vortex is extremely small, measuring about 15 to 25 ft in diameter and, according to small local variations in atmospheric conditions, it may climb or descend. That means it is no easy matter to be sure you can go straight through the centre. A number of flights were needed to settle these issues. This was further complicated by the fact that at that time there were not too many slots available for A380s that were already overbooked for development and certification tests.

It was finally decided to stick to normal law allowing the aircraft to be stabilised and thus automatically counter any external disturbance. Destabilisation by vortex is sufficiently strong as to allow for a comparative analysis, even with a rapid response of the aircraft flight controls. Penetrations are made with a side angle of about 10 degrees. This value was not based on experience, but on estimations made from models and showing that it corresponds to the most critical case. In the end, unable to make a final decision, we decided to conduct half the tests with the stick released before entry into the vortex and the other half piloting to try to reduce the vortex effect.

Early in June 2006, we were ready for these comparison flights. The idea was to do a general rehearsal with the A380 flying ahead with our A340-600 and going into vortices with an A318. We had to wait for good weather conditions, but finally there turned out to be a day when contrails were long enough, meaning about 20 NM and with little turbulence. This opportunity was provided on 23 June. There were organisational difficulties because, to perform good measurements, we wanted to be along the axis of the wind. In addition, the whole set-up lacked flexibility with an A380 and an A340-600 flying ahead and then behind, at distances varying from 5 NM to 20 NM, an A318 manoeuvring between 200 and 2,000 ft below. Finally, at 2,000 to 3,000 ft above the big airplanes, at a distance that tended to change, there was the Falcon 20 from the DLR (the German Research Centre) equipped with a LIDAR. To perform the tests correctly, there was a need for at least 100 NM of leeway or if not more at the start of each series of trials. It was extremely difficult to move such a set-up around the skies with all the civil aviation traffic. During the initial briefing with the controllers, there were a few misunderstandings, leading to difficulties during that first flight, but this proved to be a useful rehearsal.

The 'real' flight was that with the A380 and the B747-400. It wasn't that easy to organise as we had to negotiate with Lufthansa to hire a B747 on the day weather conditions were favourable with contrails 20 NM long.

Sunday 25 June 2006 looked good. Air traffic problems had all been settled thanks to a new briefing with the controllers plus a discussion with the Director General of Civil Aviation. We had insisted on the issues for air traffic managers with the need to prepare future spacing between other aircraft and the A380. We took off very early with the A318 to take advantage of the various possible test zones. On board, on the left, there was Gene Arnold, the FAA certification pilot for the A380, with whom we got on really well. We had asked him to come along to recount what he saw. Hugues Van der Stichel was on the right and I was between them on the jump seat. On the Atlantic side, no contrails! So off to the Mediterranean. Weather conditions weren't perfect but, considering the problems getting everyone together, we decided to make a go of it. Using satellite telephone, we had the B747 take off from Frankfurt and the A380 and Falcon from Toulouse. The rendez-vous was set over Marseille.

The first pass for the A318 involved sliding in 5 NM behind the two airplanes, up to 15 to 18 NM to observe the rates of descent of the vortices. A first success was notched up as we saw exactly the same result for both aircraft. The B747 vortices did indeed descend by 1,000 ft, contrary to what the FAA had claimed.

Five other passes were conducted to penetrate the vortices of the two aircraft at distances varying between 5 and 15 NM. A hundred such encounters were conducted. The results, as seen by the crew, were clear: there was no perceivable difference between the two aircraft. The FAA pilot agreed with us. After a long flight lasting six hours in the A318, we had reason to celebrate: the demonstration was incontrovertible.

Having got back to base and opened a few bottles of Champagne in the flight tests cafeteria to celebrate this success, I sent an e-mail to a number of members of Airbus Management, asking them meanwhile not to disseminate the information as we now had to go through the claws of the working group.

Matters got more complicated when it came to the meeting of the working group in Toulouse in August 2006. We put everything on the table, including the videos from the camera installed at the top of the vertical stabiliser of the A318, so the participants had all they needed to make comparisons. There was a great deal of scatter in the results, but it was clear that, on average, the vortices of the two aircraft were similar and the report from the Chief Pilot of the FAA testified to that. However, during a very small number of tests behind the Boeing B747, the A318 went very briefly through a negative load factor, meaning that, in operation, passengers not having fastened their seat belts would have been projected up to the ceiling with a risk of injury, as in extremely strong turbulence. All this even though the A318 was 1,000 ft below the B747 and about 15 NM behind, that is three times the minimum separation. Two days and interminable discussions were needed before we managed to explain that, for safety reasons, if the same separation as for the B747 was not granted to us, then the same problem had to be taken into account for that aircraft already in service. Admittedly, such a scenario is extremely improbable in real life and moreover, to my knowledge, has never been encountered since jumbo jets have been brought into service. However, we did show that it remained possible in extreme conditions.

One month later, the result came in: no specific limitation for the A380 in cruise. A first victory! This was corroborated in the ICAO 'State Letter' of 8 July 2008.

* * *

Our success in reducing separations in cruise at high altitude thanks to the proactive strategy of making encounters gave us some ideas. This technique should also give good results to reduce the penalties on approach obtained by using a LIDAR.

One of the difficulties with LIDAR is that it can only measure tangential air speeds at more than just over 16 ft from the centre of the vortex. Knowing that the active part of the vortex has a radius of less than 15 ft, it should seem obvious that LIDAR is not a perfectly reliable instrument for such evaluations: calculation of the force is performed without measuring the speeds in the most active zone. Above all, what counts is not the estimated force of the vortex, but its effect on the following aircraft. We thus needed to check the physical consequences on the airplane entering the swirling masses of air. We had successfully used this technique at high altitude, so we also took a bet it would work for the approach.

Considering the reduced size of vortices, we had to be able to spot them and to do so we needed a system of smoke generation. By pure chance, during the summer of 2006, we had an A340-300 equipped with smoke pods for an aerodynamics research programme. We therefore conducted a few preliminary tests with an A320 flying behind that A340-300. The clear answer was that this technique could be used with a certain number of adaptations. We needed more powerful smoke pods so that the vortex would be visible up to 8 NM. They also had to be capable of functioning longer and adapting to the A340-600 and A380 so as to be able to conduct comparative tests in the same atmospheric conditions as obtained between these two aircraft. In other words, new smoke pods had to be made. A request was made for them to be manufactured, with the response being that this would require a six to eight month lead-time and be accompanied by an exorbitant price tag. So our DIY team got to work to take just one month to devise a really simple system involving a huge oil tank, pipes leading to the outboard engines and a small oil injection bar to spray oil into the flux of each of these engines. Pumps and left and right side valves also had to be added to ensure a variable flow rate. The injection of oil into the hot exhaust gases created smoke. In fact, there was nothing really new here as the device was similar to that installed on the airplanes of military aerobatics teams or on certain light aerobatics aircraft you can see attending meetings (I used it on my aerobatic team aircraft when younger!). During the first test on the ground, we covered the airfield with so much smoke that we were admonished by the control tower! But we were far from the costs and lead-times initially announced.

The priority was to validate the test techniques and resume what we had already done at high altitude. Indeed, vortices are much less powerful on approach, largely because the minimum separation time is much longer - more than two minutes on approach for an aircraft of the A320 type, but around thirty-five seconds in cruise. The first tests with normal control law showed that the control surfaces countered part of the vortex effect themselves and thus the deviation was sometimes reduced, making comparisons difficult. After various tests, we chose a specific control law better adapted to this type of measurement: alternate law. The aircraft was stabilised on the longitudinal axis, with roll control direct (meaning that the ailerons only deflect in response to an order from the pilot), and the yaw damper active. Unlike in cruise, movements in the vertical plane on penetrating a vortex are too small to be able to make a valid comparison. We focused only on roll accelerations and roll rate.

We then had to choose the aircraft to represent the Heavy and Medium categories, since the separations were different. For the former, we used an A300B, the third Airbus built and that was in service for microgravity tests using parabolic flights. This aircraft had the advantage of being partially instrumented. In addition, it was theoretically the most critical airplane from the vortex sensitivity perspective as it was at the lower weight limit for that category. For Medium, we chose A320 # 1 equipped with a considerable array of measurement equipment. This was not the smallest in its category, but this was not a fundamental issue as the aim was to make comparative measurements.

In 2007, after a number of flights to finalise the test techniques, two comparative flights were made on 18 July and 12 December with the A320 entering alternatively into the wakes of the A380 and the A340-600, and an identical flight on 16 October 2007 with the A300B. Each time, more than a hundred encounters were made. Behind the A380, we had decided to conduct tests only with zero and one nautical mile penalty in relation to the Heavy separations so as to recover the maximum number of measurement points for a given flight. Very quickly, we noted that for the crew of the following aircraft and for the same distance, there was no apparent difference to be observed behind these two types of airplanes. In total, we performed 874 encounters in the approach configuration, including about 400 during these comparative flights. This gave us a good set of statistics, although some specialists might have wanted several thousands of them! It's worth noting that with an A380, we entered the wake of another A380 at a distance of less than one nautical mile and the aircraft remained controllable.

In our view, after 110 hours of tests in the approach configuration, the results were convincing: there was no visible difference between the effect of a vortex from an A380 and that of an A340-600. But we had to convince the authorities and above all the FAA. We did not receive an express veto, but the list of conditions stated was such that it was difficult to respect them all. Almost tantamount to a refusal.

* * *

We were now in quite a strange situation. After five years of intensive analyses, the 'official' conclusion was that there was an equivalent level of safety between a B747 flying 6 NM behind an A380 and the smallest of light aircraft 6 NM behind a B747!

While still in this near stalemate condition, at a meeting in Montreal in June 2008, the steering group entrusted a small team to look into the feasibility of this analysis using encounters. These were diplomatic precautions since, on the Airbus side, the work had long been completed and it was not going to lead to months more work.

The small team that went under the name of the Core Working Group (CWG) started work over the summer of 2008 with a remit to hand in a report by the end of the year. It included two to three people from each organisation. Questions and comments soon flooded in and what follows is just a small sample. What was the effect of the slope in approach descent as the tests were conducted in level flight? What was the effect of weather conditions (the tests were pursued in perfectly calm air)? The number of tests was not enough to conclude! This was the daily lot of the three Airbus engineers taking part in the CWG, Florent Laporte, Gordon Höhne and Andreas Reinke. They had to show tremendous patience. All these questions need to be put into context. Remember that the aim was in no way to perform a comprehensive analysis of safety behind the A380, but rather to make a comparison with existing aircraft. If we were to find no differences in given flight conditions, they were unlikely to change much where such conditions varied slightly

Seeing that the situation was going to remain blocked, Frank Ogilvie and I (the two Airbus members of the steering group) sent an e-mail to the group's two co-chairmen on 1 October 2008 to forewarn them. We noted in particular that the FAA made much of six encounter tests made by NASA in 1974, whereas Airbus had conducted 874 of them with instrumentation way in advance of what was available 30 years before. They had some nerve!

At a meeting in San Francisco in November 2008, an agreement seemed to have been reached with a trade-off that hurt us, but that we were prepared to accept it to make progress: "The Core Working Group has concluded that it is not yet clear that the existing encounters can be used to further refine the A380 wake vortex separations..." Putting that on paper represented a sound compromise as we were sure, with all the available results, that such refinement was indeed possible. But even that was not enough as some participants reckoned that the wording should be even more negative. So we backtracked to express a different opinion that, put simply, stated that Airbus considered the tests performed, maybe with some additional elements, could be used to reduce separations. However, all the other organisations failed to agree. This shows, at the time, the failure of the European authorities to provide support, whereas it had been proven by all these tests that there was no flight safety issue.

After many exchanges between the different participants and a whole series of changes in the details, the report remained basically the same, that is noting a disagreement between Airbus and the others. The presentation was made to the steering group in Paris on 7 January 2009.

Right from the start, the lack of support from the European authorities had constituted a major difficulty for us, even while we were in permanent discussion with the FAA, often on a 'five against five' basis, without any representative from the European authorities (JAA and EASA). Eurocontrol's support was limited considering its links with the FAA on so many projects. Due to our isolation, we were extremely concerned at the decisions that could be made at the steering group meeting. I therefore contacted, Patrick Goudou, EASA Executive Director early in December 2008, then went to see him in Cologne the day before the steering group meeting. Having given a short presentation on the tests performed and the results obtained, I asked him for his support as if we were unable to resolve matters between us, subjects were liable to go upstairs to the political level considering the issues at stake. He ensured me of his support and it is true to say that from that date on, the agency really invested in this dossier.

That 7 January meeting came as a pleasant surprise for us. Indeed, the two co-chairmen belonging to the two European authorities, Paul Wilson for Eurocontrol and Alain Leroy for EASA (who replaced the

JAA representative), distanced themselves from the conclusions of the non-Airbus CWG members and thus from their own experts. They thus decided to pursue studies by creating three different working groups: one for the technical aspect (the technical working group), one for flight tests (the flight test working group) and the last one in charge of drawing up the final report showing that the proposed distances ensured the same level of safety as for other aircraft. This time, the two FAA representatives found themselves in a minority and they protested vehemently: the recommendations of the Core Working Group (in other words, theirs) were not being followed. The two co-chairmen had visibly understood the underlying injustice of the situation where Airbus was left with no chance.

That day came as a real relief as hope was restored of handling the matter on a purely technical level. But, as we shall see, we weren't out of the woods yet.

* * *

Meanwhile, at the end of 2007, we had reckoned that it may be useful to discuss the issue in private with Boeing. The first A380 had been delivered and they were going to bring the B748, heavier than the B747-400, into service. This was no doubt in our common interest, with Airbus seeking to minimise penalties on approach and Boeing looking for savings by being dispensed from performing their own tests thanks to our results. Moreover, while we were discussing matters internally, an FAA manager had formulated the idea. Obviously, the decision could not be made without approval at the highest corporate level. Tom Enders accepted this initiative in January 2008.

Nothing is ever simple when it comes to working with the Americans as you have to get through the filter of their lawyers. I contacted Boeing's Chief Pilot, Frank Santoni, for a first face to face meeting. A few weeks later, the time it took for him to gain corporate approval, a meeting was agreed to. I met him on 8 February 2008 in Minneapolis to see what we could do together on the matter of vortices for large aircraft and especially the A380 and B748. Following a few more weeks of internal discussions at Boeing, we finally got the green light for a detailed discussion.

We organised an incognito meeting at a London-Heathrow hotel on 14 April 2008. The room was booked by Airbus and outside there was no way of knowing it was an Airbus-Boeing meeting. At this stage, the utmost discretion was required. There were experts on vortices for each side in the room, pilots and, obviously, a lawyer. We first of all presented all of our results, not hiding anything. Our concluding message was a call to work together and if we managed to get the penalties for the A380 lifted and get that aircraft into the Heavy category, there would be no further need for such testing on the B748 that would automatically become a Heavy with the dual advantage of reducing development costs and ensuring its classification (that tests failed to guarantee). Boeing then went on to set out their test programme for the B748, which was simply an analysis using LIDAR passes.

Discussions ensued. Boeing's scientists contested our approach in the strongest of terms but without pulling out a really convincing argument. The pilots kept silent. Finally, nothing was decided. All we said to each other, without much conviction, was that each party would continue thinking about the matter.

* * *

In 2009, setting up of the A380 working groups, as decided on by the steering group, was particularly laborious. Their respective tasks had to be defined precisely before deciding on the participants. We expressly requested that people from outside the usual group be brought in so that the meetings would not again be reduced to an FAA-Airbus stand-off. Some pilots from customer airlines, pilots' unions and specialists in human factors were to take part in the discussions. Now, all these people were essentially members of the flight tests group and were therefore not involved in the initial discussions.

The only group active right from the start was that responsible for technical issues. The participants were the same as in the initial working group and, very quickly, the atmosphere turned sour. Immediately, we had a barrage of questions from all sides, but above all from the FAA. It all looked much like what happens in some parliaments when the opposition tables an inordinate number of amendments to stonewall legislation! The list included more than 90 pages, each of which contained a number of questions.

In September 2009, there was a further meeting of the technical group in Cologne with six people from the FAA and just three of us.

Understandably, no progress was made. We didn't come to single matter of agreement with them. One of the topics on the agenda was the influence of thrust and the aircraft slope. Indeed, all our comparative tests had been performed in level flight in order to avoid any change in atmosphere that could have led to doubts as to the validity of the comparison. We had seen no difference between encounters behind the A380 and the A340-600. On the basis of mere common sense, it can be considered that, in conditions where there is no difference in level, if two aircraft go into descent, the variation in the vortex effect must be similar behind each aircraft, and thus the difference between the two in descent can only be of the second order and so negligible. But that wasn't enough for the FAA and an incontrovertible demonstration had to be provided. Seeing that no progress was being made, I asked how they settled such a problem within the scope of the recategorisation programme, or RECAT. RECAT was supposed to classify all aircraft more logically by increasing the number of categories. But this new categorisation was simply implemented using mathematical models and LIDAR validation. When I put this question, that seemed natural enough, I was well and truly laid into by the FAA who told me point blank that the two issues were completely unconnected. But the bottom line is that when there is a concern like this for one aircraft, it can also affect others. Coffee break discussions were heated!

A few days later, in mid-October 2009, I turned up at a seminar on vortices in Washington. I complained of our difficulties to a number of FAA people, as also to the Eurocontrol co-chairman, Paul Wilson. He understood how serious the situation was and immediately contemplated the idea of cutting the FAA out to make for a solely European working group. The Americans first seemed to accept this, then changed their minds. Then the next week, the meeting of the working group went well and the FAA accepted the principle of the analytical method developed by our engineers. It was even recognised as being 'state of the art'! It looked like a major step forward.

But questions and criticisms again came flooding in. Another crazy example related to the fact that we used the same type of inertial reference unit in our test installations for the development and certification of all types of aircraft, without the authorities (FAA included) ever having issued criticisms. The following demand came in on this matter: "The inertial reference unit includes a 3 Hertz, second order, Butterworth filter. Demonstrate that it does not affect the roll acceleration measurements" When you look at the scatter of the results, this question is clearly of no interest. What is more, as comparative measurements are involved, the said filter acts exactly in the same manner on all aircraft. This was clearly just another delaying tactic. A year and a half were needed to settle the issue!

Even though the principle for the method had been accepted, the FAA's response as to applicability to tests on the A380 was systematically negative: too much scatter, results confirm those obtained by LIDAR, etc. Come mid-March, most of our arguments in Miami went down well and the atmosphere was good, but once again, the FAA postponed its judgement as to the method's acceptability. The deadline was now set for May 2010.

Early that month, fearing impending decisions that could have dire consequences for us on RECAT, Frank Ogilvie and I sent off an e-mail to the two co-chairmen to warn them of our determination. It was inconceivable that the RECAT programme make decisions that fly in the face of all the tests conducted by Airbus. In the aviation world, no-one would have understood such an attitude, with the consensus being that calculations on models can in no event replace measurements made in flight. Admittedly we were using a form of blackmail but, given the methods adopted by some other parties, there was no point being overscrupulous. We had a private discussion with the two co-chairmen at the steering group meeting on 18 May in Toulouse. They were disapproving of our correspondence, but they understood our position. Their idea was to ensure the earliest possible transfer of the dossier to the flight tests group where resolution of the conflict should be easier. Basically, they were right, but we remained wary of further pitfalls.

What's more, a bad atmosphere prevailed at the round of technical group and steering group meetings that took place in May. Firstly, the FAA made a presentation (without putting its name to the document, as is wont in the secret services community!) in which it was shown that only three encounters respected the criteria. From then on it was clearly impossible to work out statistics, and so (QED) the method was unusable. The next day, I was only able to get to the meeting once half the various parties' positions had been recapitulated. At first, I kept quiet while my Airbus colleagues gave me a scribbled outline of the previous decisions. I then put a few questions before the co-chairman of Eurocontrol cut in to tell me that before I had arrived, the atmosphere had been extremely positive and my presence had spoiled things. That may well have been true, but when I asked that a summary of the positions adopted in my absence be projected on screen it became clear that absolutely everything was negative for Airbus, with not a single point being accepted by the FAA. That's not what I call a positive atmosphere when one of the participants systematically throws out everything, even though there were no raised voices.

Seeing that we were getting bogged down here, we turned to A380 programme management and suggested they organise a presentation of the results for our customers. Indeed, all the tests had been conducted more than two and a half years ago and we had still not shown them or given them anything as we worked in secret so that nothing that could embarrass or hinder the working group got out over the Internet. But after all that time, we were duty bound to give an account to the airlines. That presentation was organised for 6 July 2010. Not all of our customers were represented, but those who turned up appreciated the results we showed them. Obviously, no mention was made of the difficulties we were facing from other quarters.

* * *

Late in June 2010, there were rumours going round to the effect that the FAA was preparing to publish the document justifying the whole method used for the RECAT programme while completely ignoring the use of encounters. Unfortunately, we couldn't get our hands on it as it was still going the rounds for approval.

We would have liked to have been sure that Eurocontrol, being bound to the FAA under the RECAT project, would not blindly march in step to the dictates of that administration acting as a repository for US interests, but would also make efforts to look to European aeronautics' needs. This was nothing to do with asking for favourable treatment for Airbus. We simply wanted to ensure that our tests' scientific results would be assessed equitably.

Early in September in Miami, we held a series of meetings, firstly of the technical working group, then the steering group and finally the flight test working group. The results of the technical working group were finalised. To summarise, Eurocontrol and EASA agreed with all the results presented and above all to the fact that the method proposed by Airbus was viable. Conversely, the FAA refused to give a favourable view as to a certain number of points and asserted that there was no guarantee that an acceptable demonstration could be made. The two co-chairmen of the steering group then made a very good decision. They decided to temporarily call a halt to the technical working group and put everything in the hands of flight tests. A new test campaign with the authorities taking part was to be organised.

As I write these lines, I really should congratulate Vincent Trève, a brilliant engineer from Eurocontrol, who chaired the technical working group in extremely difficult conditions considering the diverging views of the various parties, for the tremendous work he did.

* * *

In parallel with this, the flight tests group started to meet under the chairmanship of François Fabre, an EASA test pilot. At the start, we were a group of pilots and test specialists and all went well. But as from the following meeting early in October 2010, two engineers from the FAA who were members of the technical working group replaced the pilots and they came back to the questions already discussed in the technical group. Luckily, we managed to sidestep all their requests thanks to support from EASA. Indeed, we were pressed for time as we had to simultaneously prepare four test aircraft: an A380 and an A340-600 with the smoke generation systems and two aircraft to penetrate the vortices, an A320 representing the Medium category and an A340-300 for Heavy. It was extremely difficult to synchronise all of this but we managed to get things up and ready for two weeks at the end of November and beginning of December 2010. The FAA wanted to delay the tests as not everything was clear to them, but it was impossible to find another slot with these four aircraft available considering the various schedules. Early in November, the American representatives announced to us that they would not be taking part in the tests. We were worried. What should we do? Both co-chairmen of the steering group immediately reassured us, saying that they would in any case use the results to draw up the report demonstrating safety equivalence.

Initially, three flights were planned. One was to compare the behaviour of a Heavy (the A340-300) behind the A340-600 and the A380. Another was to do the same exercise with an A320 (Medium category). Meanwhile, the third flight was intended to make a comparison between the severity of encounters in level flight and on the approach slope. This was to respond to a specific request from the FAA who reckoned that the thrust was different in level flight and in descent, and that this could have an influence on the force of the vortex and that it therefore had to be checked out. There was no getting round it. At Airbus, we believed this to be a loss of time (and money!): if there was no difference between an A380 and an A340-600 in level flight in given conditions, the difference between the two on a glide slope of a mere three degrees could only be infinitesimal. The second reason was the difficulty involved in execution. In level flight, the generator aircraft fly side by side and it is easy to sequence encounters behind each of them. For tests in descent, you need near perfect weather conditions across a wide range of altitudes, which limits the number of favourable days. Furthermore, with tests following on from each other at different altitudes, there could be further challenges in relation to the effect of the altitude.

On top of these three flights, I asked programme management to authorise two additional ones. The first to duplicate tests with the A320 as I feared a controversy due to the fact that there was only a limited number of tests, from 80 to 100 per flight, with the risk of not being able to generate satisfactory statistics. The second was intended to pursue encounters of the A320 behind the A340-600, in other words with a Medium behind a Heavy, with a normal control law. I wanted to check the influence of the separation for existing and already classified aircraft. This test did not therefore form part of the 'official' campaign.

The tests started on 24 November 2010, with participation from the EASA test team and a pilot from a pilots' union. At that time of the year, we feared having to cancel a large number of flights, but weather conditions were pretty favourable and most of the flights were made over the Mediterranean. After each flight, a late afternoon telephone debriefing was organised with the FAA. Note that these flights were especially long since the A320 left with full fuel tanks and landed with the minimum level, making six flying hours (seven hours with the A340-300). In total, the aircraft made about 80 flying hours and 450 encounters.

Now was the moment to proceed with a detailed analysis of the tests and above all have the results accepted by our usual contacts. Once again, this looked like being a tough task.

* * *

The supporting reports for the RECAT programme were published mid-February 2011. Apart from the issue of separations relating to the A380, the final result was not unacceptable and there was no safety problem with the new distances proposed. However, we were stupefied by the demonstration method. Without going into too much detail, it assumed a linear decrease in the force of the vortices over time and in identical fashion for all aircraft. This was merely a hypothesis, but it was not backed up and we already knew, thanks to our tests, that it was incorrect. Measurements had been made on various airfields using LIDARs or equivalent systems, but without knowing precisely the models and weights of the aircraft coming in to approach. As the table giving all the characteristics of the aircraft included errors, the results were hardly credible. How could the FAA draw up such a document, without sound justifications, while calling on us to provide an irrefutable argument? This remained a mystery to us.

So what was to be done? March 2011 in Miami, at a face-to-face meeting with the two co-chairmen of our steering group, I asked them not to treat the A380 independently of RECAT since we now had all the raw data results and all we needed were a few weeks to obtain the definitive values. The next day, the co-chairman from Eurocontrol brought together all the Europeans. He explained to us that the best solution was to accept the RECAT as it was and that otherwise we would run the risk of not seeing an outcome for the A380 dossier, bearing in mind that it would be easy to handle afterwards. He was probably right, but we decided to sleep on it! The next day, we all agreed: we shouldn't get caught up in this game, with the risk of losing all margin for manoeuvre facing the FAA.

At the second meeting of the ICAO wake vortex group in April 2011 in Montréal, the RECAT project was criticised by various participants because of its approximations. It was agreed that a new version would be presented later. It came in a little more than a month later. However, the calculation assumptions remained unchanged, and thus not demonstrated, and some of the aircraft characteristics were still inaccurate whereas it would have been easy to take a look at the manufacturers' websites to find the correct values!

Some people at the ICAO considered that the project was mature and that, just through organising a teleconference, RECAT could be accepted. I immediately opposed this idea, arguing that it was not acceptable for Airbus to ratify the penalties on the A380 obtained by LIDAR measurements in ICAO regulations as likely to apply for some decades to come. I asked that a small-scale meeting be scheduled before any decision on RECAT in order to decide on integrating the results obtained by flight tests for the A380. The principle was to negotiate a compromise: the separations obtained following tests were to be accepted and Airbus would not comment the methodology used for the RECAT project as long as no safety issue is involved. Otherwise, Airbus would make its comments. Considering those of the other administrations, especially the European ones, RECAT risked not being accepted as the ICAO could not allow regulations that had been criticised by a number of entities to be adopted. An FAA–Europe agreement was, without doubt, possible to resolve all these problems.

We received strong backing from several organisations, especially the French DGAC (Direction Générale de l'Aviation Civile). Similarly, the French representatives on the ICAO, headed by Michel Wachenheim, provided truly precious support.

We had the impression we would be able to take a major step forward at that meeting on 8 September 2011 in Cologne: either a compromise would be accepted, or be rejected and the RECAT project developed by the FAA would in turn be buried, failing an agreement on separation between the A380 and other aircraft.

We were disappointed. The FAA representatives did not want the two projects to be connected. However, it was obvious that there was no other solution. Independently of all the remarks on the methodology, we would have to set a value for separation behind the A380 that would be placed in category A and the other large aircraft placed in category B. Failing an agreement on this figure between the Europeans and the FAA, the RECAT project could not be accepted by the ICAO.

* * *

Looking at our raw results, it was clear that, for the aircraft in the Heavy category, there shouldn't be a specific penalty behind an A380. For Medium aircraft, the penalty was between 0.5 and 1 NM.

Detailed analysis of the recordings was pursued with the entire Airbus team and the Authorities, first with individual validation of each encounter, then processing of the parameters of those that had been selected thanks to the method developed by the young Design Office engineers. In mid-October 2011, at a meeting held in Denver, the first results were presented by Airbus, Eurocontrol and EASA. To summarise, for the Medium category, Airbus estimated the separation penalty compared to the A340-600 at 0.5 NM. EASA made the same analysis. Eurocontrol had not specifically looked at 0.5 NM but confirmed that 1 NM was acceptable. All these analyses were perfectly consistent. As far as the Heavy category was concerned, the raw comparative results were excellent, even showing that the A380 was very slightly less 'severe' than the A340-600, but there was some technical difficulty in correcting the results to make an incontrovertible demonstration. We needed to resort to a complementary method.

* * *

Mid-November 2011, the third meeting of the ICAO working group on vortices was held in Montréal. The main topic was RECAT project review. A large number of comments had been prepared by the various organisations in the form of 'Study Notes'. As usual, the FAA came in force with seven representatives. We all made our comments: we criticised an inadequately validated methodology, significant differences with our many LIDAR results, a certain number of errors and the fact that the effect on the following aircraft was not established by the type of measurements used. EASA, the DGAC and NATS (the British National Air Traffic Services) had comments of the same nature, as well as Boeing though to a lesser extent.

Another criticism was that this breakdown was drawn up on the basis of the aircraft population of the past but gave no consideration to forecasts for the years 2020. Thus, one of the categories, known as C, only included types that were soon to be lost off the radar screen like the B767, an old aircraft, produced in small numbers over the last years. Furthermore, improvements in capacity at the main European airfields were overestimated.

To summarise, the FAA had no support as Eurocontrol, although associated in the project, was bound to defend the interests of its member countries. During the first two days, the FAA refused to answer individual questions, simply repeating that its project was serious and presented all necessary guarantees to ensure safety. The DGAC would have liked to very slightly reduce the separation for Medium big aircraft such as the A320 or B737 type behind Heavies, just as did the RECAT project, cleverly, within the Heavy category separated into B and C. But there was systematic obstruction from the FAA for its classification to be reviewed, with them explaining that everything had been closely examined and that it was impossible to do any better.

At the end of the second day, the Chairman logically proposed that we set up a small group entrusted with answering all questions. The FAA representatives were unhappy with this and asked for the session to be suspended until the next day. That group was finally created with a remit to take questions, to answer them and, in addition, find a solution to the DGAC's request. That was the only possible channel for the RECAT project to have a chance of a positive outcome at the ICAO.

All in all, at that stage, the main protagonists were against the project as put forward. Even those who were initially rather more in favour of it started harbouring doubts as all the questions came in.

* * *

We had an appointment in Washington the day after these meetings in Montréal to examine the results of the A380 tests as interpreted by the FAA. In practice, nothing had been done. Their representatives were happy enough just to adopt a critical stance towards the test programme that had not been followed perfectly and place an objection that the meteorological conditions were not satisfactory.

Another meeting was therefore organised in Cologne on 13 December 2011. The FAA had nine representatives (despite claims that they were short of cash and available time!) as against four Airbus people. Six presentations were made one after another, not for questions to be asked, but to assert that, considering the statistics, separations behind the A380 had to keep a penalty of 2 NM for all categories, as compared with Heavy aircraft.

To put it short, for the FAA representatives, a very small aircraft could fly in complete safety at 6 NM behind a B747, but a B747 had to stay 6 NM behind an A380. They failed overtly to recognise the absurdity of this situation. They were sure their computations were right!

The next morning, I met the leader of the FAA face to face. I wanted to get across to him the message that we were heading for a stalemate situation. By blocking the A380, he was quite simply obstructing the ICAO's approval of RECAT. There was a real need for an agreement between Europe and the FAA on separations behind an A380 to be injected into this project, otherwise everything would come to nought. An agreement remained possible without in any way impairing flight safety.

The steering group met the next day (afternoon of 14 December) to decide on what to do next. First they listened to the report from the Europeans whose three entities (EASA, Eurocontrol and Airbus) showed uniform results making it clear that a significant reduction in separations behind the A380 was possible. Then the FAA 'demonstrated' that its calculations prevented any further reduction and that the 'penalty' should remain at 2 NM. This stalemate led to the consultation session being interrupted, after which the two co-chairmen, Paul Wilson and Alain Leroy, decided to pursue the study with a European group entrusted with formulating conclusions by the end of June 2012. The FAA could have observer status and possibly approve the results.

The next day's meeting was entirely devoted to the RECAT project and was subject to some turmoil. A last attempt was made to make progress. But, over the first days of January 2012, the FAA decided to withdraw the case from the ICAO. Discussions nevertheless continued as there remained a consensus that improvements are absolutely essential for the future.

As I write the final pages of this testimony, there is still no definitive conclusion on the matter of separations but there is a determination among the Europeans to settle these issues without having to refer back to the FAA. Reduced separations should soon be approved in Europe for the A380 and some other types of aircraft. I'm sure that, considering forthcoming safety demonstrations, other countries will adopt them. Just a little more time is needed.

* * *

What conclusions can we draw from this whole wearying story?

First of all, having spent forty-one years in the flight tests world, I had never come across such a situation. I saw disagreements and awkward negotiations, but, in the end, with all our counterparts, we always managed to find the best compromise solution even with other FAA departments without safety levels ever being compromised.

In 1962, John Fitzgerald Kennedy wanted the astronauts to land on the moon and less than seven years later - mission accomplished! With the A380, we started flight tests on vortices in 2005 and a conclusion for us Europeans is only scheduled for the end of 2012. About seven years to resolve this 'simple' problem.

It would be impossible to certify any aircraft using the same methodology as that used for vortices. The bureaucracy imposes highly unlikely demonstration methods. Thus, we made about 1,300 encounters on approach. All pilots agreed that there was no difference felt in severity behind the A380 or another Heavy aircraft. With such overwhelming experience, the obvious conclusion would be that there is no aircraft more 'dangerous' than another in this respect and that it would therefore be best to put them all into the same category. But we still had to provide the mathematical and statistical demonstrations to prove it, with the most difficult thing being to be absolutely sure that we are in the worst case scenario imaginable, otherwise the demonstration would not be considered a convincing enough example. Meanwhile, when it comes to certification, delicate issues of this type are often settled thanks to the judgement of experienced pilots working for the manufacturer and the authorities who together made a decision based on their intimate convictions. Safety is the byword, but there's no point throwing up false problems.

To round off on a positive note, I would like to mention the unfailing level of cooperation that marked our small team, with Frank Ogilvie, the aerodynamics specialist on the A380, a member like myself of the steering group, Florent Laporte, also an aerodynamics specialist, who has been working since the end of 2008 on the A350, Gordon Höhne and Andreas Reinke from the Design Office, who developed many analysis methods qualified even by the FAA as 'state of the art', and Terry Lutz, the American test pilot I had hired when I was still Head of flight tests. Other engineers also gave us tremendous support in our task but it would be difficult to name them all here.

CABIN COMPLETION

Delivering an aircraft with an irreproachable cabin is obviously one of the major priorities the manufacturer has in mind. The airplane can be perfectly safe, have excellent reliability, be easy to operate, but if passengers don't like it, it will never be successful. Conversely, if it is acclaimed, travellers will choose the plane rather than the airline and this will lead to some carriers winning over others. The airline that loses passengers will then seek to buy the same aircraft. And so it all becomes a major issue for manufacturers and airlines alike.

Success will depend on the aircraft's intrinsic qualities: noise level, vibrations, how it copes with turbulence, etc. and also, more directly, pressurisation, air conditioning systems, cabin fittings, the choice of seats, etc. The passenger has to feel comfortable, with the ideal temperature and good ventilation. Everything will depend on the manufacturer, but also partly on the airline, that decides on the general fitting out, with the choice of seats, spacing, specific equipment, etc. This was especially important for the A380, the prestige aircraft for our airline customers. In the long-haul aircraft field, finalising a customised cabin required dedicated testing. These trials are conducted on the first production aircraft of the airline.

As from flight # 5, I firmly believed the A380 was going to be a huge success. Just walking through the bare passenger cabin in cruise conditions had convinced me of its superiority (see the DISCOVERING THE AIRCRAFT chapter).

Development of the flight controls gave further improvements to comfort as tests progressed. Admittedly, the aircraft still moved a little in turbulence but, after all, even ocean liners are not completely stable when the sea is rough.

* * *

Development of the cabin was on the critical path for quick delivery of the first aircraft (see THE FLIGHT TEST SCHEDULE chapter). It should be remembered that we had a handicap as compared with other programmes as we were unable to install the cabin before the plane came off the assembly line as the aircraft had first to be readied for flight and delivered to Hamburg for that operation. This led to precious time being lost. So to follow the schedule, aircraft # 2 had to be readied for flight very shortly after the first take-off of # 1 and, installation of the cabin had to follow on as rapidly as possible.

In reality, this first aircraft to be fitted with a complete cabin only commenced flights on 3 November 2005, in other words a little more than four months late based on the initial schedule as revised for the date of aircraft # 1's first flight. We then brought it over to Hamburg on its third flight on 8 November 2005.

In the flight tests team, we knew very little about what was happening during installation of the cabin. But our specialists on the job were witness to some strange events. For example, to signal the end of a particular task (putting a cross in the box), all the covering panels sometimes had to be installed but then taken out the next day for another operation. Another example was when the aircraft had to be placed on jacks to make final settings on the flaps (following what had been seen during the first three flights of aircraft # 2). Works commenced, but to stick to the planned schedule, the aircraft was again set on its wheels before the end of the operation and setting had to be started again from scratch. We weren't aware of all the desiderata but it is likely that more realistic organisation would have allowed a few days to be saved on the installation work.

The fourth flight of the aircraft with equipped cabin was made on 22 May 2006 after the 'cabin virtual first flight' described later. We took advantage of this to bring it back to Toulouse as the length of the Hamburg runway at that time made it impossible to make landings with a heavy aircraft and so continue cabin tests on site.

The schedule was even more critical for aircraft # 7, the second to be equipped with a cabin. It had first of all to be prepared to conduct evacuation on the ground with a high density cabin, this being a demonstration required for certification (see the EMERGENCY EVACUATION OF THE CABIN chapter). There was no need for this test for the galleys to be fitted or for In Flight Entertainment (IFE) to be operational, but the entire on-board communication system had to be to certification standard, as also everything relating to safety including the slides, floor escape path lighting, lighting, etc. After that demonstration, everything had to be removed again and the cabin installed as appropriate to the next tests, bearing in mind that the same aircraft would also be used as a showcase for a number of commercial rounds. It was also to be used for endurance flights just before certification.

We had it fly on 19 February 2006 and it was ferried to Hamburg on 22 February for its second flight. On 26 March, the evacuation demonstration was successfully completed. Then, cabin installation really took longer than expected. As from early August 2006, it appeared to us to be obvious that this aircraft, that was to perform endurance flights, would not be ready on time and that it would have to be replaced by # 2. Now, the schedule had to be respected, and so no decision was made to do so! As time went by, everyone had to face the truth. A/C # 7 only flew again on 8 December to be brought back to Toulouse, that is after those endurance flights. And even so, there were many problems of all kinds concerning the aircraft itself as also cabin installation.

To summarise, all basic certification tests relating to the cabin, except for the evacuation demonstration, were conducted on # 2, then both aircraft were used to finalise the various fittings and options.

* * *

All the cabin equipment was initially tested on the appropriate test benches. Then, as with the aircraft that was first tested on the simulator, integration took place on 'cabin zero'. This full-size mock-up only represents a part of the aircraft but it allows the great majority of systems it has to be correctly evaluated. Such integration tests had been performed on previous aircraft, but they were pushed still further on the A380 in order to ensure the highest possible level of maturity as from first flights.

On the aircraft, many of the cabin development tests are conducted on the ground. That applies to most of the systems relating to communications, passenger entertainment, manoeuvring and setting the doors, loading and unloading the cargo holds, etc. A certain number of other tests required considerable preparation on the ground, including ventilation and air conditioning, galleys, etc. This can reasonably be estimated to come to about 70% of the volume of such tests, with the rest of them being covered by validation in flight.

So what are those tests to be conducted in flight? There are plenty of them as variations in pressure, humidity related to the air conditioning, heat exchanges with the very cold atmosphere at cruise altitude, even slight deformation of the fuselage, etc. can create difficulties. Many development flights are thus needed to obtain a result to answer to customers' expectations. This is all the more true for a prestigious aircraft like the A380.

* * *

Development concerns a host of different specialities to cover air conditioning, pressurisation, noise, firefighting, individual computers, toilet maintenance, etc. Good global coordination is essential as during all these flights, that are generally very long, a large number of tests are conducted in parallel. The team entrusted with the work, the 'Integrated Flight Test Team', included more members than the other groups working on a single subject like the autopilot or flight controls. It was headed by Jean Piatek who had considerable experience in flight testing of all sorts. Wolfgang Brüggemann, also a flight test engineer, assisted him.

* * *

Have you ever complained about the temperature in an aircraft? The chances are you will have felt too hot, too cold or been bothered by draughts. Getting on board, it's cold and you put on a pullover then, after take-off, you are too hot. You may also feel your legs are cold. The human body is extremely sensitive to small variations in temperature and that's why development of the entire air conditioning system is so difficult. It is essential to have a uniform temperature throughout the cabin. It must not be colder in the side areas, close to the windows or the doors.

The air used for air conditioning can come from a number of different sources: carts on the ground, the APU or the engines, and this very diversity is the first reason for the complexity involved. We had to develop two air conditioning systems known as 'packs' and then work on distribution of the air throughout the aircraft until a satisfactory solution was found. This distribution very much depends on how the cabin is fitted out and the relative position of the galleys, the doors and the toilets influences the final results. That's why on each first production aircraft of an airline we conduct an analysis of the air flows so as to adjust them accurately for each zone using appropriate restrictor systems. It is worth noting that the only time we had decided not do this work, on an A340, where the fittings were similar to those on another already delivered airplane, the aircraft was sent back to us by the airline as there had been a series of complaints about the air conditioning. We thus had to rework all the settings. In principle, the situation is much the same as in a car when outside temperatures are either very hot or very cold and it sometimes proves difficult to ensure everybody's comfort. On the A380, optimisation was made especially complex due to the number of zones to be considered in the cabin, whence a great many tests and flights to obtain satisfactory results.

For all settings, you have to ensure a sufficient air flow is maintained in all the zones otherwise passenger comfort will suffer. Tests on the ground normally allow such adaptations to be made.

The mean outside temperature at cruise altitude is -56°C but can go down well below, even as low as -70°C to -75°C in extreme cases. Maintaining an even temperature over the entire cabin runs up against a number of difficulties. Firstly, the temperature of the aircraft's skin varies considerably between the part exposed to the sun and that in the shade. On some types of aircraft, you can test this yourself by touching the windows in different areas in such conditions. Thermal insulation eliminates much of these differences, but a perceptible effect can still remain. Temperature regulation sensors have to be carefully placed and tests are sometimes needed to find the appropriate locations. On the A380, we didn't have to move them around, but their environment was modified to ensure correct air circulation in the zone where they were installed.

Another major difficulty related to the doors as locally there is a significant difference in thermal insulation. Thus, if nothing is done to counter it, you can feel cold in their vicinity, just like at home next to doors and windows in winter. However, the contrast between the inside and outside temperatures of the aircraft is huge. This cold bridge is extremely localised and boosting the temperature in this zone by just distributing hotter air would lead to an overall imbalance throughout the cabin. So, the favoured solution is to have a heating carpet, much like underfloor heating in a house, although here it is electrical and is installed in the door stiles.

Considering how difficult it is to get the air to circulate in the galley zone without creating strong local air currents, these are also equipped with electrical heating carpets similar to those for the doors.

To fine tune the air conditioning, we have to use small sources of heat representative of that released by the actual passengers and crew. Dummies that generate heat by a system of tubes and that fit in with the shape of the seat are thus installed in one or more zones, on all the seats. You can then set the heat each of them releases to conduct a detailed study of this parameter's influence. Our flight crew humour had it that a sleeping dummy would release 50 watts, 75 watts when eating and 100 when seated to an attractive member of the opposite sex!

The air is partially recycled in all aircraft air conditioning systems, being mixed with that taken on the engines and treated by the packs as on certain cars that are equipped with an automatic recirculation system. However, one of the A380's special features is the circulation of air between the two decks. This difficulty (which was new for Airbus) considerably complicated development.

Many flights, but also tests on the ground, are needed to adjust these air flows and ensure control over the adjustment techniques. The flights concerned are generally very long ones as you have to wait for thermal stabilisation conditions. Globally, volume-wise, they represent a major share of the cabin development tests, even though some other verifications can be conducted at the same time. In flight, either the APU or the engines can provide sources of air. The failure of one pack must not degrade the quality of temperature regulation and you have to check this. Development of the system involves both cold weather (-30°C) and hot weather (+45°C) test campaigns on an aircraft equipped with its cabin.

Surprises are often in store as long as all the outside conditions have not yet been explored, especially with extreme temperatures. Thus, during specific air conditioning tests on one of the hot weather campaigns, at Al-Ain, near Abu Dhabi, late in July 2006, we noticed that cooling was insufficient with just one pack in operation. This was unsatisfactory as it is always possible for one pack to fail. There is no risk for the aircraft, but passenger comfort must not be impaired.

Even with normal outside temperatures, we sometimes discover oddities. Thus, while we were satisfied with air conditioning on aircraft # 2, during flight # 15 of aircraft # 7, performed on 20 February 2007 from Toulouse with heating dummies, it emerged that some zones weren't being cooled as they should have been while, conversely, others were not adequately heated. In other words, the setting methods applied on aircraft # 2 failed to give good results on # 7. Everything had to start again from scratch due to the different position of the galleys and further work had to be commenced on our pre-setting models.

But we were still not out of the woods. In August 2007, just some months before entry into service, as we were thinking that we had practically completed these air conditioning tests, there was a new hiccup. During a sequence of flights in Dubai conducted on request from Emirates to check the boarding and disembarking times at their new terminal and also provide something of a treat for their personnel, it emerged that on the ground, the temperatures to the rear of the upper deck remained much too high, at about 29°C, during the stopover. Cooling with the air taken on the APU was insufficient. Modifications on regulation of the packs had to be made before delivery.

Once all these difficulties had been overcome, the final result was extremely satisfactory and there was not the least criticism from the airlines.

Setting of the air conditioning also has to be conducted for the cockpit, a particularly complex matter due to the number of surfaces exposed to variable temperatures, the need to have adequate air circulation and the draughts of various origins. We have to proceed by trial and error to ensure sealing of some zones, sending air at the right temperature and ensuring it is distributed correctly. Considering the cockpit's glazed surface area, sunlight plays a key role and validation has to be conducted for both day-time and night-time.

We had to make the same settings for the two pilot rest areas situated behind the cockpit as well as for that used by the cabin crew. Where the latter is situated is an airline choice. It can be either to the rear of the cabin, on the upper deck or at cargo hold level with various possible fittings and access by a stairway in the middle of the main deck.

Finally, we have to check the temperatures in the cargo holds, remembering that the transport of certain (e.g. perishable) goods may require a fairly low temperature, while animals may demand a milder temperature.

Overall, while difficulties were encountered finalising the A380, we reckon that this was all par for the course in developing a new aircraft. Certainly, a great number of tests and flights had to be made to optimise all settings, and the novelty of having circulation of air between the two decks was a major contributory factor. In this field, as in so many others, computer models have their limits and nothing can replace the common sense of the specialists!

* * *

Pressurisation caused concern for us as from flight # 2 (see the DISCOVERING THE AIRCRAFT chapter). Certain principles had to be reviewed and improvements were made over the fourteen months following the first flight. One of the difficulties in developing this system was to eliminate, or at least minimise, pressure variations related to changes in engine thrusts. Indeed, when the thrust is modified, for example when commencing a descent as the engines are set to idle, the pressure of the air taken from the engines to ensure pressurisation can be lower. On older aircraft, there is then a characteristic change in noise and a variation in pressure is sometimes felt in the ears. On the last generation of aircraft, design office engineers are called on to devise systems so that this unpleasant feature can be eliminated. Then there is just the gradual change in cabin altitude that can affect the ears, especially in descent and when you have a cold to boot. But this is inevitable as the cabin is usually at a greater altitude than the airfields of departure or arrival. The rate of climb of the cabin is normally limited to 500 ft/min and, in descent, it comes to 300 ft/min with a higher maximum value if the aircraft has to descend rapidly. Note too that on the A380, unlike aircraft of older design, with the standby system used in exceptional circumstances, the crew now commands a rate of climb or descent rather than having to open and close valves to handle the situation that can lead to uncomfortable cabin altitude variations.

The final standard of the pressurisation system was initially to be installed on aircraft # 1 but, clearly, it was much more interesting to have it on # 2 equipped with the cabin. After some discussion, in April 2006, programme management finally acquiesced to our request and we thus rapidly completed development so as to have a fully operational system for the first flights with passengers.

* * *

Well before the first flight, the doors were the subject of ongoing discussions between the crews and the designers. For the first time, we had electrically operated doors. Our short-lived CEO, Christian Streiff, expressed his surprise at this during the first meeting with all the leading execs. He reckoned that the mechanical system traditionally used was perfectly adequate and that this innovation was risibly superfluous. Sour smiles and mildly suppressed laughter all round! But his conclusion was no doubt a little too hasty, especially for the doors on the upper deck.

A new rule required the doors to remain locked until the pressure inside the cabin was the same as the pressure outside. The aim was to avoid untimely opening with a pressure difference that could even suddenly project people out of the aircraft. This type of accident had occurred a number of times, despite red warning lights being triggered on the doors when attempts were made to open them in such conditions. To reinforce these warnings, various modifications had been implemented and some had been installed in retrofit. But EASA had the additional requirement that unlocking must in no event be possible as long as a pressure difference subsisted. The drawback was that, in the event of an accident on the ground, there was no guarantee that unlocking can take place, meaning that all the doors could even remain blocked. There had been two tragic and fatal accidents at discotheques early in the 2000s with many people dying due to the emergency exit doors remaining blocked. So we were ready to contemplate the worst case scenario where an aircraft suffers an accident. For the flight test team, this was unacceptable and we made a commitment never to have passengers carried in such conditions. Since a number of members of the administration stuck to their position, talks on the matter were arduous before we finally hit on a compromise. We insisted, in vain, on the fact that there were many existent modifications to provide an equivalent level of safety to that sought. Although at Airbus we were unanimous on the principle, some engineers were unwilling to take the risk of seeing the programme schedule slide due to a dispute with the certification authorities. It is interesting to note that the EASA crews backed us up. Discussions were heated before it was finally decided that (to simplify matters) any electrical cut-off in the various zones of the aircraft would lead to the door safety systems being unlocked.

Another debate concerned door numbering. It was initially planned to give numbers 1 to 9, followed by L or R for Left or Right. Door 6 didn't exist on the version in construction of the A380 as this number was reserved for the door furthest to the rear of the main deck of a longer version, making 7 the first of the upper deck. The flight tests crew wanted numbering starting with a letter to tell you the deck (M or U for Main or Upper). Our principal argument was essentially for better understanding for any outside parties having to intervene (fire-fighters, for example) who would not necessarily be aware of the position of door 7 and possible confusion with the position of door 6. Most of the airlines supported our position for operational reasons and we gained satisfaction. So now, door U2L means the 2nd door of the upper deck starting from the front on the left hand side.

Throughout the first part of tests on the development aircraft, we had had many warnings on doors on departure on each flight, for example for failure to lock on closure. Of course a check was made before starting the engines but, on more recent aircraft, there has to be a device to prevent pressurisation if all the doors are not declared closed and locked (this safety will no longer be active if there is a warning after take-off). The first type of Airbus to be equipped with this system was the A340-600 in 2001. To avoid loss of time or even seeing test flights cancelled due to false alarms, which are generally pretty frequent at the start of a programme, before flights have even commenced, we had installed a system on the airplanes to inhibit this safety feature. On the development A380s, a similar device was installed. It was nevertheless designed so that causes of the malfunction or its display could still be analysed for lessons to be drawn.

* * *

There are 16 emergency evacuation slides on the A380, with 10 for the main deck and the remaining 6 for the upper deck. Each slide has two runs so that two passengers can be evacuated at a time and each of them has a short 'deceleration strip' providing greater friction as you get closer to the ground to slow you down before landing (scrubbing the seat of your pants in the process!). In the middle of the main deck, the two emergency exits open out onto the wing and slides lead down to the rear. The slides on the two forward doors of the main deck (M1L and M1R) are different in having an 'extension' that is deployed if the aircraft is nose up (with more than 2.5 degrees of pitch). The aim is to facilitate evacuation if the aircraft is resting on the main landing gear and the rear fuselage as these slides, located well to the front, will then be too short to reach the ground with a normal slope.

The slides of the upper deck were a big challenge for their manufacturer Goodrich, obviously because of their length, but also because they have to remain in position even in strong winds (to specified values in the regulations) whereas their sheer size makes them vulnerable. Changes had to be made a number of times during development as initial certification tests met with failure. In practice, fairly high side edges, inflatable walls, were first installed on the slides to stiffen the assembly. They also have the advantage of reassuring passengers when facing such a long slide thanks to a tunnel effect masking the side drops as they get ready to go.

When stowed, the upper deck slides occupy a considerable amount of space and, rather than being placed in the lower part of the door as is usual, they have their own recesses in the main deck ceiling with a dedicated opening to the outside.

Certification requires slide deployment tests to demonstrate their reliability. First, five consecutive deployments for each door and then all doors at once, sixteen slides in all. This demonstration is independent of the evacuation test itself.

As we came to delivery of the first aircraft, fairly frequent inspections of the slides were made. A series of modifications were enough to remove any remaining these.

* * *

Study of the noise in the cabin is fundamental in development of an airliner. No test is prescribed by the regulations but efforts have to be made to detect all parasite noise that could reduce comfort. There are many possible sources and identifying them is often difficult as they are sometimes to be found outside the aircraft.

Imagine you live in an apartment block and someone starts up a power drill. The noise is persistent and pervasive and you are utterly unable to identify where its origin is. It may be coming from a flat above or below your own. You won't know until you have gone to the door of the apartment where works are going on. It's much the same on an aircraft but with the major difference of being unable to reconnoitre outside your flat!

As from the first flights of aircraft # 1, we had been intrigued by a fairly strong noise to the rear of the cabin. Given its special tone, we had called it the 'turboprop' noise as it sounded much like such aircraft engines equipped with propellers. A number of straightforward searches were made, for example by cutting off the air conditioning, varying speed and altitude, or manoeuvring the aircraft in tight turns, but we were unable to determine the cause. For flight # 14 on 27 May 2005, we set up a dedicated test installation intended to make recordings and analyse the frequencies to try and pick up the trail, but this proved unsuccessful. We then had the idea that the noise could be coming from vibration of the side plates to be found at the horizontal stabiliser root, just against the fuselage. Various tests showed us that this was not the cause. Another hypothesis was posited suggesting it could be coming from the fairing at the bottom of the fin. Some additional checks were made that confirmed that we were on the wrong track. Finally, in autumn 2005, we discovered that the seals placed on the attachment of the leading edges of the horizontal stabilisers in the zone close to the fuselage were too flexible and that the structure vibrated locally. A temporary correction was made for the November round of flights to South-East Asia and Australia. The final modification was installed early in 2006. This provides a good illustration of what our work is all about – much searching around before being able to identify the origin of the problem.

One of the commonest sources of noise is around the doors. We often have to rework this area after the planes have come off the assembly line following the first flight due to whistling or rumbling noises. Most of the time, the settings of the door hooks and stops have to be adjusted. On a development aircraft, you also need to make sure there is no design problem. These noises are often due to quite simple issues like the setting of locking stops or hooks, flexibility of the fuselage, speed or pressure difference between the interior and exterior of the aircraft. Few tests were conducted on aircraft # 1 and 4; as they had limited sound insulation, a thorough evaluation would have been difficult and, furthermore, their test programme was a particularly busy one. We therefore only really started this work on aircraft # 2 with the cabin installed.

The most frequent cause of noise close to the doors is due to the cover plate. This is a plate secured to the upper part of each door that is intended to ensure good airtightness around it. On the A380, it includes a flexible composite material seal that must mate perfectly with the structure of the fuselage when the door is closed. Even when all seems perfect on the ground, the situation may well prove to be different in flight due to dilation or slight deformations of the structure and the seal may fail to stick to the fuselage. The plate can then vibrate and generate a whirring noise. Setting of the doors or the cover plate then has to be adjusted according to the results in flight to ensure perfect mating. On series aircraft you want to avoid having to make such settings after the first flight as this can lead to having to proceed with an additional flight for verification before delivery to the customer. This all means extra work and time spent on development aircraft. One of the new features on the A380 was to have this part made of composite material rather than being metallic as on other Airbus planes. That previously meant a good metalworker could perform retouches on return from the flight to eliminate the noise. This was no longer possible on the A380 where you have to make sure the door is a perfect fit. Many modifications were made to the cover plate before it was finally brought in to service, and even a few other changes shortly afterwards.

On 31 May 2007, on aircraft # 7, one of the forward doors was leaking and making noise. A setting made previously had not proved effective. Our cabin specialists had already conducted experiments on similar situations. They had let a liquid flow at the bottom of the door and, with a sufficient pressure differential between the interior and exterior adequate, it flowed outward and froze on exiting, thus plugging the leak and eliminating the noise. The specialists even told me that they had used a bottle of wine left over from a VIP flight to validate the technique. I never checked out whether this was a joke or not, but the result was there! That day, we absolutely had to find a solution as a VIP flight from Paris-Charles-de-Gaulle was planned for the next day.

So, a special installation was prepared to have liquid gently and discreetly flow there and thus eliminate the noise. A do-it-yourself solution again, but it worked!

Among the other sources of noise you regularly find on all types of aircraft, there are the seals of the belly fairings (fairings located between the fuselage and the wing as explained above). These fairings do not sustain any other structural loads than those of the aerodynamic forces working on them. Rubber seals ensure contact between the belly fairings and the wing on one side and the fuselage on the other. If a seal breaks, it can locally generate noise or even vibrations that can be unpleasant for the passengers. Although these are secondary parts, great care needs to be devoted to their design. On the A380, as from the first flights, we realised that there were anomalies, firstly on the fairings and secondly on the seals. The fairings themselves proved to be insufficiently rigid (see the FLUTTER chapter) and the seals couldn't stand the strain. All the first flights were conducted with aluminium tape being applied to ensure these seals held. I have no idea of the number of rolls of this tape used, but there must have been quite a few as after most flights, big lengths had to be replaced. With the reinforced fairings, the seals still failed to hold and tended to come apart. The material used was therefore changed as well as the shape to obtain a satisfactory solution. Meanwhile, so as not to lose time, we continued flying with our aluminium tape (a basic material for mechanics working to develop aircraft, however big they came!) Note that this development was implemented on aircraft # 1 and # 4 as we needed to ensure as quickly as possible that there was no anomaly in principle on the belly fairings and the seals that could have been the cause of the noise.

Abnormal noises in a cabin can derive from other sources that need to be tracked down as soon as possible. To do so, microphones were set over the entire test aircraft. Analysis of the frequencies allowed the origin of most of these to be detected and study possible improvements. For the A380, these measurements were mainly conducted on # 2 during the various flights. Some of our customers are extremely sensitive to the slightest unusual noise when they take delivery of an aircraft. So before handing over the airplane to them we have to ensure that everything is nominal. To limit the work searching for such anomalies as the plane comes off the production line, it is best to identify possible sources as from the development phase.

Measurements also have to be made in the cockpit on the captain's and copilot's seats to reproduce the noises in the simulators. The regulations give the precise list, covering all flight sequences from power up through to start-up of the APU, closure of the cargo doors, etc. and then complete shut-down of the engines and the APU, and electrical cut-off.

* * *

Putting out a fire in the cargo can only be done remotely. The principle is to extinguish it by saturating the atmosphere with a neutral gas, halon. We have to ensure that the concentration of this gas is adequate for it to be effective for four hours in cruise. The aircraft must have the time to reach a diversion airfield when flying over regions where there is no close lying landing area as, for example, on the route going close to the North Pole. Usually, discharge occurs in two different stages to obtain a concentration that remains reasonably constant over time and not have a gradual reduction. To do so, after the instantaneous discharge of a first canister, a second gradually releases halon to maintain a sufficient quantity of the gas.

Test flights on this system always take a long time. While the concentration is normally adequate at the start, what really counts is how things stand at the end of the period – will we remain within acceptable limits right through to the very last minutes? We have to go through to the end of the four hours to see whether the test is successful or not, unless the concentration drops more rapidly.

Measurements of the concentration must be made at different positions in each hold. In practice, there are about sixteen check points as we have to ensure that there is no local deficiency that would allow the fire to be kept alive. Success lies in positioning the discharge systems and in sealing of the cargo. It's rare to get things right first time, and tests have to be conducted on all the holds. For the A380, that meant checking the forward and rear cargo as also the cabin crew's rest quarters where these are situated under the main deck.

For the A380, no particular difficulty was encountered during these tests.

* * *

One of the most complex systems on modern aircraft nowadays is the In Flight Entertainment (or IFE) system that provides video, music, games, etc. on demand. This is not subject to certification but we still need to ensure that it answers to all the regulations concerning airborne systems and, in particular, that it does not create interference that can affect the avionics. If the IFE fails to work on a long-haul flight, the passengers are likely to remember it and the aircraft will be discredited in their eyes rather than the system itself. We had our work cut out when IFEs were first brought into service in the mid-1990s due to their disastrous lack of reliability. It was rather like your laptop cutting out suddenly for no apparent reason. But much progress has been made since that time and the failure rate now has become extremely low on the latest aircraft brought into service. All passengers can bear witness to this. However, upgrades still lead to new bugs emerging from time to time and these need to be pinned down before the start of commercial operations.

On the A380, two systems are offered the customer, one manufactured by Thales and the other by Panasonic. For their development, they were each installed on one of our cabin equipped aircraft, with Thales on # 2 and Panasonic on # 7. Test benches are used to check out these systems but nothing can replace the simultaneous operation of several hundred systems installed on an aircraft. So, during flights with many guests aboard, we would ask them all to use their IFE to the full with all functions.

A lot of modifications were made overall but, on entry into service, the IFEs worked perfectly.

* * *

The Cabin Intercommunication Data System (CIDS) manages all communications between cabin crew, with the cockpit, announcements for the passengers, lighting, air conditioning, and so on. Clearly, it would be unthinkable to make a flight with the CIDS in a situation of total failure considering the consequences on flight management. That explains why the system is duplicated. In addition, some functions have special protections due to their importance in terms of safety, for example those to trigger an emergency evacuation.

The CIDS was mainly developed on the ground, but validation of all the functions is pursued in flight. Indeed, it is an essential feature as it is used to control a large number of settings, including the temperature in the various areas, the lighting, etc. It is also needed for information to be exchanged between test crew members, pilots included.

We encountered some difficulties during development, especially sudden shut-downs (again as can happen with personal computers). Restart too was just the same as a reboot!

* * *

The A380 has a centralised supplemental cooling system for the galleys. Indeed, on previous generations of aircraft, these galleys had one or more refrigerators installed. A study showed that these could be done away with to make weight savings with this system distributing cold fluid to the galleys. The principle is similar to centralised air conditioning in a building.

The first application of this system was on the A340-600. During development we had many failures and leakages and development was pretty laborious. The centralised cooling system lived up to its reputation on the latest aircraft in the family and it caused us a lot of trouble. Tests on the ground were not enough to detect all the anomalies and so some of the verifications were made in flight in realistic conditions, in parallel with other cabin tests. It proved tough going and on 7 July 2006 we even interrupted the first flight mainly devoted to supplemental cooling due to multiple failures. Furthermore, during certain rounds and during endurance flights it remained unserviceable for long periods of time and we had to take large blocks of ice on board just like in the old days! This was the case for the second series of endurance flights to Hong Kong and Tokyo where the system conked out completely. But we finally got the system to work correctly!

* * *

A number of oxygen supply systems can be used in the event of depressurisation. The best known and as now currently installed on most airliners is with the generator placed just above the seats. All you have to do is pull on one of the masks. As it comes down, it triggers a chemical reaction that provides oxygen for the time needed to make an emergency descent to an altitude where you can breathe normally, that is towards 10,000 ft.

On the A380, to reduce weight, we installed a centralised oxygen supply system with carbon fibre canisters each holding 3,340 litres at atmospheric pressure, but with the gas stored at a pressure of 128 bars. The customer chooses how many are to be installed according to the planned routes and the number of seats installed. Thus, if high mountains are not flown over for long periods and the aircraft is not in a high density configuration, you can just have six canisters, giving about fifteen minutes autonomy so you can make an emergency descent on most routes. As an option, you can install up to thirty canisters as, on the most critical route between Asia and Europe that includes flight over the Himalayas, you may have to fly for at least forty-five minutes before being able to go down to 10,000 ft, choosing a path that takes you rapidly away from the mountains. The main advantage of this centralised system is that it can be adapted to the routes actually operated by the airline to considerably reduce weight.

Considering the schedule for our test aircraft, we had initially accepted just having fifteen minutes autonomy on aircraft # 2, with six canisters installed, as very few trips were planned, with endurance flights being performed with # 7. But on this latter aircraft, we wanted to have more than fifteen minutes available so as not to have any limitation on routes flying over the Himalayas. It turned out however that the endurance flights were actually made on # 2. In practice, the route restrictions on this type of flight were not critical since, one way or another, we had to perform a minimum of flying hours for these tests. The situation is fundamentally different for an airline where you absolutely have to seek reductions in flying time and fuel consumed.

The oxygen tests were limited to making sure the system worked correctly, especially activation, the masks coming down normally and providing a correct flow.

* * *

Fitting out of the cabin was done at the Hamburg-Finkenwerder plant. Now, at the time the A380 was being finalised, test flights could not be conducted from this airfield as the runway had not yet been lengthened completely and there were landing restrictions. However, we wanted to assess installation of the cabin before bringing aircraft # 2 to Toulouse. If there had been too many retouches to be done, it would have been more difficult to do them in Toulouse as the competent personnel were in Germany. We therefore decided to make a virtual flight on the ground (the 'cabin virtual first flight') to get all the cabin systems working. Some people were against this, arguing that we were going to be losing a number of days on the schedule with this dedicated test. We even had to argue our case to first perform calibration of air circulation in the cabin despite the fact that it was obviously useful as we were going to get a first impression of how the air conditioning worked with 500 people on board, even though all this was on the ground.

For the first time, a very large number of people were going to be taken on board an A380 and we had to ensure that all these 'passengers' could get out of the aircraft in a reasonable time in the event of smoke or fire. At the time when the cabin virtual first flight was organised, the evacuation test had not yet been done and we obviously remained uncertain as to its success. What is more, due to lead-time problems, we couldn't bring the aircraft into conformance with the configuration for that test. We therefore had to find another solution and prepare a rapid exit of the passengers by other means. As a result, we decided to install ten stairways, five on each side, with eight for the main deck doors and two to the rear of the upper deck.

That 'flight' took place on 9 May 2006 with about 500 passengers, including many engineers and technicians, and with participation from Lufthansa cabin staff. It lasted three and a half hours. The engines were not running and electricity was supplied by APU generators. The conditioned air came from packs using air taken from the APU. There were a few failures, for example with the conditioning that stopped practically at the end of the test, and on the CIDS. The IFE failed to give total satisfaction. Finally, we diagnosed a number of anomalies in terms of temperature distribution. To summarise, it was an extremely useful exercise and a success at this preliminary stage in cabin completion. In particular, it allowed us to perform an effective evaluation of its general condition. It also provided a demonstration that cabin flight tests could be started without further ado. However, we still had to complete preparation of the aircraft to get it back in flight after this long period of work. This was thus commenced immediately and, on 22 May 2006, the aircraft was ferried to Toulouse to kick off testing.

* * *

One amusing example of development involved the toilets. As always, most of the work is done on the ground but, in flight, validations are needed. During the A330's endurance flights in 1993, we had had leaks from these toilets on a number of occasions. That was pretty embarrassing! There had been minimal verifications on the ground, but we had not done anything specific in flight other than using them on long trips. For the A380, it was out of the question that we should find ourselves in such a situation, whence a test campaign over several flights. These flights had a number of goals and the toilets were just one issue among others.

A number of toilets to be tested were equipped with a small counter that set the following sequences: deposition of a small sample 'product'(!) followed by a few leaves of toilet paper, then a little rinsing liquid and, finally, flush. A new sequence started again a few minutes later. This toilet inspection shows how thorough we were in all our development work. Such tests were genuinely necessary as the in-flight environment is different than that on the ground, with the effect of altitude (even if the cabin is pressurised), the attitude of the aircraft (often slightly nose up), turbulence, etc. In addition, we had to check the entire system for recovery of waste from the sewage tanks.

* * *

Just like in a house, the evacuation of waste water is a serious matter. The waste from the toilets is stored in tanks that are emptied when the aircraft comes onto the apron. Water from the washbasins meanwhile is evacuated outside during flight and immediately vaporises with the speed of the aircraft. The evacuation drains and sometimes part of the pipes have to be heated as the formation of ice would block any water from exiting and make all the washbasins unusable. During development, we therefore have to ensure perfect draining in all conditions, especially in extreme temperatures. We did indeed encounter icing of these pipes on 26 August 2006 during a long flight to the North Pole intended especially to prepare the aircraft for the first flights with a large number of passengers. A makeshift repair was made during the flight and, on return, we were able to proceed rapidly with the necessary modifications.

* * *

Not all problems can be seen until you have flown with a large number of passengers on board and for a sufficiently long time. That's why we organised a series of cabin Early Long Flights devoted to development with about 500 people (see the EARLY LONG FLIGHTS chapter).

* * *

Finally, let's just browse through a few examples of issues that have to be considered during development from among a host of others that can't be covered exhaustively in the present publication.

Cabin lighting is important when it comes to passenger comfort. On long-haul flights, a wide range of modes exist, with different arrangements and more or less gradual changes in intensity as you turn off or on. This is a strong feature in customising the aircraft for each airline. We therefore seek to develop a range of possibilities rather than a single solution. In addition, we had to validate our own choice for our two cabin equipped aircraft.

The passenger announcement system requires special care. It has to be clearly audible, with the right level of volume and the position of the loud-speakers has to be adapted to the arrangement each customer wants.

Some airlines asked for installation of a trolley lift to carry food from one deck to the other. We had to check out how it worked and so it was installed on aircraft # 7 to be tested. This led to a number of modifications.

Apart from temperature settings as already mentioned, how the rest areas for the pilots and cabin crews are fitted out had to be validated. There may be differences between an often incomplete model tested and the real installation and any anomalies have to be picked out as soon as possible so they can be corrected. Maybe we can boast about a certain success in soundproofing. It turned out that the ambient noise in the aircraft was so quiet that for one of our customers having chosen to place the cabin crew rest compartment on the upper deck, we soon had to proceed with internal sound-proofing improvements on entry into service as the crew were bothered by noise from the passengers (with the usual chatter not being crowded out by the noise of the engines as on other aircraft!)

Yet another really simple example. A toilet or cabinet door may work correctly on the ground but unexpectedly jam in flight due to very slight deformations of the aircraft structure.

Customers sometimes want specific options. Emirates for example wanted showers to be installed for first class. According to the nature of modifications needed on our development aircraft and the foreseeable difficulties in finalising such an option, this feature can either be validated on our own airplanes or on the first in the customer's series. It was this latter solution that was chosen for the shower as this installation had already been introduced on aircraft we had delivered to private operators and States.

To round off this chapter on a lighter note, we sometimes had to schedule dedicated flights to settle the bubble problem, something encountered on pretty much all types of aircraft. This concerns pockets of air that become trapped under the carpet when it is laid. In cruise, this leads to a bump being created on the floor as the air expands. So we fly with a mechanic on board holding a syringe. We increase the cabin altitude, reducing pressure slightly so the bubbles trapped at the higher pressure on the ground expand, and the mechanic then punctures the bubble and draws off the air into the syringe. High-tech stuff in modern aeronautics!

EMERGENCY EVACUATION OF THE CABIN

In the event of an accident, the passengers have to be able to get out of the aircraft immediately. Also, during certain incidents, the crew may decide to have the airplane evacuated rapidly if they consider there is a risk in staying on board (for example if smoke is detected). For certification, the regulations therefore require that a demonstration be made of emergency evacuation in extremely stringent conditions.

To ensure the required safety level, the aircraft's design, the layout of the cabin and installation of various systems must respect a number of rules. This applies to the doors, emergency exits and slides, for example. A number of these issues are mentioned in the CABIN COMPLETION chapter. The maximum distance between the emergency exit doors, as set by the regulations, has serious consequences both on the aircraft structure and how the cabin is fitted out.

As for the evacuation demonstration itself, the conditions for performance were defined extremely precisely by the FAA long ago. No passenger taking part must be familiar with the aircraft or have already taken part in an emergency evacuation. This means the manufacturer cannot resort to using volunteers from among its employees. All participants have to be recruited from the general public and the cabin crew employed (who should have no prior experience of this type of operation) traditionally come from a customer airline's staff who are first qualified on the aircraft type.

The exercise has to be conducted at night, with only half of the emergency exits being operational. Which ones are used is left to the judgement of the certification authorities and the cabin crew only discovers that some of them do not open once the exercise has commenced. Evacuation of the passengers and crew has to be performed in less than 90 seconds. If not all the occupants manage to get out, the evacuated passenger count at the end of the 90 second interval is simply used to set the maximum number of passengers that will be authorised for the aircraft.

* * *

The European and American authorities are well aware of the serious dangers this type of demonstration represents. Many years ago, after such a test with McDonnell Douglas (now Boeing), a woman was injured and became hemiplegic. There have been plenty of other more or less serious injuries. Broken legs are a fairly common occurrence. Personally, I find it difficult to understand why the administration accepts the idea of having non-professionals take such serious risks with very little prior information on the real nature of the danger. Such risks should only really be taken by people capable of assessing the issues, in other words, professionals. There are numerous examples in flight tests where it has been accepted that certification procedures be reviewed so as to make them less hazardous without in any way detracting from their demonstrative value. So why not do the same for emergency evacuation?

We never managed to convince the authorities to use equivalent but less risky means. A simple method would be to conduct a first evacuation onto a platform placed at the level of the aircraft doors without the slides. Then, a flow of passengers through each type of emergency exit door and slide could be measured with a reduced number of people. Analysis of the results from these two tests would allow the aircraft capacity satisfying the regulations to be determined accurately with a minimum of risk. Unfortunately, as is often the case, we shall have to wait for a serious accident to occur during this type of test before deciding to adopt a wiser demonstration method.

Initially, Airbus had suggested doing the test in two parts – first the main deck and then the upper deck. The advantage with this was to reduce bottlenecks on the ground and thus minimise risks for the participants. Organisation would also have been facilitated. But yet again, this was refused by the FAA some months before the demonstration whereas EASA, the primary certification authority, had accepted the principle. We therefore had to re-organise the entire test for simultaneous evacuation of both decks.

We managed to obtain two amendments to the regulations. First of all the test was not conducted in complete darkness but in twilight conditions, so as to facilitate the rapid egress of the passengers towards the hall exit. It was also accepted that the upper deck slides be deployed before the start of the demonstration. The reason was that they were not totally reliable at the time and the fact that failure of just one of these slides to come out would have required the entire test to be repeated with a thousand new volunteers and a new airline cabin crew for reasons explained above. In the end, everything was organised for a second attempt to be launched a week later as there were other potential reasons for failure. In any case, specific demonstrations of the upper deck slides were conducted after that test.

* * *

A scale 1 wooden model of part of the cabin was built. Here, the aim was to finalise the technique to get the participants out of the way as quickly as possible when they get to the bottom of their slide so there would be no collisions. The slides had two runs and we placed eight people at the bottom, four for each side, to take care of the 'passengers' as they arrived and make room for the following ones. This model was used to train nearly a hundred assistants for this task.

In so far as was possible, protections had been placed on the ground, on each side of the slides, to reduce the risk of injury if someone came over the side wall on each slide.

Aircraft # 7 was to be put into the 'high density' configuration with the maximum number of passengers for whom the A380 had been designed, i.e. 853. This installation was relatively quick, as the galleys and toilets were replaced by wooden partitions and there was no oxygen system. The only things that had to be in perfect condition were the systems relating to evacuation, as with the computers managing the signals and public announcements, illuminated escape path, doors and slides.

Lufthansa provided us with the cabin crew. We prepared fifty people to cover the eventuality of the test having to be repeated the following week. In reality, there was not supposed to be any special training for this exercise, as the regulations forbade it. All we did was qualify them on the A380. The list of staff selected had even to be approved by the administration!

Recruiting the volunteers turned out to be a major task. We needed 1,000 people, substitutes included, for each of the two trials planned, 2,000 in all. By mid-November 2005, for a test scheduled for the end of February, we had 11,500, so everything seemed fine. But the authorities had their requirements and we had to have a certain percentage of people over fifty, as also from each sex. In addition, the selection was accompanied by other criteria, especially medical, and about half the files were rejected. Finally, we were short of candidates, both men and women of more than 50 years of age. So new recruitment campaigns for that age group were initiated. All the participants received a small compensation of 60 Euros for the operation, but I'd bet many of them would have paid to go through the adventure. After all, going back to your childhood whizzing down such a big slide was well worth the cost of a trip to Disneyland®!

* * *

Several months before that demonstration, flight tests had asked that Airbus aircrew be able to observe in the cabin, without intervening. We proposed that these people be hidden in wooden cabinets simulating the galleys. They would only have got involved to stop the test in the event of serious danger, for example if someone was trampled on.

Despite our insistence, this was refused by the authorities and we were not supported by our management. It has to be said that the flight tests team was not responsible for organising this demonstration and I am not sure that all those opposing our wishes had correctly weighed up the risks. Indeed, just saying that nothing serious had gone wrong until then in the cabin during all the previous evacuations was not a convincing argument as accidents can so easily happen. But there was nothing we could do.

At the last moment, some EASA certification managers wanted to 'hide' on board as observers, an idea that made sense. Once representatives of the authorities were admitted to the cabin, there was no further reason to refuse Airbus personnel, and that's how, finally, we obtained satisfaction: four of our specialists were able to watch the evacuation from inside the plane.

• See photo n°32 – page 245

For the record, in order to make things look more authentic, the representatives of the authorities distributed dolls to simulate babies that also had to be evacuated. They also placed a few cardboard boxes and cushions in the aisles to hinder the passage and make the scenario of a disaster landing more realistic, just as required by the regulations.

* * *

On Sunday 26 February, the crowds were out at the Hamburg site for this evacuation, starting with the 1,000 'passengers' in addition to whom there were the other participants, including the cabin crew, the evacuation assistants alongside the slides, medical aids, firefighters, the technical staff, the organisers, the certification authorities, and so on.

We were all extremely worried for a number of reasons. First of all, I'm sure that few of us would have wagered on the test being a success, meaning the evacuation of 853 passengers and 20 crew members in the 90 seconds as required. The second, and the main concern, was that we feared there would be serious injuries as, statistically, considering the number of participants and experiments conducted on other aircraft, that certainly could have arisen.

For the organisers of this exercise, there was another unknown. On the upper deck, the first forward door was located some way behind the passengers seated in the first rows. How would those passengers react? Would they head for that door or, on the contrary, go down the stairs to go out by the forward door of the main deck, which could have created congestion locally? We were not entitled to give them any instructions.

The authorities finally decided that the doors to be opened would all be on the same side. That facilitated observation but had the disadvantage of concentrating all the passengers on the ground in the same area and thus making it more difficult for them to move away after having left the aircraft. The certifiers no doubt considered that this situation was the most critical for the flow of passengers away from the plane and so the exercise was to be conducted in those conditions. Indeed, the slide that is deployed to the rear of the wing comes down below certain of the four slides to the rear of the aircraft, rapidly leading to a major concentration of evacuated passengers in this zone.

The traditional pre-flight briefing that, as usual, included evacuation instructions, was given by the cabin crew, obviously avoiding the provision of specific information for the exercise.

The pilot-in-command gave the signal, evacuation commenced and the stopwatch started. The five main deck slides were deployed (those of the upper deck already being so). A first stage had been completed! Indeed, failure of any of the slides to extend would have compromised the test as a whole. Exit of the passengers started immediately through the eight doors. The pace was good and everything went well. The videos showed the pilot leaving last, some time after the passengers. In a real accident, it would have been normal for him to take some time to inspect the cabin and make sure nobody needed help. In the present case, this was to no purpose, and an extra couple of seconds were in the offing on the time!

The final score, as validated by the certification authorities, was thus for an evacuation of 873 people, passengers and crew members, in 78 seconds. Considering this excellent figure, it has been planned, after analysis, to slightly increase the maximum number of passengers, without having to conduct the test again.

Finally, contrary to what we had feared, there was just one broken leg and a few light injuries. That's what you can call good luck!

Success all down the line.

EARLY LONG FLIGHTS (ELF)

ELF used to be the name of a major French oil company. The choice of the term and its acronym (well before the A380 adventure) was a wink in the direction of one of our major suppliers.

What do these flights involve and why are they essential to effective development?

Their origin goes back to the A340-300, the first real Airbus long-haul aircraft, being brought into service. The cabin for this type of airplane has to remain perfectly operational on very long flights, whereas you can be much more tolerant on A300 type medium-haul airliners built in the early days of the Airbus adventure. For the A340, development of the cabin had been reduced to the strict minimum and we only had one flight test engineer responsible for all verifications. And happened what was bound to happen on entry into service with a host on anomalies in the cabin. Technically, the aircraft worked perfectly, but to the rear of the cockpit, there was failure after failure.

Just to recall some personal anecdotes, I remember having made two round trips to the United States and Mexico in the early days of the A340-300's entry into service, in principle to advise Air France pilots. I hardly helped them at all, as they were perfectly familiar with their aircraft and were well trained. However, on request from them, I intervened many times in the cabin to repair videos, toilets, seats, and so on. Usually this just involved a simple circuit breaker reset (much like a personal computer reboot!). Cabin crew were not authorised to do this, as a minimum of technical knowledge of the systems was required. I even remember a return flight from Washington to Paris when all the toilets failed over Newfoundland, that is before even starting to cross the Atlantic. The passengers were sleeping. What was to be done? Divert the flight? Continue? Finally, the crew decided to head on to Paris, only waking the passengers as we began the descent and depriving them of breakfast. The flight arrived early and it turned out to be the best option. However, such an incident shortly after departure, on a very long haul and in day-time, would clearly have led to a diversion.

For the A340-600 programme we had thus decided to make considerable efforts to finalise the cabin. Clearly, as already seen, most of the development can be pursued without passengers, but you need to fly in conditions that are representative of airline operation to be certain of detecting all problems. That means some flights with the full complement of passengers, or at least with a sufficient number, on long-duration flights. Thus, sometimes condensation accumulates at the start of descent and gently rains down a little in the cabin or even in the cockpit. This cannot be detected without passengers on board as they are the cause of the humidity. In addition, just installing temperature sensors to validate the air conditioning is not enough as there are sometimes draughts felt by the passengers that are not encountered on initial development as the way the air circulates in the cabin will be different when you have the complement of passengers on board. Making flights in these conditions also means using all the equipment made available to the cabin crew, and the latter's comments are obviously extremely useful.

Preliminary test flights are thus not enough for development and you have to make later ones with passengers so as to garner as much information as possible. That's why we launched a series of ELF flights for the first time on the A340-600 programme.

* * *

Why, apart from the passing reference to the former oil company, the ELF (Early Long Flight) title? This acronym summarises perfectly what we wanted to do: long and even very long flights, as soon as possible in the programme and with passengers.

Before the start of the A340-600 programme, we had hired Anne Kerrien, Chief Cabin Attendant with Air France, who had vast operational experience on long-haul flights, at flight tests. There was a dual goal here: firstly to actively prepare such operations with Marie-France Perrin's team that was coordinating the organisation of all events, and secondly to take part right from the start in defining the A380 cabin.

One of Anne's first missions was to organise A340-600 ELF flights for the cabin part.

For the A340-600, we had decided to conduct flights lasting different durations, from seven to fifteen hours, including a night flight where the behaviour of the passengers is different and operation of the air conditioning is not the same as the temperatures on the skin of the aircraft are on average lower since not affected by solar radiation. We also had to choose the passengers. We had decided to have only Airbus personnel on board and above all nobody from the media. So, those seats not reserved for the hundred or so people working on board and engineers we wished to reward, became lottery prizes, with the winners being drawn by lots at the various Airbus sites. Flights were thus organised from Toulouse, Nantes, Madrid, East Midlands and Hamburg.

These flights provided us with a considerable amount of information and one can only regret that the said data weren't used to the full before entry into service.

* * *

While these flights are a key element in development of the cabin, they create difficulties with the schedule. Indeed, to take such a large number of passengers on board, the test crews have to have the intimate conviction that the aircraft is sufficiently reliable. For the A340-600, these flights had been conducted about seven months after the first flight. We reckoned that, considering all the new features installed on board, we needed a little more time for the A380 and we initially worked out an eight month lead-time, on condition, of course, that the test aircraft come out at the planned rate. In addition, we needed a minimum of two months of tests on the aircraft used for the ELF flights in order to check for correct operation of all the safety systems. Development of the cabin also had to be sufficiently advanced as to be able to make those flights worthwhile.

Delays in aircraft # 2 coming off the production line and above all its being fitted out (that took longer than planned) totally disrupted the initial schedule. Finally, these ELF flights could only get started early September 2006.

* * *

Our experience on the A340-600 had shown us that it was not necessary to make more than four long flights, with the chosen durations being 7 hours, 10 hours, 12 hours (night flight) and 15 hours. Furthermore, with an aircraft that was even bigger than the A340-600, it was hard to entertain the idea of flights from the various European airports for logistics reasons and so we focused completely on the Toulouse departure.

The issue of knowing where exactly to embark and disembark was the subject of lengthy discussions: should we use the boarding bridges in the airport terminal, board at the Lagardère plant, where the aircraft are assembled, or at Saint-Martin, the base for flight tests? Finally, we opted for the terminal for reasons of easier registration with a large number of passengers. However, disembarking was proceeded with by bus close to the terminal.

Here again, to fill the aircraft, we chose the method of drawing lots from the (very long!) list of volunteers. For each flight, we had 460 places and about 150 were reserved for engineers and technicians with a job to do on board and for our own list of guests (people we wished to reward for various reasons). Distribution between the different classes was pretty much by chance.

Clearly, as the aircraft was not to final certification standard, on Friday, just a few hours before the first ELF flight planned for the following Monday, we saw a long list of limitations come in from the Design Office. Needless to say, this annoyed us considerably, all the more so in so far as almost all of them concerned truly minor issues that had no impact on safety. The Permit to Fly signed by EASA and authorising us to conduct that flight nevertheless came in on Friday afternoon. About time too!

To check that the aircraft was in perfect condition for long stages, on 26 August we organised a test flight to the North Pole. This also gave us an opportunity to validate navigation when flying over that quite special area as anomalies are always a possibility there.

The first ELF flight on 4 September 2006 lasted 7 hours, the pilots being Frank Chapman, Jacques Rosay and an instructor from the training centre. A large number of journalists had come to be there for the departure at the terminal. The CEO, Christian Streiff, was present to meet passengers and journalists as this flight, the first with so many people on board, was obviously highly mediatised with many interviews before, after and (for those having remained on the ground) during it. Boarding of the 460 passengers went really quickly, in about 16 minutes, with only two boarding bridges. This was an extremely good result, although not completely representative of a standard flight as the passengers were carrying less luggage than usual. Take-off took place two minutes ahead of the planned time. The flight went well and the passengers were delighted.

This was the first opportunity to discover possible problems relating to carrying such a large number of passengers. At the end of the flight, throughout the descent, some unlucky ones had a few drops of water fall on them. This was due to condensation as the humidity due to the presence of so many people, combined with the rapid change in outside temperature in descent, led to this localised phenomenon and this light 'shower' in areas that were not so well drained. Even the pilots weren't spared. This obviously re-occurred on all the following flights and modifications were clearly needed.

The second flight, which I took part in, was a little longer, taking 10 hours. Charles Champion was there, discreetly seated in the business class. He had just changed position in the new organisation and when I announced to all the passengers that he was on board he was warmly applauded.

There were only some minor problems during this second flight. However, difficulties relating to the balance between temperatures in the different zones were more clearly highlighted as, in each of them, it was colder further forward than to the rear. The absence of the heated carpets that are usually on the sides of the doors to heat up these areas sensitive to thermal exchanges with the exterior did nothing to improve matters.

The third flight lasted 12 hours and involved flying at night. Our experience on the A340-600 had shown us that you could draw useful conclusions as to the air conditioning as explained above.

During this flight, we encountered a fuel transfer problem. The huge tanks failed to transfer automatically into those supplying the engines without the failure being announced. Thus, the engine supply tanks dropped to a fairly low level (4 tonnes each) before the warning was activated. This was clearly unacceptable for certification and correction was essential.

The fourth and last flight lasted 15 hours in day-time. The flight went very well and nothing new was reported, but there was still condensation and again there was the fuel system anomaly.

We took advantage of this last flight to make cabin noise measurements in realistic conditions, meaning with the seats occupied by passengers. On the main deck, the value was lower by 2 decibels than the values aimed at by Boeing for the new B787. On the upper deck the difference was 3 decibels. Quite remarkable!

* * *

During all these flights, the atmosphere on board was extraordinary as the passengers were delighted to take part in the adventure. Unlike what happens during an airline flight where you stay in your ticket class, there was complete freedom to move around the entire aircraft. Obviously, people crowded in to try out the first class seats! We had authorised access to the cockpit, but with so many passengers, the visits had to be kept relatively short. A huge number of photos were taken. Also, on each flight, four lucky people were able to follow take-off or landing in the cockpit on the jump seats.

We had asked Lufthansa to provide us with the cabin crew. We often worked with them for this type of operation and it was a great pleasure to do so. They were accompanied by Anne Kerrien as well as by the Airbus cabin specialists usually in charge of production aircraft in the team managed by Pierre Buzy-Vignau. For the Lufthansa personnel, this was a good opportunity to get used to the aircraft, to get a first idea of their future working environment and tell us what needed to be improved. Service being made complicated by the fact that all the passengers were constantly wandering around the aircraft, we had chosen high quality meal trays, the same being served for all classes.

* * *

The outcome of all these flights was extremely positive. Generally speaking, all the passengers appreciated the comfort of the cabin and the low noise level as compared with other aircraft. The Lufthansa crew also told us how satisfied they were with the working conditions on board.

But in our job we are not so much there to linger on what is positive as to look at what needs to be improved. The only really bothersome issue that looked like being difficult to correct we identified through these flights was the condensation that led, in some areas, to water droplets falling at the end of descent. Improvements in drainage were therefore immediately started up.

Other corrections had to be made relating to the temperature distribution, the fuel transfer system and some other minor issues.

* * *

In 2007, before entry into service, many modifications were made on all the systems installed in the cabin and to validate a certain number of them, there had to be passengers on board. Furthermore, on aircraft # 7, the way the cabin was fitted out was fairly different from that on aircraft # 2, used for the first ELF flights. Finally, many new options were introduced, for example IFE. This led us to launch a second series of ELF type flights in order to make all the necessary checks. The first took place on 9 February 2007. This was the twelfth flight of aircraft # 7. The rules for the choice of passengers were still the same, with a few guests and Airbus employees drawn by lot.

We found anomalies that we had already encountered, as with condensation, proof that the supposed solutions applied to aircraft # 2 had not settled matters as we had hoped. The CIDS (the interface used by cabin crew for communication with all systems) was not up to the task and here too there was plenty of work ahead before delivery!

There was a second flight on the same airplane on 28 June in the same conditions for final validation of the solutions applied. For cost reasons, our President Louis Gallois had asked us to make the flight at night. Quite rightly, he wanted workers from all the Airbus plants to be able to take part. But with a day-time flight, passengers living far from Toulouse would have had to have arrived the day before for an early morning take-off before going back home the next day (due to the flight's duration). That would have meant two nights at a hotel in Toulouse for each of them. The global saving thus meant 700 to 800 room bookings, not to mention the time spent. This meant sleeping conditions in flight were tested while our thrifty President also made a few precious savings!

The results of this last flight were globally satisfactory. We were therefore optimistic about into service.

ROUTE PROVING

Route proving flights are a statutory requirement for the certification of any new aircraft or new engine. The origin of this goes back several decades with the need to demonstrate the reliability of the engines, that were then not up to scratch on entry into service. This is no longer the case, but the rule has remained as these flights give you a good idea of the aircraft's overall reliability.

According to the regulations, route proving must be conducted in operating conditions similar to those of an airline. Initial preparation thus starts with the choice of one or two partner airlines, usually from among the first operators of the new airplane. They first of all provide technical crews working in twos (or threes if the flight is long) for the entire flight, except for the take-off and landing phases where an Airbus test pilot occupies the co-pilot's seat while acting as captain, since the aircraft is still under the manufacturer's responsibility. The cabin crews are from the airlines. This doesn't prevent having our cabin crew members on board to help, but above all, check operation of all the systems, repair if necessary and take note of anything that can help improve the aircraft. Routine maintenance can also be conducted by the airlines, obviously under our supervision. For a carrier who is going to start operating a new aircraft, taking part in these endurance flights is an extremely valuable plus as pilots, cabin crews and maintenance teams have the opportunity to familiarise themselves with procedures and special features in favourable conditions so that they are then fully operational the day of entry into service.

Route proving flights are really the last step towards certification and, often enough, the Airworthiness Certificate is awarded some days after their completion, the time it takes to address any minor anomalies that may have been encountered.

The regulations require 300 flying hours be performed in the conditions stated above. In practice, to reduce costs while remaining true to the spirit of the standard, the certification authorities accept that 150 hours correspond to the various ferry flights, and that at least 150 hours be made on the same aircraft as if it was being operated by an airline.

In so far as is possible, these flights are organised with the full complement of passengers. In the case of the A380, this was actually quite a problem as it meant there were 500 people to accommodate at the route stops, generally at the expense of Airbus or of the airline taking part in the flight and the cost involved on a number of trips became prohibitive. As will be seen, other factors led to choosing a different form of organisation.

We made two series of route proving flights with the A380: the first in November 2006 for certification of the aircraft with Rolls-Royce engines and a second one, a year later, to validate the version equipped with Engine Alliance engines.

* * *

Route proving flights on the A380 were to be pursued with aircraft # 7. It soon became apparent in the summer of 2006, that it wasn't going to be ready on time for these flights scheduled in November, with a certification date in December. Moreover, the date announced for its return to flight with the cabin installed was September and, with the usual slippage in the schedule, we suspected this would not be kept to. What's more, we had demanded it be flight-ready at least two months before the start of route proving. This was the time necessary to finalise the cabin, with its new options and safety systems. The aircraft that was going to be presented to our customers and their first guests had to be well nigh impeccable and two months preparation would certainly be needed. The obvious conclusion was that the most reasonable option was to use aircraft # 2.

There was much heated debate among us as to the issues. There were those who wanted to stick to what had been planned at all costs and who wished to use aircraft # 7. The others, including ourselves, reckoned that this was unrealistic and that we shouldn't launch aircraft # 7 into this operation while guessing that it would inevitably be ready late. What happened subsequently bore us out since it finally flew only after the end of route proving.

As the sales team did not want to show and operate aircraft # 2 with our customers, we conducted the route proving flights among ourselves, with only a few passengers for the major flights, just some technicians monitoring the systems, and distinguished guests.

Sales Chief John Leahy had decided to prohibit any visit of the aircraft at stopovers. But how can you imagine the crew refusing access on board to customers at a stopover on one of their main bases? The situation was untenable. With Fernando Alonso, we therefore jointly signed a letter for the new Programme Manager, Mario Heinen, to let him know we disagreed. If there was no change, and so as not to create difficulties for our crews, the choice of itinerary would include airfields that are not essential for our A380 customers like Iqualuit, Timbuktu and Chateauroux (though I have nothing against Chateauroux as I was born there!). Finally, John Leahy adopted a more flexible stance and authorised visits from airline VIPs, but the media and photos were barred.

* * *

What destinations to choose?

Traditionally, these flights provide the opportunity to visit our customers and organise a local VIP flight with the airline executives and other personalities, but here too, in principle, that was out of the question.

One destination had been eliminated as from 2004 during the first discussions on the matter - the United States. Indeed, I had been outraged by the attitude of some people in their administration during A340-600 route proving flights in April 2002. What happened was this. We flew in to San Francisco from Frankfurt with a Lufthansa crew where we disembarked the passengers, but our technicians still had some work to do on board. I stayed with them and was in uniform. That's when an immigration official ordered me to "Get out" in threatening tones. All the technicians came out without even having the time to cut off power on the aircraft. So 45 minutes or so went by, with the aircraft energised, until the following crew had time to get on board. This situation was potentially dangerous as, if there had been smoke, a short-circuit or any other problem, there would have been no-one on board capable of intervening. The next day, as I boarded the plane after it had made a round trip taking in Canada, I found no-one from the Airbus team that had just come in on board. They too had been thrown out. However, there were plenty of American visitors on the airplane who had no business being there.

Having stated at the earliest opportunity that for safety reasons I didn't want to go to the United States during these route proving flights, I was immediately supported by Charles Champion and that put an end to the discussion.

As we had no customers in either North or South America and future operator airlines were in Asia, Australia and Europe, tours were re-focused on Asia.

* * *

On the pilots side, as the customer airlines were not taking part (unlike in all previous programmes), we thought we would have the training centre instructors take advantage of this so that they could rapidly obtain their type rating and that of instructor. Indeed, well before entry into service of the first aircraft, they were to start training Singapore Airlines crews. In this situation, these instructors make the entire flight, with the sole exception of take-offs and landings where a test pilot is seated on the right. During these phases of the flight itself that pilot only looks after radio communications and ensures responsibility for the flight as the aircraft was not yet certified.

Then it was, just six days from the first departure that the EASA test crews demanded to be allowed on board for practically all the flights. We tried to dissuade them from this. What could be the advantage for these test pilots of making some of these endurance flights when they had already clocked up several hundreds of hours on the A380 and also taken part in a good number of long flights during the certification campaign? The whole matter went back up to EASA certification management level with a number of teleconferences and exchanges of e-mails. But finally, we just had to bow to pressure although not an iota of justification was provided, except to say that it was necessary for certification! We merely obtained a reduction in the number of flights in which they could take part.

We were pretty annoyed by this requirement for two reasons. Firstly because that took away some flights that would have been useful for our future pilot instructors. The second reason was that the participation of some EASA crews was practically systematic as soon as there was the slightest certification test and this was not even open to discussion, whereas we were delegated to conduct most of the actions needed for certification (DOA).

* * *

We finally decided to perform four series of flights departing from Toulouse with various stopovers.

The first tour started on Monday 13 November 2006. This involved four flights to Singapore then Seoul and return back to Toulouse by the same route in reverse. At each stopover, there was a press conference, and a visit of the aircraft by VIPs. Photos were forbidden, as expressly requested by John Leahy. However, you can imagine how difficult it was to have this instruction respected with miniature digital cameras and cell phones. We'd have needed a minder behind every visitor to succeed in this. But we did our best.

We had a few failures during this first series of flights: orientation of the nose wheel, fans, pressurisation and many false alarms on door locking.

The aircraft left for the second series of flights on Friday 17 November going to Hong Kong and Tokyo before returning directly to Toulouse. The main problem here was failure of supplemental cooling.

Tuesday 21 November saw the departure of the third series for China with a VIP flight to Canton (the official position had been moderated somewhat!), Beijing and Shanghai. Nothing new. The A380 was well received everywhere it went.

* * *

The fourth and last circuit was the real treat for some of us as we did a world tour via the Poles. We had a few prestige guests on board, including Joël Rault, Director for Civil Aviation South (DAC Sud), based in Toulouse, who had been stalwart in helping us solve problems as they arose on the platform. We also had on board a lot of EASA ground engineers who were working towards certification and were our guests on this exceptional tour.

We had already flown over the North Pole a number of times as it is important to check that the navigation systems continue to operate correctly when you put them through such rough treatment, i.e. instantaneously switching from a northerly heading to a southerly one. We knew that the North Pole was not problematic, but checking is not usually done on the South Pole as no route goes over that point. There are no cities in the southern hemisphere involving great circle (or orthodromic) routes, meaning the shortest across the globe that comes close to the South Pole. In any case, we needed to accumulate flying hours to obtain the required minimum, so this was a good option to go and see what would happen.

This world tour included three stopovers: Johannesburg, Sydney and Vancouver, and thus four flight segments. We had decided to keep one of them for the FAA Chief Pilot, with whom we got on really well.

The first stopover was Johannesburg. Firstly, looking at things from Toulouse, Johannesburg lies on the route for the South Pole. Secondly, South African Airways is a good Airbus customer, and even though they were no doubt not interested in buying the A380, a potential market for the A350 was on the horizon.

We therefore set off from Toulouse for a night flight on Saturday 25 November 2006. The next morning, I made a fairly short flight with Hugues Van der Stichel to validate automatic landing on this rather special runway (see THE AUTOPILOT chapter).

The afternoon's proceedings were devoted to commercial activities with visits to the aircraft. The next morning, we set off for Sydney via the South Pole. This was to be a long flight, lasting some 15 hours 45 minutes. As we weighed in pretty heavy at about 556 tonnes, and Johannesburg is high in altitude, we were limited by the length of the runway required for take-off. Also, take-off performances diminish as the temperature increases since the engine thrusts are lower. We therefore decided to take off at 8 a.m.

We arrived over Antarctica in clear blue skies over the immense expanses of ice covered terrain below. There was absolutely no air traffic around us, this being easy to see from the total absence of contrails in the sky. We were supposed to contact various air traffic organisations via HF radio but all we met was total silence on all frequencies. It was as if we were aboard a small private airplane off for a few hours up in the air.

The big moment approached as we prepared to fly exactly over the Pole. I was at the controls alongside Jacques Rosay together with fifteen or so people crowding the cockpit to be there for the event. A bit like getting on the tube in rush hour, but with such a great atmosphere! Suddenly, we tilted from a southerly to a northerly heading. As we really were alone, we took advantage of this to turn back some twenty miles later to make another pass.

• See photo n°33 – page 245

Now we had to take a northerly heading. The options from the South Pole were unique in covering a full 360°! In reality all we had to do was validate the navigation programmed by the computer. Ever wary, Jacques had thought about all of this in advance. The position of the Sun was to tilt through about 120 degrees in relation to the path of arrival. But there was an even easier way to check we were on the right path in relation to the time and Sydney's position: the Sun was to be directly behind us. Everything became elementary: all you had to do was look at the video taken by the camera of the fin and check that its shadow was in the middle of the fuselage. So that's the sort of improvised solution that was used to confirm navigation! But in fact, there was no need to watch the video. All we had to do was observe the shadow of the aircraft and its contrails on the ground to know where we should be heading. We just followed the shadow!

This flight over the South Pole remained an unforgettable moment for all concerned. Indeed, it made me want to see Antarctica again in closer quarters, and this I did some years later when I went on a cruise there with my wife.

We got to Sydney next morning for the usual press conference. One of the participants was Nancy Bird (91 years of age!), the first ever Australian woman aviator. It was planned to name the first Qantas A380 after her.

We set out for Vancouver on Wednesday 29 November. EASA pilots were at the controls for take-off and landing as demanded by that organisation. In so far as was possible we wanted each of the ground engineers from the authorities to get the chance to see a take-off or a landing from the cockpit. I took responsibility for naming them as some of the EASA flight crew members wanted to monopolise these 'good seats' with no consideration for their colleagues who were flying for the first time on this magic aircraft. They could, in any case, continue to follow all the pilots had to say from the flight engineers station at the centre of the aircraft.

It was extremely cold (-9°C) in Vancouver when we arrived on the afternoon of the very same day (we had crossed the date-line). The Canadians hadn't seen such conditions at that aerodrome for a number of years. The two Airbus pilots who'd made the flight, Guy Magrin and Frank Chapman, took part in the traditional press conference and the others went off to the hotel for some rest to be ready for the next stage.

As we were getting ready to head back to Toulouse via the North Pole, snow started to fall. We tried to have the time of departure moved forward by an hour, but the aircraft had to be de-iced and we were in fourteenth position in the queue. De-icing an aircraft involves spraying, while on the ground, the entire surface, but mainly the wings and the empennages, with a product that prevents ice from forming during take-off and up to initial climb. Finally, after discussion, we were granted preferential treatment and four vehicles came up to perform the de-icing operation. The delay on the initially chosen slot was just 20 minutes.

For the departure, the two pilots were our guest Gene Arnold, FAA Chief Pilot, and Hugues Van der Stichel. They were used to team work. Shortly after take-off, a red warning indicator came on "Engine 4: turbine overheat". This only lasted three seconds and was in fact a false alarm due to a difference between two sensors. Gene quite rightly qualified this as being unacceptable. But the modification was already being implemented for engines coming out of the factory.

We passed over the North Pole at night and much less excitement was aroused than when going over the South Pole.

During the last part of the flight, Gene and Hugues went off for a rest so as to prepare for landing. We therefore settled in up front with Jacques. As there wasn't much to do, we devised a little game: guessing the time the Sun would rise. For a given geographical position, the calculation is simple because all you have to do is take a look at the tables we have available that give the time in relation to the date, the longitude and the latitude, and interpolate. But in a moving aircraft, longitude and latitude change constantly and estimation is distinctly more complex. Jacques is a real pro for this sort of exercise and once more he gave the best answer just a few minutes out.

On arrival in Toulouse, a crowd of journalists were waiting for a final cocktail. We were relieved and certification flights were over. The few problems they had pinpointed were in process of being resolved. To summarise, there was still the matter of the false alarm relating to the engine mentioned above and condensation of water at the end of the flight that led to mild showers in some parts of the aircraft, including in the cockpit. Work still needed to be done on drainage.

* * *

For the Engine Alliance alternative route proving campaign, everything was easier as there was only a single test aircraft equipped with that engine, # 9, and it had no cabin, just thirty business class armchairs that had been scavenged, for all the technicians on board.

This time, the pilots flying with us were those from the first two airlines that were going to bring the aircraft into service with that type of engine: Emirates in mid-2008 and Air France in 2009.

Reticence about going to the United States had been allayed. The procedure to be followed had been clearly identified during demonstration flights conducted in March 2007 to New York and Los Angeles. Moreover, considering the equipment on the aircraft and the small number of people on board, these flights could in no event be treated like commercial flights. This gave us some leeway when it came to managing disembarkation and avoiding the sort of unpleasantness encountered during flights with the A340-600. There were, however, other drawbacks, with the need for each passenger to have a visa for example.

Before this official route proving campaign, a number of world tours had been made with that aircraft. The engines still had not had all the modifications needed for certification applied to them, but the differences were minimal and we set off with a backlog of positive experience for this series of flights.

They started on 26 September 2007 with a round trip to Bogota. Why Bogota? Because, although we didn't yet have any customers in South America, this airport, at an altitude of 8,300 ft availed us of the possibility to certify a higher altitude for autolands (see THE AUTOPILOT chapter).

The second series of flights, run with Air France, started with the United States as destination. This first of all involved a visit to the two engine manufacturers who had teamed up to build this new engine. On 2 October, we were at Bradley Hartford, Pratt and Whitney's base. Then the aircraft left the next day for Cincinnati, where General Electric's were located. The return included a stopover in San Francisco.

The third tour was made by pilots from Emirates and included Dubai, Melbourne, Manila, Clark and return with a stopover in Dubai. The fourth series was a small world tour taking in Bogota, Vancouver and Osaka before ending on 18 October 2007.

As a few flying hours were still needed to have the required minimum of 150 hours, we made a trip out to the Dubai Air Show with a return early in November.

No problems were encountered with the engines although we did have a minor problem with the autopilot (see THE AUTOPILOT chapter).

CERTIFICATION

With the A380, EASA (European Air Safety Agency) that had just been set up, managed a certification for the first time. It replaced the JAA (Joint Aviation Authorities) that had previously brought together the specialised agencies of the various European countries so as to avoid duplicating tests and dossiers. In the United States, we had our usual FAA correspondents.

Certification of a new aircraft is a complex process that takes a number of years. In practice, discussions with the authorities start well before the airplane is officially launched so as to determine the applicable regulations. There are basic rules valid for all transport aircraft know under the term JAR 25 (Joint Aviation Requirements) in Europe and FAR 25 (Federal Aviation Regulations) in the United States. These two sets of regulations are identical apart from a few details as, over time, harmonisation between the two continents prevailed. Sometimes, the rules turn out no longer to be adequate to the specific features of the aircraft to be certified. For example, for the A320, we had fly-by-wire flight controls on a transport aircraft for the first time (apart from Concorde). The existing regulations therefore had to be adapted and new rules had to be devised. It also arises that the authorities imposed more stringent rules on us considering the experience acquired on other airplanes or because a change in the regulation will soon come into effect and it is desirable for the new aircraft to satisfy it. Finally, the manufacturer may, of his own volition, ask for a rule that is not yet compulsory to be applied. All these situations are encountered on each certification of a new aircraft.

The rules must normally be frozen five years before the date planned for certification. The reason for this is obvious: you can't ask the manufacturer to change the design of the structure or systems at the last minute. Modifications to the basic JAR 25 regulations are made through documents known as CRIs (Certification Review Items) where the authorities and the manufacturer take it in turn to set forth their arguments until a compromise is found. Sometimes there is immediate agreement, but in other instances, discussions are more difficult. Sometimes too, the five year rule is not adhered to strictly as reaching an agreement can take more time than foreseen.

All the aircraft manufacturers have Design Organisation Approval (DOA) as issued by EASA, recognising that they are capable of designing aircraft. This implies that, for certification, the authorities can trust us and, unless there is a particular problem, they won't check out all the details of the structural design file, operation of the systems or computers. EASA has a general role to monitor things, meaning they have to check working methods, analyse the dossiers submitted in a common sense manner and ask the right questions to ensure that we had not forgotten anything. The final objective for this agency is flight safety. An analogous document, the POA (Production Organisation Approval), exists for the production of aircraft that are to be delivered to customers.

Aircraft certification covers a wide range of domains, including structural design, operation of systems, electromagnetic protections, etc., and not all can be covered here. We shall simply concentrate on flight tests in this chapter.

* * *

Flights are organised in such a way that there is no break between development tests and those devoted to certification, meaning there is no single moment where we say that the aircraft is completed and that certification tests can start. Many tasks are run in parallel in line with progress in development. For some validations, when we are extremely confident as to the result, we sometimes proceed directly with the certification flight, without prior verification, to save time.

For the A380 certification flights, test data sheets were prepared as usual by the flight tests crew with assistance from some specialists. Common sense is required here as, for a given type of validation, it is clearly impossible to cover all flight conditions: altitude, speed, weight, centre of gravity position, quantity of fuel in the wings, etc. We therefore have to choose a broad range of situations as being representative of all flight conditions, above all covering those that may prove to be critical.

For each type of test, it is stated whether the presence of an EASA pilot or engineer is needed. The engineer simply observes the test, while the pilot conducts the flight. In theory, considering the DOA, we could perform all tests ourselves. But, clearly, it is quite normal that the certification authorities intervene to check a certain number of results, for example handling of the aircraft in such and such an instance of failure. For the A380, the proposal for EASA pilots and engineers to be present or not was formulated according to the custom that had emerged on other programmes whereby we went ahead with the flight without them where they were not available on some tests, while for others we postponed it so they could attend.

* * *

At the time of A380 certification, EASA, still a new organisation, did not have all the resources it needed. In the flight tests domain in particular, it was unable to cater for all the current programmes as Dassault Aviation too was to almost simultaneously have its Falcon7X, the first of its business jets with fly-by-wire flight controls, certified. EASA therefore resorted to the usual resources of the JAA authorities to perform these tasks. On the French side, there was the CEV (Centre d'Essais en Vol - Test Flight Centre) and, for the British, the CAA (Civil Aviation Authority). Germany never had a strong team and, over the last decade, practically none of its representatives took part in Airbus certification flights.

The Chief pilot from CEV on the A380 was François Fabre. He had taken part in all Airbus certifications since 1994 and was thus particularly well trained to perform the most delicate of tests. The CEV introduced a second pilot, Didier Poisson, who was younger but through work on Airbus certifications had nevertheless acquired a level of experience many test pilots might envy. But being on the second line, apart from a few exceptions, he didn't take part in the most difficult tests, especially those relating to the flight controls.

The CEV flight engineers were led by Jean-Pierre Sarrato. Michel Cartier followed engine testing very closely and Thien An Deville focused on systems.

Two members of aircraft crew represented the CAA. The pilot, Dan

Griffith, had more recent experience on our Airbus planes, and had taken part in certification of the A340-600. He is an extremely skilful pilot, capable of performing tests with great precision. He also exuded confidence, while in flight tests the general rule is to remain wary. The only British engineer who frequently took part in flights was John Matthews.

* * *

The first flight by François Fabre took place on 25 May 2005, one month after the first take-off. This was one of the last series of rate of climb measurement tests with an engine shut down to decide on the flaps deflections (flight # 12). This was not a certification flight, but it is usual, in each programme, to get the pilots, as also the engineers, from the authorities to fly really early. The advantage lies in getting them to know the aircraft, train them in the same way as we train our pilots, so that, when the time comes, they are ready to perform even the most difficult certification tests.

Dan Griffith flew a little later on 11 July 2005. This was flight # 38 on aircraft # 1 intended to determine the aerodynamics to reset the models used in the development simulators.

Apart from these two flights in which flight engineers also took part, the latter were also present in the aircraft on 13 and 14 July 2005 for VMU tests as it remained possible that certain measurements be retained for certification.

The first series of flights with near systematic participation of the pilots involved checking rates of climb with an engine shut down over the last months of 2005 on aircraft # 4, whose engines were equipped with the entire test installation to determine the thrust extremely precisely. These were flights to make measurements so as to establish performances for operating manuals, so presence of certification pilots was in no way essential but, once more, here was another opportunity to get used to the aircraft.

Early in 2006, the CEV and CAA took part in two campaigns. The CEV went out to Medellin for the first tests on a high altitude airfield, and the CAA to Iqualuit for the first series of cold weather checks.

Over the second quarter of 2006, the pace of certification flights intensified in a variety of domains, including braking, fuel, natural icing, flight characteristics with ice shapes, hydraulic failures, navigation systems, take-off performances, and so on.

Then in August 2006, flight certification activity was further increased and it became a common occurrence for several flights to take place on the same day with participation of EASA pilots or engineers.

* * *

Early in 2006, the atmosphere between Airbus and the CEV turned sour. The CAA remained outside ongoing discussions.

The first signs appeared early in January. The CEV complained that they were not taking part in enough flights. Indeed, at that time there were few certification flights going on so there was nothing illogical about it. Complaints continued to flow in and on 10 July 2006, during a certification meeting in Cologne, we were again taken to task on the matter. Hugues Van der Stichel, the Airbus certification pilot, was able to show that to date, there had been 266 certification flights and that EASA had taken part in 270 development and certification flights! We were in no way against their taking part in flights outside certification, but this way of portraying matters with EASA management stuck in the throat.

On 1er August 2006, we discovered that in July, some engine certification flights involving simple measurements originally classified as 'W' for Witness, had been unilaterally re-classified as 'P' meaning that the presence of an EASA pilot was mandatory. For the flight scheduled for the next day, there was no way out and the pilot from the authorities just had to be present. I then asked Jacques Rosay to assign this flight to me. During the preliminary briefing, I explained that, in principle, I would myself perform all manoeuvres not related to certification, and especially take-off and landing, whereas the tradition has it that the EASA pilot does the entire flight or possibly that repetitive tests be shared. A friendly and professional atmosphere nevertheless prevailed on board.

One day, an engineer refused to ratify a measurement test simply because he hadn't been on board. Fair game! Our flight engineer prepared a new certification data sheet with a similar test but that was conducted on a day when that engineer from the administration had absented himself!

On 7 November 2006, we again had a difficulty with the requirement that certain crew from the authorities should take part in route proving flights that were to start on 13 November. No such request had ever been made on previous programmes. But we just had to accept this (see the ROUTE PROVING chapter).

To summarise, the Airbus flight tests crew was a little disappointed by this rather cavalier attitude. This had nothing to do with the technical skills and capabilities of the pilots concerned. In that respect, we had every confidence. For example, out of the six landings made with a crosswind reaching more than 55 kt in gusts, the CEV pilot did four while we Airbus pilots only did two. Similarly, we regularly let them land the airplane at weights greater by almost 200 tonnes than the maximum landing weight, something we would have been perfectly entitled to refuse.

On technical issues, it is not unusual to have differences of opinion between the aircraft manufacturer and the certification authorities, but these were actually rare occurrences on the A380.

* * *

As soon as the decision to certify the aircraft in the last quarter of 2006 was made, we flew a round trip on 11 April 2006 out to Cologne where EASA's headquarters are with the Chief Engineer and Head of Engineering to ensure we had their support in all areas so as to keep to that objective. Patrick Goudou, their Director, confirmed that they were going to do their utmost to help us. In mid-September, another meeting took place to review matters and make sure everything was ready on the EASA side for certification in December. All went well.

Then on 6 November 2006, it emerged that two major issues could jeopardise certification: fire resistance of the APU support and an issue concerning flutter with the reinforced fin. On the latter question, the EASA specialist was dissatisfied with the model based differential computation method used and wanted an additional test in flight. However, considering the various other modifications, we would have had to conduct the test on an aircraft without test instrumentation, this being difficult to entertain for a flutter validation. Two independent experts were appointed and the problem was finally resolved with no further test taking place. The other issue was also settled rapidly.

A Type Board Meeting was held on 16 November 2006 to report back on progress on the certification dossiers. For some reason unknown to me, Airbus flight tests were not invited, but we were accused of endangering the schedule as, for the flight aspect, there were said to be about 300 still open issues. Having checked this out, all the data sheets corresponding to these problems were put on the table for closure as solutions had already been found and we were simply awaiting final validation by the EASA flight panel. It turned out simply that for one reason or another, the responses had not found their way to the office centralising them. Four days later, there were just four open issues remaining. They were settled at the last certification meeting on 5 December 2006.

For certification, a small number of limitations remained. One of them was a restriction on speed to 330 kt rather than 340 kt between altitudes of 20,000 ft and 30,000 ft when the aircraft weight is greater than 550 tonnes. This limitation had been adopted due to the discovery of a very slightly inadequate margin (2%) on the calculation of wing bending forces in extreme cases. It was to have been lifted within one to two months, either by confirmation that a small structural reinforcement (already under production) was adequate, or, possibly, by a minor adaptation of the flight controls. Some people at Airbus wanted to modify programming of the computers for this limitation to be shown on displays. Others, including flight tests, were against this so as to avoid taking the risk of introducing a bug into a hastily modified computer, also bearing in mind that that solution made no real contribution as the first aircraft was only to be delivered a year later, well after validation of the real modification.

Other certification tests were scheduled in January 2007 as integration of the electrical and fuel systems had not been examined in the final version (see THE FUEL SYSTEM chapter). In the end, autoland worked well, but certification tests had not been completed.

Finally, there were only a very few operational limitations remaining for certification and, from a strictly technical perspective, apart from cabin installation, about three extra months would have been enough to put the A380 into a satisfactory configuration for operation by an airline.

* * *

One subject that arouses lively debate in the flight tests community is as to whether to use a fixed base simulator or a full flight simulator (with motion) to qualify the seriousness of failures. As the same discussions arise in the field of training, I'd like to develop a few ideas here. In other words, is a motion simulator better adapted to demonstrating certain unusual situations to pilots?

The answer might obviously seem to be 'yes', motion will be better. But in reality, such movement induces illusory sensations. Let's take an example: the aircraft can bank by more than 60°, whereas the simulator cannot. So how does it react? As soon as it reaches its motion stop, it has to return to its neutral position, otherwise the pilot would have the impression of flying with sideslip. The inner ear, that ensures one's sense of balance, is no longer in tune with the real movement of the aircraft and with what the pilot can see, either on his instruments, or on displays giving outside vision. The crew can then sustain a form of disorientation that can impair their judgement.

It was also demonstrated, during an accident, that training on the simulator can have a negative effect as it will never be able to reproduce extremely violent accelerations. In 2001, an American Airlines Airbus A300-600 took off from New York. It entered the wake vortex of another airplane, leading to a slight lateral movement. The pilot had been trained on the simulator to counter this type of jolt in roll using the rudder pedals, just by giving a kick on the pedals. But, in the aircraft, this action led to a lateral acceleration that was surprisingly violent. The simulator was much gentler. Surprised by this, the pilot made a sequence of right and left yaw commands until rupture of the fin thus causing the accident. Training on the simulator can lead, in some cases, to the opposite effect to that sought as, by their very nature, the aircraft's motions cannot always be faithfully replicated. That's what we call negative training.

That being said, I had asked for us to get equipped with a development simulator endowed with motion as I was sure that EASA would ask us to do so. This was a dead cert, all the more so in so far as Boeing used one. So it was that EASA made this demand at a certification meeting in mid-July 2006.

Experiments conducted on a number of failures that can be reproduced in flight showed that analysis on a fixed base simulator is closer to that conducted in flight than that performed on a full flight simulator. The result may even be reversed between the two categories of simulators. We therefore decided not to use motion for our own tests. The difficulty remained for certification with EASA who wanted motion. Nevertheless, whenever there was a doubt, we managed, as in the past, to find trade-offs when classifying failures. All in all, we even sometimes overestimated the seriousness of some of them so as to move on and close the case.

This experience is neither specific to the A380, nor to tests and the debate is no doubt still far from finding an outcome.

* * *

Certification of workload or human factors has always been a source of endless discussions. The human factors dimension here essentially concerns studying the behaviour of the pilots to validate a system's ergonomics or design. This issue was subject to considerable emphasis at the time the flight engineer function was taken out of the picture due to the automation of so many tasks. Old-timers will no doubt remember the endless discussions with some French trades unions. The first Airbus to have been flown with a two-man crew was the A310, brought into service towards 1980. Faced with violent opposition from some quarters, a special additional seat was fitted out but only on Air France A310s to keep the 'third man' in the cockpit. Later, for the A320, the battle raged on, but only in France. On this aircraft, it was clearly out of the question to fit out the cockpit for three, while the A320's direct competitor, the B737, was flown by just two people.

Validating the workload in operations and the ergonomics in any new cockpit as with that for the A320 is perfectly logical. The human factors specialists threw themselves into it with enthusiasm. How many pilots had to fly with a battery of electrodes on their heads to record so many mysterious parameters? In fact, what counted most was to analyse concepts, as with the sidestick, the thrust levers with detents, the presentation of failure procedures, how the cockpit is fitted out, etc. Much time was spent in discussions on the simulator, not only with the certification authorities, but also with customers to validate new features and the entire installation. All this was extremely fruitful.

For the A340, everything had to start from scratch. The cockpit was identical to that on the A320 with some improvements resulting from the lessons learnt from operating that airplane. You might have thought going through it all again was futile. But, given we were talking about a long-haul aircraft, some people thought that everything needed to be looked at again. Analysis was thus resumed. We had to make long and short flights simulating a certain number of failures. Faults and errors made by the authorities' crews were counted by another pilot from the administration. This led to squabbling between the pilots as some considered too many mistakes were being counted against them while they flew and were quick to let themselves go when counting for the pilots flying after them. We Airbus people found this pretty amusing! I retain a few fine memories and amusing stories.

So how to proceed to validate the A380? According to Airbus, as the way the cockpit was to be fitted out derived from that on the A340 and the certification pilots having accumulated considerable in flight experience, the conclusion appeared obvious. Meanwhile some people at EASA reckoned that some, even minor, differences, justified a specific evaluation. Just before starting these evaluations, the three EASA pilots counted a total of 360 flying hours on A380, to which we can add many sessions on the simulator. You would have thought that such a wealth of experience, with so many delicate tests that included various failures or abnormal situations, would have been enough. But this was not to be and we therefore had to proceed with a formal evaluation.

In order to minimise lost time, EASA accepted the idea of conducting the various evaluations relating to the workload on ferry flights for the various campaigns and during a route proving flight from Singapore to Toulouse. This was summarised succinctly with dry British humour in just one lapidary sentence by Airbus pilot-in-command, Ed Strongman, during analysis prior to take-off by EASA when he said “I would never have thought it could take so many words to take-off an airplane!”

Obviously, all this evaluation was extremely formal and no noteworthy remarks emerged (unless you count Ed's!).

* * *

U.S. certification by the FAA is normally automatic and stems from that granted by EASA. In our field, it is customary to keep our American opposite numbers, specialists in flight tests, informed of progress in development. The form of organisation usually retained is firstly a week of evaluation during the development programme with a few meetings for presentation of the aircraft and flights, then, one to two months before certification, a progress review as well as other flights to go over the topics of special interest to them.

For the A380, the team mainly included a test pilot, Gene Arnold, and a test engineer. Gene was highly experienced, full of common sense and knew how to distinguish between the major issues and the details, a fundamental quality in a test pilot.

This time, considering the aircraft's novelty, the FAA team came over on three occasions. The initial visit, during the first week of December 2005, was to make contact, mainly to 'get to know the aircraft' and discuss potential certification problems. A first dedicated flight had unfortunately to be cut short as the crosswind on landing was greater than what we had provisionally demonstrated. The FAA team also took part in a rate of climb measurement flight with one engine shut down to pursue familiarisation with the plane.

Early in June 2006, there was a new visit from the FAA, and this time, given the state of progress of the programme, this really involved an evaluation. Hugues, who was normally to have flown with them, was busy with other matters, so I made two flights with Gene. He was extremely interested in lateral control. He was probably concerned by possible difficulties due to inertia in roll. He also wanted to get an idea of manoeuvrability on the ground. The aircraft is equipped with two cameras, one under the fuselage, to see the nose wheels, and the other on the top of the fin, giving an overall view, as well as the position of the wing landing gears. The first flight took us to Tarbes for a landing facing the south. Take-off has to be facing north due to the proximity to the south of the Pyrenees. After landing, there are two successive turns, left then right, on narrow taxiways to get to the take-off runway, with very little lateral margin, so this was good training! To complete evaluation on the ground, on return from the flight, we performed tight half-turns (as have to be made at the far end of the runway in some airfields where there is no other solution than to backtrack to go into take-off position) on our ground run area. To do so, we had plotted standardised markings out on the ground to validate these manoeuvres. Gene thus made the manoeuvre once with the cameras and once again without them. He was impressed by the good handling on the ground and concluded jokingly that he could do without the cameras. We took his word for it and the next flight was without them!

The second flight mainly included an evaluation at high altitude. Finally, the FAA was satisfied. The only reserve stated concerned handling in roll that they wished to see again in turbulence and with crosswind.

The last stage was final evaluation between 16 and 20 October 2006. A second pilot joined our crew. This was Guy Thiel, whom we knew well as he had taken part in tests on a number of programmes. Four flights were made. One of the concerns was the 150 ft wide runway despite the validation procedures adopted. For this, they wanted to conduct a night flight, firstly to see normal procedures for use of the aircraft, and secondly to land on an outside airport with a 150 ft wide runway. We chose Lyon-Satolas as our destination. No announcements were made but the rumour got out and Lyon people came out in large numbers to see the aircraft. They were crowding round the edges of the airfield even though it was already 10 p.m. and cameras were flashing non-stop. This was a surprise for all of us and especially Gene Arnold who in substance came out with the phrase "We're getting a rock stars' reception!"

All these flights went really well and satisfied the FAA team members. They reckoned approval for operations on 150 ft wide runways would be a mere formality. However, they did want EASA pilots to participate in crosswind validation. But this was already planned.

Note that outside these official evaluations, Gene Arnold took part in two additional flights, one to make a comparison of high altitude wake vortices between the A380 and the B747 (see the WAKE VORTICES chapter) and the last route proving flight between Vancouver and Toulouse via the North Pole (see the ROUTE PROVING chapter).

* * *

European and American Airworthiness Certificates were handed over to Louis Gallois on 12 December 2006 by EASA Managing Director, Patrick Goudou, and FAA Administrator, Marion Blakey.

RUNWAYS

Discussions on the runways commenced extremely early for this programme since the decision as to where the assembly was to be located depended in part on the characteristics of the airfield we would be using for take-off and landing. In November 1996, while we still didn't know much about the aircraft's performances, I wrote a note giving some recommendations on the runway for an assembly site. My assumption was quite simply that the A3XX as it was still known then, would have performances similar to those of the A340. My conclusion was that if delivery flights were not made at this location, a length of 9,000 ft was adequate, otherwise we would need to aim at 10,000 ft. I also insisted on the environment, stating that I preferred 9,000 ft in a quiet area to 10,000 ft with too high a population density around the airfield.

In 1998, DASA (Daimler Chrysler Aerospace Airbus) published a document explaining that everything should be done at Hamburg-Finkenwerder. In their view, this was the ideal location and there was no problem extending the runway to the required length. In passing, this memo also argued that installing the production line in Toulouse was unviable for many reasons!

The response from Aerospatiale, defending Toulouse, was a swift one. The report came out in November 1998 and consisted of twelve dense pages to demonstrate that the Hamburg site was absolutely inappropriate and that the runway was totally inadequate!

Discussions then raged on to recover the assembly line and the length of the Hamburg runway was one of the major deciding factors. How many memos we must have prepared over the years 1998 and 1999 to back up our position concerning Finkenwerder? We regularly explained that while we had no detailed knowledge of the future aircraft's performances, so as to avoid taking risks, we ought to stick to a length of about 10,000 ft. Finally, the decision on the share out of work between Toulouse and Hamburg was made at a political level in year 2000 between Prime Minister Lionel Jospin and Chancellor Gerhard Schröder: the assembly line was to be in Toulouse and installation of the cabin in Hamburg. Deliveries were to be shared between these two aerodromes.

* * *

The width, length and condition of some landing strips posed various problems for us over the programme, first of all on the airfields where aircraft are assembled and equipped, but also on those used for development and certification flights.

Let's take a look at some dimensions. The runways are generally supposed to have a shoulder of 25 ft on each side. These shoulders are there to avoid pebbles being projected by the blast of the engines or sucked into them. This takes the overall width of concrete or tarmac to 200 ft for the runways of about 150 ft in width and 250 ft for those measuring 200 ft as found at most major airports.

On the A380, the distance between the axes of the two outboard engines is almost 170 ft. In addition, the distance between the outer sides of the outboard wheels of the wing landing gears is greater by 20 inches than that of the Boeing 747.

* * *

During flight tests, we damage runways more than other users (see the chapter on the development of braking and the problem of rubber deposits being left). We therefore have to regularly clean up the runway used.

Other damage can occur during tests given the special conditions prevailing during the manoeuvres performed, for example during VMU tests (see the OPTIMISING THE AERODYNAMIC CONFIGURATION chapter), when the skid installed on the rear part of the aircraft scrapes the runway. What's more, in this situation, as the aircraft is in a high nose-up position, the blast from the engines is oriented towards the ground. This can damage the markers and panels along the side of the runway through projections of pebbles. There are even greater risks of causing damage during VMCG tests (see the MINIMUM CONTROL SPEEDS chapter) as the aircraft is aligned 30 ft from the central axis. One of the outboard engines may, at maximum thrust, blast the edge of the shoulders or even beyond, leading to many projections of debris.

* * *

The longest Toulouse-Blagnac runway, 'Runway 2', was built in the mid-1960s to cater for the needs of the Concorde programme. It is almost 11,500 ft long – adequate for most of our tests, remembering we can also resort to the more than 16,400 ft at Istres for more critical manoeuvres.

It was to be refurbished due to its age and the damage sustained in certain areas. Works were conducted in two stages. During the summer of 2003, it was neutralised for more than one month to stabilise the concrete slabs. The following year, it was enlarged and asphalted over the summer, being modified to have a number of special features. It is 150 ft wide, but has two shoulders on each side, one measuring 25 ft allows the maximum weight of the A380 on the ground to be borne (but not wheel touchdown) while the other, more to the exterior, also 25 ft wide, provides protection against the absorption of foreign bodies by the engines and avoids possible projections of pebbles. The markings were recessed. We wanted the great majority of tests on the A380 to be conducted on a runway almost 150 ft in width, and this proved to be the case. However, for a certain number of manoeuvres that have nothing to do with airline operations, we were not to be aligned with the axis and sometimes even made alternating turns to left and right, whence a risk of colliding with the runway lights (which had happened sometimes previously on various other programmes). That explains why we asked for embedded lights and the possibility to egress from the 150 ft for some specific tests on the ground.

Financing for all these works was shared between the Chamber of Commerce and Industry of Toulouse (CCIT) and Airbus. In its broader outlines, reinforcement of the runway due to aging was the responsibility of the CCIT, while Airbus financed its specific requirements for the A380 (which was logical).

At Toulouse, a single runway is authorised for autoland with extremely reduced visibility ('category 3' operations), the one known as

'14R', which is runway 2 facing south. Toulouse-Blagnac airport sometimes goes for years practically without any fog and then years where it is a frequent occurrence, all this being highly unpredictable. Initially, with the A380, we were theoretically barred from taking off in foggy conditions on runway 14R as we couldn't taxi along the taxiway on the western side due to lack of lighting approved for operations in category 3. We were also prohibited from going round the aerodrome to join the taxiways taken by other airliners as, in a small area, there was too short a distance between the wing tip and the airfield fences with the road behind them. Later, works were conducted to ensure the spacing required by the regulations.

Every day of testing lost costs big money on a development programme. During the winter of 2005-2006, during days with foggy conditions, we were in theory condemned to stay on the ground. Luckily, the air traffic controllers took pity on us and authorised us on a number of occasions to take off in the opposite direction to normal traffic (on runway 32L). This led to a few late arrivals for commercial operators, but share and share alike! In the end, not a single day was lost due to fog. I'd like to express a special word of thanks here to those controllers.

The following year, we couldn't continue to take advantage of their favourable disposition towards us. Indeed, we could count on their help for a small number of aircraft but with five A380s undergoing tests at the same time, that would have been unrealistic. We therefore decided to equip our taxiway, on the western side, with all the lighting systems needed for it to be used in category 3. The cost came to about 0.5 million Euros, but we had no cause for regret as, the following winter, we had our fair share of fog. So it was a good investment!

* * *

Development of Hamburg-Finkenwerder was much written and talked about. In the 2000s, the length of the runway was 7,615 ft, following extension works in 1997 that involved filling in a small yacht harbour on the Elbe. One of the difficulties here was that landing was most often on runway '23', that is towards the south-west, due to the prevailing winds, and there were only 6,152 ft that could be used. Indeed, in this direction, the point the wheels touch down is shifted further along as, on approach, you mustn't fly too low over the hill on the other side of the Elbe.

To ensure operation of the A380, there were a number of situations to be covered. First of all landing on runway 23 in rain, and thus with reduced braking, for an aircraft coming in from Toulouse just after having come off the assembly line. This was not the most critical case as the aircraft is light, weighing in at around 280 tonnes, without the extra 30 tonnes of the cabin installation. But, to deliver aircraft from Hamburg, we had to be able to make flights with customers, with the cabin installed and enough fuel to check the transfer systems. This led to landing at a weight of about 350 tonnes, and it was really not sensible to ask customers, who were not necessarily familiar with the A380, to go through a short runway landing. Delivery of aircraft thus required us to lengthen the runway to allow for comfortable landing facing south-west. There was also an additional safety issue since, in the same direction, the strip ended a few feet away from an embankment. In the event of runway excursion, there was bound to be a dramatic accident whereas, practically everywhere worldwide, airports have a sufficiently extensive area at the end of each runway to allow the aircraft to slow down before encountering obstacles.

For take-off, the situation was different. Taking off from runway 23 posed no problem as long as the aircraft was not too heavy. However, on runway 05, facing north-east, we had to go over the hills on the other side of the Elbe at sufficient altitude to get by even with an engine failure. In these conditions, not all the length was usable, meaning the weight on take-off was considerably reduced.

In October 1999, a meeting was held in Hamburg in the presence of Noël Forgeard and the DASA executives, including Gustav Humbert. On the Airbus side, I was accompanied by the Chief of the Large Aircraft Division who was in fact A3XX/A380 Programme Manager, Jürgen Thomas. This man is a great engineer for whom I personally have great esteem. The DASA executives complained that, for some reason unknown to us, documents sent to the Senate to obtain financing reported on a runway length shorter than our team was seeking: about 8,745 ft instead of 9,960 ft. Jürgen slapped them down by refusing to reduce the margins that are necessarily built in to this stage in the programme as we could not take a false step and realise later that the runway was too short! We nevertheless all agreed not to take extreme cases into account for sizing.

I inherited a dossier from DASA to be studied, the matter being how to increase the ILS glide slope from 3 degrees to 3.5 degrees and, as a result, be able to displace the runway threshold for landing more than 900 ft to the north for runway 23. This increase in the glide slope had been approved by the German Ministry of Transport. However, not unsurprisingly, flight crew had not taken part in drawing up this proposal and weren't even informed – a clear policy of fait accompli. DASA had in its hands a concession from the Ministry, so the issue was closed, and they didn't want to bring what they appeared to have gained into question.

I couldn't say exactly how many memos were written to explain that this was nonsensical but I can state categorically that I was confronted by managers with no operational culture and who obstinately refused to listen to what we had to say. First of all, contrary to what they maintained, slopes of more than 3 degrees for ILS are extremely rare and only exist for specific reasons. This applies especially for Marseille, when landing facing the strong northerly Mistral wind. But at Marseille the runway is very long and the strong face wind helps you. Now, at Hamburg, the usable distance was short and these managers wanted us to land a huge aircraft (whose flight characteristics still remained unknown to us) along an unusually steep slope. For the record, we did in the past adopt an ILS glide slope of 3.5 degrees for several months in Toulouse. The aim was initially to reduce noise nuisance on approach and, following this, the flight tests team insisted we keep that ILS so as to test flight controls and the autopilot to the limits. The Air France pilots finished by protesting and demanded we go back to a normal 3 degree slope. But the aircraft concerned was the relatively small A320 and they had more than 9,800 ft before them to come to a stop. Here we were talking about a big and as yet unknown aircraft facing a very short runway.

DASA also wanted to shift the point of touchdown to the north. Logically, as the glide slope was steeper, having gone over obstacles at the same altitude, you should be able to displace the touchdown point. In fact, they had cheated a bit by having an aircraft much bigger than the A321 and A319 (which the local inhabitants had got used to) fly even lower over the hill. Trouble ahead! All my notes proving that the change in ILS glide slope and displacement of the touchdown point were poor, and even potentially dangerous solutions, due to possible difficulties on landing changed nothing. That's when I had the idea of asking one of the engineers, Vincent Rouquière, who also did great pencil sketches to do a scale drawing of an A380 over the hill before the Elbe, first on the glide slope and then a little further down as can occur with dispersion (see drawing). The visual impact was so striking that the discussion finally came to an end with those solutions being abandoned towards 2001. Sometimes a good sketch is worth more than a lot of blather!

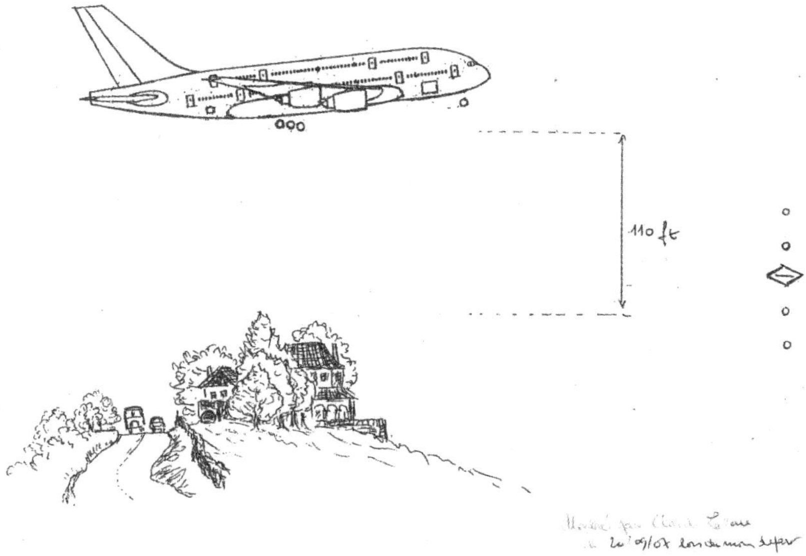

Artist's impression of the A380 coming into approach at Hamburg over the hills along the Elbe – © Vincent Rouquière

To avoid any later polemic, Jürgen Thomas and I signed a joint memo in February 2001 giving the distances required for take-off and landing according to aircraft weight on runways 05 and 23, in both dry and wet conditions. This remained our own vade mecum as clearly you couldn't possibly contravene a document issued by the Head of Programme and the Head of Flight Tests. These distances were re-adjusted later in line with technical changes and then also our knowledge of performances. The margins used were removed and the necessary lengths reduced shortly after the first flight, as the landing speed was much lower than forecast. Finally, various changes were introduced later as we became confident enough in the tuning of the braking system. However, in all discussions, we considered that wet runway conditions had to be covered as, firstly, operations should not be stopped in such circumstances considering the cost of a day's delay, and secondly, there should not be a diversion in the event of unexpected rainfall.

What then were the difficulties involved in extending that runway?

First of all, it should be said that whereas in Toulouse, flight tests representatives took part in the major technical meetings concerning the airfield, even sometimes to the extent of being the initiators or drivers, in Hamburg everything was shrouded in secrecy. Decisions were made without us being informed, although we were the first people to be concerned by issues like the available runway length. Even the (German) local Chief Pilot didn't know what was going on!

A past history of flooding had led to embankments being erected around the plant to protect it. These would now have to be moved and their stabilisation would then take a number of years. Clearly, all these works were hardly compatible with the aircraft's development schedule. But oddly enough, shortly afterwards, it was discovered that the logic behind the embankments was no longer valid and people had been wrong for all those years. So that was settled then!

To the south, there was farming land and some of the owners refused stubbornly to have their fields taken away from them. These farmers got together under an association to protect their interests as they feared an increase in noise pollution over the small village of Neuenfelde. The situation remained deadlocked for some time due to expropriation difficulties. In 2003, a small extension by about 180 ft was made, just by taking the end of the runway up closer to the embankment. But the distance available for landing facing south-west was just 6,329 ft, woefully inadequate for flights with the cabin installed.

Huge works saw the runway extended north towards the Elbe by over 1,000 ft in 2004. This improved the situation for take-off on runway 23 but did nothing to settle the problem of landing in the same direction and take-off in the opposite direction. That's why we were surprised by the decision to undertake such major works, whereas extension to the south seemed the most reasonable solution. Indeed, I remember arriving in Hamburg with Noël Forgeard and telling him, as we discovered these works on the Elbe, that they were extending the runway on the wrong side! As might be expected, I hardly received a warm welcome from our friends at Airbus Deutschland (formerly DASA)! I only understood the real reasoning behind this extension a few years later when the southern extension had been completed and I had a bit more data to chew on. Frankly, it would have been much simpler to have drawn up the strategy together by spilling all the beans from the start.

Early in 2004, thanks to better knowledge of the aircraft, the Design Office decided to do away with four brakes, thus making for considerable weight savings without any serious operational consequence considering the runways normally used. However, this was bound to have an effect on landing at Finkenwerder, as the distance on a wet runway for the simple ferry flight went from nearly 6,040 ft to 6,385 ft. Despite the small runway extension, we were theoretically unable to land on runway 23 in rain.

To improve braking performances on wet ground, you can resort to a special coating on the pavement or cutting grooves to drain the water away. Some roads have the same features. The local managers had decided to choose an asphalt coating, once again without discussing matters with the Design Office and flight engineers who had considerable experience of such braking problems. Indeed, we were sometimes called on by customers to assess the improvements in performance that some runway characteristics can contribute and our specialists knew the good and the bad solutions. The Finkenwerder runway was thus 'resurfaced' over the summer of 2004. After a few A319 and A321 production flights, it emerged that after each flight the tyres suffered considerable damage and had to be replaced for delivery flights.

We decided to conduct braking tests on this runway with our test instrumented A320. The initial idea was to see the improvement provided in rainy weather. We thus set out on a day it was going to rain, but the forecast was wrong. So we made three or four acceleration stops and landings, firstly on dry ground. Then we asked the firefighters to spread water on the runway to do tests on wet ground. The result was disastrous. Firstly, the tyres were heavily damaged and their condition was only just good enough for us to fly back to Toulouse. Performance was notably poor, as there was a loss in braking distance of 28% on dry ground and 20% when wet. Finally, to make matters worse, minor damage to the runway led to projections of gravel that could damage the flaps. The pavement coating simply had to be done over again – this time with advice from specialists. We also called in the specialists from ADP (Aéroports De Paris) to help. These works were completed very quickly from the end of September 2004 as we had to minimise impact on the A319 and A321 assembly line.

It was also planned to make grooves in the runway to improve braking in wet conditions. This could help compensate the increase in landing distance after removal of the four brakes. But having dithered about the type of grooving to be chosen, nothing was done.

Other difficulties awaited us in 2004. Indeed, expropriation of land needed to extend the runway to the south had been refused after lengthy legal proceedings. The extension requested included part of the basic length of 1,024 ft and another additional component of 909 ft to compensate for the refusal to increase the ILS glide slope and displacement of the point of landing on runway 23. The extension had been justified to the local authorities only on the basis of the existence of a heavier cargo version. This was a poorly argued dossier as there was only a very small number of cargo aircraft, hardly enough to justify an extension. Although we were not involved in preparing the document, we did mention to the DASA executives at an early stage that they should insist on the need to conduct development flights with the aircraft loaded with fuel and the cabin installed as well as flights of the same nature with customers. As usual, they didn't deign to listen to our concerns! I have to say I didn't push the issue as, personally, I was behind Toulouse so that operations would be simpler for us with a single delivery centre instead of two for the A380. A first judgement in April 2004 accepted the expropriation. However, the opponents contested this and Airbus Deutschland was forbidden from commencing works during the appeal proceedings. Various judgements then followed until August 2004 and they all went in the same direction. It was considered that the expropriation was unjustified as the community would not benefit much since the delivery of a very small number of cargo aircraft was not going to lead to safeguarding a significant number of jobs. I'm convinced that a well argued case file based on all aircraft and the fact that lack of an extension to the runway would prevent deliveries to Hamburg would have had a much greater impact with the likelihood of victory.

As the expropriation procedure failed to work, Airbus Deutschland launched negotiations with all the landowners to the south of the plant. To complicate their job, ten people had become owners of a plot of around 1,000 square feet only! I was obviously not privy to the details of these secrete negotiations, but, finally, Airbus bought all the land so as to able to extend the runway. The last phase could start with the destruction of the embankments and re-routing the roads. The first earthworks started over the summer of 2006 and the runway was opened a year later.

In its final configuration, the runway was 10,442 ft long, i.e. an extension of 1,640 ft. Meanwhile, the cargo version of the A380 had been abandoned and better knowledge of the aircraft's characteristics meant unnecessary margins could be dispensed with. We kept to the certified values for take-off and landing performances. Arriving facing southwards, the displaced threshold remained unchanged and the distance available for landing was 8,261 ft. Facing north, to avoid flying over the small village, the runway threshold was displaced by 2,256 ft. There were hardly any remaining limitations, except if an aircraft has to land in excess weight conditions.

As, facing north, I landed for the first time on the new runway with displaced threshold, I understood the probable reasons for the huge extension works towards the Elbe. Negotiations must have started at an early stage of the programme for the altitudes when flying over the village of Neuenfelde and its fine church not to be too low in the event of runway extension. So, to have adequate lengths for landing, the extension towards the Elbe was indispensable. A pity we hadn't been informed about all this earlier on.

* * *

A number of other runways were used regularly for tests. The main one was that on the Istres military air base. Its test history began well before the beginning of our story as it had been used to develop many military aircraft prototypes after the Second World War. Right at the start of the 1990s, Airbus had financed its extension to take its length from 11,470 ft to 16,400 ft in order to perform emergency stop tests with maximum energy on the A340-300. For the A340-600, longer by 43 ft than the A340-300, parking on the Flight Tests Centre apron was problematic. Therefore, it was decided to build a new area with the idea of accommodating not only the A340-600, but also the A380-800 and even lengthened versions.

On 13 July 2005, we went to Istres to perform the first VMU tests. We quickly realised that these manoeuvres threw up a lot of pebbles. Given the number of development and certification tests that were still to be conducted on this airfield, improvements were obviously needed.

The idea was to widen the shoulders to take the protected width to 250 ft. Clearly, we wanted a decision and quick implementation so as to pursue the tests without hindering the rest of the air base's activity. All these additional works to widen the blast pad were executed between 22 August and 5 September 2005. This really was record progress despite it being during the summer holiday period. The Airbus France teams, and especially Patrick Libralesso, had clearly anticipated these needs. Thus, in co-ordination with the A380 programme, they were able to launch all operations, including administrative authorisations, contracts, performance of the works, etc. Without such responsiveness, we would no doubt have had real operating problems on some days at Istres. The whole team deserves a word of thanks here. The price tag for this work was about one million Euros.

* * *

But our troubles were far from over yet as there were other runways we had to use for tests and that could have posed problems. We soon came to the conclusion that with grass covered runway sides, there should be no real problem. So we landed at nearby Tarbes to allay any lingering doubts. This airfield is regularly used by Airbus for the development of automatic landings and braking tests. As a precaution, sweepers were put on standby. Everything went well, thus showing that with a 150 ft wide runway, the usual shoulders measuring 25 ft on each side with grass beyond, there was no aircraft operation limitation. This validation at Tarbes proved that we could unrestrictedly used Brest airfield, which has lush greenery alongside the runways.

* * *

It soon emerged that there was one runway that was going to be a source of problems: Perpignan. Perpignan is the only airfield close to Toulouse where you can conduct a large number of autoland tests in very strong head wind conditions as needed for development. The strong, dry northerly Tramontane in the region blew regularly face on at 30 kt and this was important so as not to lose precious days on the programme waiting for the right conditions. Despite having the Mistral, the north-facing Marseille runway is not suitable due to the excessively steep ILS glide slope. Not being able to use Perpignan would have cost us in flying hours and lead-time. Indeed, the other airfield where the desired type of wind was available was at Keflavik, Iceland, but about four hours away.

Knowing the Perpignan runway well, and after our experience at Istres, works would obviously have to be done before it could be used by the A380. Indeed, beyond the usual shoulders alongside the 150 ft wide strip, there was a considerable amount of stones and gravel. We would thus have run the risk, on each takeoff, of throwing up stones onto the runway, which was hardly acceptable for regular use.

We made a proposal to the airport managers to enlarge the shoulders, at our expense, to take them to 25 ft on each side over two thirds of the length of the runway facing north. This was to cost around 0.5 million Euros. In return, we would be granted reductions on landing fees for the A380. Over the long term, this was a break even financial operation for us, but it meant huge time savings.

* * *

Apart from these runways used for tests, we never encountered any particular problem anywhere we went around the world. However, wariness towards the aircraft was widespread, especially in France.

The A380 landed for the first time at Paris-Charles-de-Gaulle early in June 2007 at the end of a VIP flight before taking off on a commercial round performed with Air France crews. We then returned on 26 June 2007 for the inauguration by the French President of the new terminal 2E. Early in the morning, after radar guidance, we were in final approach on one of the two parallel southern runways. Shortly before landing, I received the order to go around and take a northerly heading on the pretext that the runway hadn't been cleaned. As this had been said over the radio, all crews listening in could have got the impression that a special cleaning operation was needed for an A380 to be able to land, which was disastrous for the airplane's image. I therefore protested on the same frequency against this absurd form of discrimination so that the pilots would at least understand that such treatment was abnormal. We landed on one of the two parallel northern runways (27R). I didn't ask whether it had been swept or not! Then, just after landing, three cars drove back up our runway to inspect it. What a welcome!

At the reception following that inauguration, I had the opportunity to meet a number of airport managers and members of the Direction Générale de l'Aviation Civile (DGAC). They explained to me that they had to make such inspections for the first ten landings of a new type of aircraft. But I fail to see the interest when our weight was around 350 tonnes and the maximum take-off weight was 560 tonnes!

* * *

In Europe, the minimum runway width on which an aircraft can operate forms part of the certification results. To avoid later discussions, as explained previously, we had seen to it that the one usually used in Toulouse had a width of 150 ft, even if part of the shoulders could bear the weight of the aircraft once on the ground. Considering the only slight difference in the track of a B747 landing gear, we were confident about certification on that issue. The judgement of EASA test pilots confirmed the matter during the many flights they made. The aircraft was thus certified in Europe right from the start for 150 ft wide runways.

For the United States, the situation was different as determining this minimum width is not a certification requirement. Instead, the matter is subsumed under aircraft operations and is thus covered by another FAA department. Hugues Van der Stichel was in charge of this issue and he had to organise a number of meetings with the FAA to determine a test protocol to ensure validation. It was agreed to record all take-offs and above all landings except those relating to particular tests and to compile statistics on deviations from the central axis. These statistics were to compare the A380 with other aircraft, and in particular the A340-600, for which, some years later, we had to analyse a large number of tests conducted during development and certification. These statistics were to cover different populations of pilots: tests, certification and airline.

The situation that can be the most critical is normally with crosswind. That's why, for this instance of flight, special statistics were drawn up for test pilots and airline pilots. In addition, a dedicated flight was organised on 5 June 2007, in Keflavik, to check performances of three airline pilots who had as yet had little exposure to the A380, with a 20 to 35 kt crosswind. The maximum deviation in relation to the axis was about 16 ft. Remarkable!

All the results were extremely good since the mean deviations in relation to the runway axis recorded for the A380 were slightly lower than those that had been obtained on the A340-600. American approval came in a few months after certification.

On a lighter note, early in December 2006, one week before certification, I had the pleasure of having Nick Sabatini, number two of the FAA, fly alongside me. At the end of the pre-flight briefing, I jokingly told him that he couldn't come on board unless he first signed the authorisation to use 150 ft wide runways. Finally, I asked him to take great care as his landing would form part of the statistics, even though he had little experience on big aircraft. He made a perfect landing that was spot on in relation to the axis and the entire Airbus crew signed the print-out from our trajectography system and awarded it to him as a memento. He had been pleasantly surprised when I ostensibly placed my right hand on the cockpit coaming rather than let it linger close to the stick ready to take over the controls. He was flattered by this display of confidence. The whole episode made an impression on him as, some years later, he reminded me of it at a meeting of the Flight Safety Foundation board.

* * *

The last issue that had to be settled with the Administration, and especially the FAA, was what we call the determination of 'minima' on landing and the margin in relation to obstacles when you arrive near the runway on landing and make a go-around.

So what's the problem? An aircraft in approach for landing has to follow a path whose accuracy depends on the means for guidance used. For each type of approach, a complex statistical calculation allows the probability of striking an obstacle to be computed, considering that the greater the deviation in relation to the nominal flight path, the greater the likelihood of collision. This probability must not exceed a certain value. There are several categories of aircraft, from A to D, with A representing light aircraft and D the biggest. In practice, the distinction is made according to the approach speed as these categories allow either the decision height or the minimum descent altitude (the difference relating to the type of approach) to be determined. On reaching that height or that altitude, pilots must be in view of the runway to pursue landing, otherwise a go-around is compulsory. The faster the aircraft, the earlier the crew must be able to see the runway to prepare for landing and the speed is thus a logical criterion in determining these values that are known as 'minima'. Thus, the latter are higher on an aircraft in category D that approaches at more than 140 kt than on an aircraft in category C with a speed of between 120 kt and 140 kt. Generally the biggest airplanes fly faster, whence the confusion between 'size' and 'minima'.

Originally, the unanimous view was that the A380 was going to go into category D due to the planned approach speed. Now, the ICAO had decided to define a new category for aircraft with a wingspan of between 214 and 262 ft. The reason for this was that, for the same approach speed, with a slightly greater wingspan, the risk of touching obstacles increased. So there was to be a new category called 'DL' (L standing for 'Large') that could, in some cases, have very slightly increased minima. Only the A380 was concerned at that time as it was the only civil aircraft to have that wingspan. Category DL had been devised for it especially.

But, after the first flights, it finally appeared that the A380 had a much lower approach speed than predicted, which brought it into category C. Should a new category CL based on the same principle as DL be created to increase minima from 10 to 20 ft in some rare cases? Reason finally prevailed over nonsense and the A380 is now simply in category C.

As far as flying over obstacles during go-around were concerned, the situation was different as margins are established in relation to the size of the aircraft. A large number of studies were made on the simulator in Berlin and Toulouse, but the aircraft was not penalised.

TYPE RATINGS

So what came first, the chicken or the egg? An interesting question for an administration to address!

Certification of the A380 took place on 12 December 2006 and, well before this, Airbus had been concerned by the procedures to be applied to train the pilots of the first customer airline: Singapore Airlines. How to go about attributing the first type rating and those for the instructors? First of all, Airbus pilots had to be qualified. But to be able to give them a type rating, you had to find an instructor for them who already had his own rating. But who?

Back in 2000, the European manufacturers had made a suggestion. Indeed, a joint proposal for the regulation of flight tests was filed with the European authorities. It was expressly provided for that test pilots having taken part in developing a new aircraft would automatically receive their type rating, as also an extension for that of instructor, if they were already instructors. That way the pump would easily have been primed so as to qualify new pilots. But, after six years, this proposal hadn't advanced by an iota and, when the A380 arrived, the authorities found themselves with their backs against the wall and the need to resolve this conundrum.

At the Airbus Training Centre, responsible for qualifying airline pilots, things had moved forward appreciably with an international organisation known as the Joint Operations Evaluation Board (JOEB) bringing together those Europeans and Americans appointed by their respective administrations and that was entrusted with validating the A380 training methods and programmes. So how had they had resolved the chicken and egg problem? The Centre had appointed one of its instructors, a Belgian, who, it should be added, not being a test pilot, had only been able to take part in a very small number of flights as a co-pilot, since the aircraft was still in the development and certification phase. The method used was to organise a flight with a test pilot as captain and the said instructor from the Training Centre in the two front seats. A third man acting as an observer entrusted with issuing qualifications was to be installed on the jump seat. So who was this third man who held in his hands the power to sign (or not to sign) the first qualification on A380? You'd never guess! That person was to be another instructor from the Training Centre, but who had never flown on an A380! Meanwhile the test pilots having developed the aircraft and clocked up more than 500 flying hours on the machine were not qualified. What buffoonery!

I therefore expressed my dissatisfaction to the Direction Générale de l'Aviation Civile (DGAC) on the use of such grotesque methods worthy of Alfred Jarry's King Ubu. Obviously, my correspondents agreed with me, but that hardly solved the problem. What could we do? An engineer from the administration then suggested a convoluted way of getting around the problem where we would be granted temporary authorisations allowing us, as we tested each other, to issue a final qualification. I found it rather ridiculous for me to test Jacques Rosay with whom I had made the first flight (unless it was the other way round). Nevertheless, we accepted the principle as being the lesser of evils.

Three months after certification, nothing had shifted on the French side despite many reminders. A tour of the United States commenced mid-March. It really would have been desirable to have had our rating affixed to our licences. Only the British test pilots' situations were in order as their administration, the CAA, having grasped how absurd the situation was, had adopted a pragmatic approach and issued both the type and instructor's qualification in consideration of their experience on the aircraft. A few days before departure, Jacques and I received a temporary authorisation, equivalent to type and instructor qualification, but that could not be inscribed on the licence. The other French pilots had nothing.

Between 16 and 18 March 2007, I made a quick round trip to Los Angeles with the A380 # 1 (see the SALES TOURS chapter). On return, 18 March, I found a photocopy of the licences for all the Belgian training pilots from the Training Centre, with the type and instructor ratings on A380 affixed, on my desk. That copy had been obtained, under the table, by my assistant, Christine Hubert, who was familiar with the affair. I saw red and immediately phoned a friend, a top civil servant with the DGAC, to ask him what they intended to do faced with such a shameful situation: on the one hand French test pilots with several hundred flying hours of tests on the aircraft (more than 500 for some), with no qualification, and on the other, Belgian co-pilots with limited experience, had been issued with European qualifications, including that of instructor. That very evening, everything was settled: we had the DGAC's approval to issue A380 qualifications to all the French test pilots and the next day we received the official documents!

By way of a conclusion, a big word of thanks to our Belgian friends for having helped the French administration settle the problem of the chicken and the egg in its aeronautical version!

AIRSHOWS

What are airshows for?

Aircraft manufacturers often sign many contracts at them. But in reality, this is more of an announcement effect as either everything has been prepared in advance, or the last details will be reviewed later. These shows do not actually get sales made, but they do provide the background for meetings between personalities from the aviation world, including aircraft and equipment manufacturers, airlines, administrations, etc. They also provide aircraft and equipment manufacturers with a prime opportunity to make publicity. For some airshows, the big question is whether or not to take part with the latest prototype as that costs huge amounts of both time and money.

As far as the A380 is concerned, we came under enormous pressure to attend. Firstly, it is difficult to avoid going to the internationally recognised Paris-Le Bourget and Farnborough Airshows. Berlin is of little economic interest, but for political reasons, you can't miss it unless you want create a diplomatic incident with our German friends. Then there are Dubai and Singapore. How could we fail to go to Dubai, the headquarters of Emirates that, at the time, had ordered more than fifty A380s? As for Singapore, our first customer is based there, so we had to go.

On average, taking part in an airshow 'costs' ten days of tests per aircraft if you add up the preparation, ferrying and actual presence. Doing five shows in two years therefore means taking an aircraft out of the test programme for about fifty days. That's huge, especially when the Top Managers are at the same time asking you to get through to certification in twelve months.

In our initial schedule, we had only included participation in the Le Bourget Airshow. On the flight tests side, we knew perfectly well that this was totally unrealistic and that we'd have to go to Dubai, Berlin, Singapore and Farnborough for the reasons given above. However, we reckoned that it wasn't up to us to plan it and fit it into the schedule.

* * *

Airbus has always had the habit of doing display flights at these airshows. Personally, I saw no problem with this as it was an exercise I enjoyed. Making a success of the act brings great satisfaction. When events in flight don't go altogether as planned, even if on the ground few people will have noticed it, you have to think about how to go about things the next time, which is always a rewarding exercise.

At the end of the 1990s, I officially asked the question: “Do we have to continue doing flying displays?” The answer was “Yes”. That suited me fine and I no longer felt any remorse in deriving pleasure from the exercise!

* * *

At that time, there were three pilots qualified for flying displays. Firstly, several were needed to cover colleagues' absences, and sometimes a number of aircraft had to be presented at the same show. In addition, it's an exercise that requires some training.

I already had considerable experience in demonstrations since, back in 1978, I had set up an aerobatic team with my best friend Denis Legrand to fly two light aircraft. This team went by the unusual name of ‘Les Porthos’. Why chose one of the Three Musketeers as a title? Quite simply because one day, we were at Brétigny (the former Flight Test Centre base in the Paris region) with the Patrouille de France for rehearsals. Their leader did some radio checks with his team mates: “Athos check radio”, “Athos 2”, “Athos 3”... So in the end, for a joke, I followed on with “Porthos check radio” and we kept that name!

Demonstrations with two aircraft are much more difficult than solo as manoeuvrability is not the same. Even when not in tight formation but when, for example, you are preparing crossovers, you need to take the wind into account to perform them facing the show centre. With two fairly slow aircraft flying in opposite directions, possibilities for correction are reduced, whereas when alone, everything is much easier.

Having arrived at Airbus with quite some experience in displays when I became Flight Tests Director, I naturally joined the two or three qualified pilots. The Chief Pilot was obviously involved, firstly Bill Wainwright then Jacques Rosay. Finally, at the time of the A380, Peter Chandler, who had large experience in the display of military aircraft, joined the team.

* * *

I gradually adapted the profile of flying displays. Since the mid-1990s, it had not changed and, up to my departure from flight tests, it remained standard on all types of civil Airbus aircraft. The fact that it remained practically identical is hardly surprising as the manoeuvring speeds are of the same order of magnitude from the small A318 to the very big A380. The thrust to weight ratio is not always the same and, generally, it is significantly greater on a twin-engined airplane. Similarly, drag is relatively smaller on an A330. All this leads to adaptations from one aircraft to another, but these remain minor.

To make a good flying display, whatever the aircraft type, you have to respect the basic principles. Thus, in so far as is possible, the aircraft must always be in movement. The spectators don't come to see it fly in a straight line at cruise speed.

It is also important always to stay at the show centre as much as possible. You sometimes need to move away from it, for example to prepare a slow pass. But decelerating astutely, you can be at low speed, in a good nose-up position just when leaving the turn once the trajectory is parallel to the axis for presentation and close to the public. If you look at the traces of radar recordings against the ground for the display of a Rafale fighter jet and that of an Airbus, as described below, you can see that the surface occupied is very much the same, apart from the turn before landing.

Above all, you have to avoid scaring the spectators. So passes at very low altitude are absolutely to be prohibited, especially for really big aircraft. For an A380, even if the lower limit authorised is 500 ft, we often fly past towards 800 or 900 ft.

* * *

Here's a short description of the display as performed in the first years of the A380.

It started with a take-off and an initially very slow climb to then pick up speed and reach the maximum attitude of 30 degrees. Then, according to the take-off point in relation to the position of the public, either the pilot makes a teardrop turn to come down towards the spectators, or two turns are made to return in parallel to the presentation axis.

The following figure is a 360 degree turn with a bank angle of 45 degrees, firstly in climb with an attitude of about 20 degrees, then in descent.

This turn is followed by a short two to three seconds flying away from the public, in deceleration, so as to get into position to come back and pass by in parallel to the presentation axis at maximum or near maximum angle of attack, of around 15 to 16 degrees. The indicated air speed will be about 110 kt. As this pass is performed facing the wind, the real speed can be considerably lower.

At the end of the pass, throttle is set to full thrust and the aircraft starts to climb and then, after stabilisation, a tight 180 degree turn is engaged (half turn), ‘stick in the corner’, or almost according to the type of plane, to continue climbing and adopt a fairly steep bank. Coming out of the turn, the pilot lets the attitude stabilise towards 25 to 30 degrees (again according to the aircraft type). Then, simultaneously, reduced power, stick full forward to dive rapidly at an angle of 15 degrees (the limit with the flight controls protections) and extension of the landing gear.

The display is almost over. Two turns are used to get into position for landing. If possible, no thrust reversers (or at minimum thrust) so as not to make too much noise and not too heavy braking to avoid smoke!

Put like that it all seems very easy. In fact, when the conditions are good, it's almost a walkover. But that can be deceptive; problems can arise so quickly and you have to react immediately if there is an engine failure, not because of performance in climb but because, in some cases, the aircraft can be distinctly below the minimum control speed with one engine out. Everyone knows what to do in such instances and the test flight engineer is especially attentive in monitoring the engines in critical phases.

The other problem situation, encountered fairly frequently, relates to respecting certain volume and axis limitations. The minimum altitude is not normally a concern, as we adopt margins, but we may have lines not to be crossed. This applies, for example, at the Le Bourget Airshow when taking off facing south-west. To get in place for the low speed pass, you have to turn around between two imagined 'barriers' making an angle of 60 degrees as, firstly you must not cross a line parallel with the runways of Paris-Charles-de-Gaulle Airport, and secondly you must not cross the axis for presentation parallel to the public. Imagine making such a manoeuvre with your car in a car-park. It's child's play when driving very slowly, but things are not so easy in an aircraft travelling at about 150 kt, possibly in strong wind. You really need to anticipate and the slightest misjudgement can have serious consequences.

Even if displays can appear easy on some days, you still need to have experienced pilots.

These demonstrations proved the agility of our aircraft and that of the A380 in particular. Such good handling characteristics are essential. Sharing the same runways with other (often very different) aircraft, you need to have the same level of manoeuvrability.

* * *

With the A380, the major novelty during these airshows was the visit of the aircraft interior, something not yet seen on other programmes. Visitors went through, sometimes uninterruptedly, from the moment the show opened right through to its closure. We knew peak periods with more than 600 visitors a day.

It was during the first Le Bourget Airshow in 2005 that we discovered the enthusiasm this visit aroused. We were then very badly organised as we hadn't seen things coming. For things to run smoothly you need a registration and filtering system, security personnel and plenty of people with good knowledge of the test aircraft layout to assist. We came to the conclusion that you needed at least ten minutes for a visit and that it should be conducted by a test crew member, with five to seven visitors at a time, following a well-established circuit. Nothing of all that had really been foreseen. We remedied that at subsequent shows and everything went like clockwork for Le Bourget 2007.

On the test aircraft, the visit started with the main deck, where our guests could see to the fore, close to the entry, big tanks that are used as ballast to adjust the weight distribution. This means you can modify the centre of gravity position in flight by rapidly transferring liquid from the forward tanks to those to the rear and vice versa, using big pipes. We jokingly informed the visitors that these tanks contained Bordeaux and Burgundy red on the main floor and white on the top floor, with Champagne being kept chilled in the cargo! Right at the start of the programme, we still had the crew's emergency evacuation door in place (closed off of course!). Going to the rear, on the engineers' station, explanations were given on the instruments the flight engineers have available to them to conduct tests and we were able to do a few demonstrations on various displays. The visit then continued to the rear with all the electronics cabinets and the storage area for spare parts we may need during the show. There is also an access to the rear cargo hold by a ladder that does not exist on a standard aircraft, but that we could use, if necessary, to conduct additional checks and listen to any unusual noises. At the foot of the rear stairway there is an ordinary armchair with an intercom to communicate with the pilot and the engineers. We explained that some flight controls settings commenced on the simulator required additional adjustments in flight, especially where passenger comfort was concerned. The pilot analysed the effects of the manoeuvres performed himself for the front of the aircraft, while test engineers did the same for the central part and another engineer seated in that seat did likewise for the rear.

You climbed up to the upper deck by a small half-winding stairway. There was nothing much to be seen on the upper deck, apart from the ballast. On one of our test aircraft, there was an extremely fine wooden table. A number of people told me (although I never had formal confirmation) that it had been made to be installed on a Caravelle, still at the prototype stage, during a trip from Rome to Lourdes by Cardinal Roncalli, the future Pope Jean XXIII, in 1955. It has since then always been placed on one of our test aircraft and, for us, it still remains the 'Pope's table'. When we are out travelling, the mechanics are delighted to use it for impromptu picnics or to play cards!

From the upper deck, there's a really fine view over the more than 9,000 square feet of the wing. Still moving forward, after crossing an area fitted out with thirty business seats, all different, recovered from other programmes, but really good for long trips, visitors arrive at the top of a majestic stairway leading you down to the main deck. You then have to climb a few steps to gain access to the cockpit.

The whole visit comes to an end with the climax of entering the cockpit. Specialists occasionally ask for some technical explanations but most visitors just want a photo souvenir of them in the pilot's seat. We still needed to keep a sharp eye out for the odd would-be pilot wanting to have a go at the controls!

* * *

The first airshow the A380 took part in was Le Bourget in June 2005. For this occasion, the Airbus chalet was a permanent building with two floors rather than just a removable chalet. Marie-France Perrin was the great organiser of all these events and her idea was to have a permanent location close to Paris where you could organise meetings, seminars, aircraft deliveries or any other event. At the time of the airshow, it had not been completely fitted out for these additional functions for lead-time, but also budgetary reasons.

At the time, we still had little experience in flight on the aircraft as we were at rehearsal flight # 20 and ferry flight # 22. We therefore had several serious limitations when it came to the presentation. Firstly, it had to take place with the landing gear extended, as the landing gear manoeuvre included an automatic test lasting two minutes, which was incompatible with the flight time. It had also been decided not to go beyond the maximum angle of attack normally authorised as some uncertainties remained as to the behaviour of the flight controls computers.

When discussing the first flight crewing we agreed with Jacques Rosay that he would be the demonstration pilot for the first Le Bourget Airshow. His co-pilot was Wolfgang Absmeier. Obviously, Gérard Desbois was flight test engineer and a number of other engineers came in in turn to support the team and monitor all the parameters, including flight controls, loads, engines, etc.

The aircraft came in on Sunday 12 June 2005, awaited by about 200 media people. The usual scenario with plenty of interviews, visits around the aircraft, demonstration flight, etc.

But the real problem lay elsewhere. The organisers wanted to impose a six minutes interval between landing of the A380 and take-off of any other aircraft. The previous evening, having arrived with the A340-600, I had commenced discussions to try and improve matters. To no avail. There were people there who were obdurate, failing to understand that we would be taking off at just 300 tonnes. Imposing a six minute separation was patently absurd considering the fact that B747s took off daily at 396 tonnes with intervals of just two to three minutes. We resumed discussions the next day. Flight Director Christophe Cail (codename 'Jupiter') was convinced by our arguments, but orders came from on high. We tried in vain to get Noël Forgeard in contact with the 'Head or Air Navigation'. This was a Sunday and we were alone in seeking a solution to this problem. What really worried us was the idea that our customers would get the impression that operating the A380 would really require such separations. The next day, just after the demonstration flight for President Chirac, we had the opportunity to meet the Cabinet Director of the DGAC. Within an hour, this spacing problem was settled (in principle). That afternoon the details were finalised during a local meeting and, to do some face-saving, we accepted a separation of two minutes between the A380 and a medium weight aircraft. Just the time to cross the runway! On Tuesday, the Airshow Director, who was a reasonable man, confirmed to us that there were some people who took things to extreme lengths but there was nothing he could do about it because they were the specialists in their field.

Another difficulty was waiting for us around the corner with the dust and pebbles thrown up on application of power for take-off on runway 03 (facing north-east). This was the strip usually used for displays, except where there was a strong westerly or southerly wind. The Sunday rehearsal had revealed the problem. Four hours were thought necessary to sweep the area. That seemed a huge amount of time, but what could we say to those actually doing the work? In reality, after the demonstration flight for the French President, one hour had been enough to clean everything up. Following the earlier afternoon clean-up, this one took just eleven minutes. The two test flight mechanics we sent out to take a look and confirm the need for sweeping expressed doubts. But the 'sweep vehicle' had the last word! Finally, an inspection was maintained for the Tuesday and then no further sweeping intervention was made after the flight of the A380. Perhaps the blast from our jet engines had cleared enough dust away from the runway shoulders!

Overall, the display flights went well considering the state of progress in development. There were a few problems, as for example with some transient losses on certain of the main computers for flight controls, and electrical and APU anomalies, but nothing to prevent a flight display being pursued in complete safety.

There were many visits from major personalities, starting with President Jacques Chirac on Monday. He even shook hands three times with the two flight test engineers: the first time when entering the aircraft, then when approaching their work console and then again on leaving the aircraft! Many Ministers and airline CEOs came by during the week. And one visit ran into another. This even became problematic as it was difficult to have some personalities stand there waiting. What caused much more trouble was a big delegation of Israeli Ministers with their heavies, wives, etc. sent in without warning by EADS. They refused to be separated into several groups to visit the aircraft in the best conditions. Close supervision was required to make sure they didn't interfere with anything. Indeed, during all these visits, everyone had to remember that the aircraft was to remain in airworthy condition.

Next Tuesday, Pierre Graff, President of Aéroports de Paris (ADP) suggested a stopover at Roissy on the Sunday after the last presentation before going back down to Toulouse. We accepted this in principle and a number of meetings took place to organise arrival at CDG. Everything looked good until, on Friday morning, a Top Manager from ADP came along to tell us that it was cancelled. Apparently, this decision was based on concerns as to insurance for the risks in docking the boarding bridge. I quickly obtained documentary proofs from Airbus showing that we were correctly insured but again to no avail. ADP had decided it was too late! I must admit that this mishap stuck in the throat and later, during our tests, we did all we could to avoid Paris-Charles-de-Gaulle airport.

* * *

The next airshow was Dubai. This too had its fair share of emotions, but for other reasons.

It all started off really badly with a truly serious diplomatic incident with our biggest customer: Emirates. The Dubai Airshow came in just at the end of the sales tour to Singapore, Australia and Malaysia (see the SALES TOURS chapter). Now, for each of these countries, we had applied stickers to the aircraft with the colours of the local airlines: Singapore Airlines, Qantas and Malaysia Airlines. But Emirates was, by far, our biggest customer. The Emirates management were outraged to see the aircraft with the other airlines' colours, even if, in reality, there had just been one other logo before having presented the aircraft in their own colours.

To try to appease Emirates and more especially their CEO, Tim Clark, Noël Forgeard and John Leahy offered to send them the two A380s undergoing tests, # 1 and # 4. But Tim Clark refused to take their calls. So much for the general atmosphere! Meanwhile at flight tests, we were not in favour of this solution as we reckoned that a single aircraft with the Emirates livery would have had more impact than two of them, with the other one bearing Airbus colours. Secretly, we were also hoping there would be just one aircraft at the airshow, so as to reduce the amount of time lost for testing. But, considering the diplomatic tangle, our point of view hardly counted. Finally, the decision came in on Thursday 17 November 2005: just one aircraft, fully covered in Emirates colours. Order of Emirates!

During the first tour in Asia and Australia, I quit aircraft # 1 in Brisbane to go direct to Dubai to settle all the details of the in-flight demonstration and organisation of a flight photo. Aircraft # 4 arrived on Friday 18 November 2005 towards 8.00 p.m., after having crossed # 1 back from the sales round above Crete in visual.

On Saturday morning, the photo flight commenced with a Learjet equipped with airborne cameras. Despite the many passes at different altitudes close to the famous seven star Burj Al Arab Hotel, there was still no good photo and we had to make do with a montage. A whole flight to no purpose! The afternoon was reserved for general rehearsals. For co-pilot I had Harry Nelson who, outside his test pilot functions, was my deputy.

We had acquired good knowledge of the aircraft and displays were now up to the standard of all other Airbus airplanes with the profile explained above. The environment in Dubai is ideal and no particular difficulties were encountered.

We still had some technical issues. On Tuesday, as we were getting started, we only had half the electrical equipment available and, despite our many attempts, we failed to recover what was lacking. So, thirty minutes from take-off, we decided to cut everything off and restart. Just like a computer reboot! The problem derived from the alignment of the three inertial navigation units that takes ten minutes and has to be done with the aircraft stationary. We had no margin as we then had to push the aircraft out, start up the engines and taxi up to the take- off point. But everything worked well. Relief! The next day, we accumulated a number of minor failures but nothing that prevented the display.

Inauguration of the airshow took place, in accordance with the tradition, on Sunday morning. A host of major personalities wanted to board the aircraft after Sheikh Mohammed Al Maktoum, the Emir of Dubai, but we had to hold them back at the foot of the stairway. A constant flow of visitors in a test aircraft, ready to fly that afternoon, requires a certain number of precautions and extremely strict surveillance. We were hardly qualified to ensure such control but, finally, all those personalities were able to make short visits in small groups.

Generally, we had retained the lessons learnt at the Le Bourget Airshow to organise visits, with candidates having to sign up in advance for a given time. But we still needed to make progress.

The Airshow finished on Thursday and it was part of the tradition to have pilots from the local airline, Emirates, fly on our new aircraft on Friday. This time, this was a major issue as they would be the first non-test pilots to land the aircraft apart from a few instructors from our training centre. We decided to strike twice in one go by organising the first part of the flight with the Patrouille de France to take photos and video sequences. The in-flight rendezvous was set at 9 hours precisely close to the Burj Al Arab Hotel. It was planned to devote 35 minutes to this exercise and air space was reserved for us. It didn't take one minute more for the accompanying Alpha Jet to take some fantastic photos of the A380 with the Patrouille de France, plus a few videos.

• See photo n°34 – page 246

The flight continued through to Ras Al-Khaimah airfield just over 60 miles away from Dubai. There was practically no traffic, meaning we could devote a quarter of an hour to each pilot to make two visual landings. And there were eleven candidates. There were supposed to be five of them to start with, but the list got longer as others wanted to have the pleasure of flying this new aircraft. There were Airbus pilots, who obviously felt at ease straight away. But there were also 'pure Boeing' pilots who experienced no difficulty. In the end, it was a great flight for all. This was important for us as getting other pilots to fly can lead to surprises or pinpoint difficulties they may have handling the airplane. This had already happened in the past. But nothing of the sort with the A380. We were now convinced that we were close to the final configuration for the flight controls.

Having refuelled, we immediately flew back to Toulouse. I must admit that I didn't take part in the take-off from Dubai. I left my seat up front and was already sleeping as we sailed up into the air, exhausted by the long morning flight and the amount of concentration it had demanded.

* * *

At the end of February 2006, we took part in the Singapore Airshow. To avoid any jealousy, the A380 had been entirely painted with the Singapore Airlines livery, as for the Dubai Airshow where it was in Emirates colours.

The aircraft was ferried out a few days in advance to Singapore for a brief test campaign in hot weather, on the ground and in flight.

It was Peter Chandler's turn to perform flight demonstrations. He was accompanied by Jean-Michel Roy, who was then Director of the Training Centre.

In Singapore, flight demonstrations are performed along the coast and thus start with an aircraft already in flight and not as from take-off, meaning a minor adaptation in profile is needed. No difficulty was encountered during any of these demonstrations.

The Singapore Airlines pilots had the opportunity to fly. However, we had to keep their number down compared with what we had done in Dubai as the traffic on the international airfield prevented us from doing such rapid circuits. A flight was also made just for the airline CEO and some of its managers.

* * *

Every other year, in May, the same year as the Farnborough Airshow, there is the Berlin Airshow. This is an international show, but in practice, there are few foreign visitors. As Airbus is every bit as German as French, it is impossible to avoid making a go of it. The main advantage lies in showing Airbus to Berliners and more generally to German personalities. In 2006, we demonstrated an A380, an A340-600 and an A318 in flight.

German airshows are a bit special as the law gives priority to commercial traffic. The Flight Director can do nothing about this. Thus, despite there being two runways at Schönefeld Airport, the timetable was never respected and there could be up to an hour's delay for demonstrations. These generally take place fairly early and as Berlin is not far, it's easy to get back to the hotel early to visit the city. What's more, I can generally take advantage of the chance to go to the opera. With La Traviata and Norma being staged in May, I had a pretty enjoyable airshow!

There were no worries for visits of the A380, with everything being in German. We had come in with a lot of young German flight engineers who were delighted to show off this fine aircraft to their compatriots.

As at Le Bourget, Jacques Rosay demonstrated the A380 with Wolfgang Absmeier, and I took care of the A340-600. Thursday afternoon, after the demonstration, the A380 left for England for a flyover of the Airbus Filton and Broughton plants, followed by a landing at London-Heathrow. The aircraft came back the next day, in time for the demonstration.

* * *

Sunday 16 July 2006 saw our departure for the Farnborough Airshow with a flight to validate the demonstration shortly after arrival. Obviously, there were two Englishmen at the controls: Peter Chandler and Ed Strongman.

Everything started off badly as we were shocked to discover, as the aircraft came out of the paint workshop the day prior to departure, that the front part of the fuselage bore the legend "Recyclable by design". We were furious: our magnificent aircraft was treated like a humble bit of packing to be put in the recycling bin! I told communication management what we thought of this in no uncertain terms, but for them this was the sign of progress: a recyclable aircraft! Decidedly, between aviation enthusiasts and some communication specialists we were worlds apart!

It was really hot that summer in Farnborough, with temperatures reaching 36°C! Once again, there were many visits, firstly before the demonstration, then during the afternoon, as the aircraft was immediately brought back to the static exhibition area after landing. The organisation was a priori good, but, unfortunately, some visitors came late or VIPs, that it was difficult to turn away, came along without warning.

Thursday morning, we called in at the flight schedule office and were astonished to see that my A340-600 demonstration flight had been taken out. As they had to find a slot for the pass of the A380 with the Red Arrows, it had been decided to do away with the A340-600's flight. Among the organisers, nobody had been polite enough to have told me. I lost my temper and told them we had had to pay a huge amount of money to take part in the show and that the organisers had no consideration for their customers. Finally, I called their bluff and announced that we would be pulling out the next day. The flight plans for a return of the A380 and A340-600 to Toulouse the next morning were filed. Obviously, I had no powers with Airbus to make such a decision based on so futile a reason. I don't know if the organisers were taken in or not. In any case, the A340-600 demonstration was put back on the programme for the afternoon, and the A380 fly past with the Red Arrows was also maintained.

* * *

For EADS, the Paris-Le Bourget 2007 Airshow took place against a background of economies. The Power8 Plan had been implemented. One of the main measures adopted affecting us was the requisition of 'our' building by EADS. We had to share it with EADS Management and Eurocopter, with overall organisation being under EADS. That hardly pleased Marie-France Perrin who had had the idea for the entire arrangement of the premises. So, to co-ordinate visits, she had created what we called the 'Gauls village' as in the Asterix series. This was something like a small shack. There was one room to register visitors and serve as a waiting area, another used as a bar and VIP area, a miniature restaurant with four large tables and then a fourth room used as an office. It was in this small building that the entire Airbus technical team met up to get away from the crowded 'chalet'.

We had come in with two A380s, # 7 that stayed permanently on the static exhibition and # 9 for in flight demonstration. The latter was parked outside the airshow as such, which greatly facilitated the flight departure as there were no other aircraft to be removed.

I was responsible for the demonstrations, Richard Monnoyer and Karl-Heinz Mai taking turns as co-pilots.

A large number of visits were scheduled but in the end, we packed even more people in. Visits were organised differently from how we had originally proceeded due to the fact that the aircraft had a real cabin, and we didn't need flight test personnel to lead people around. So, five marketing engineers presented the cabin and the flight tests representatives were in the cockpit. There were three levels of filtering with a number of security staff. It all worked extremely efficiently. During the days devoted to professionals, the number of daily visitors was estimated to come to between 600 and 700. There were also about 300 to 400 a day over the week-end, even though the aircraft was supposed to be closed!

Unlike the Le Bourget Airshow tradition, the French Prime Minister officiates over the inauguration rather than the President. I was in aircraft # 7 with Louis Gallois to welcome him. The briefing was extremely short and, as soon as it was over, I went back down by another door to join a police car that took me to # 9 that already had two engines running as there were only a few minutes between the end of the visit and the start of the inauguration demonstration flight.

The weather was pretty awful throughout the show, with mostly southerly and westerly winds. Almost all take-offs thus took place on runway 21 (facing south-west). The take-off was highly appreciated as the initial climb before the turn was directly in front of the chalet with an attitude of 30 degrees. But, as explained at the start of this chapter, execution of the demonstration in that direction involves a number of difficulties. Landing on runway 21 was forbidden and we had to land on the runway facing west, far away from the spectators.

We took our time flying back to Toulouse, making a detour to fly over the Airbus plants of Laupheim, in Germany, then Méaulte, Saint-Nazaire and Nantes in France.

* * *

The cycle of presentations was pursued at various international airshows. The Santiago de Chile show was added to the list in March-April 2008. We wanted to push sales of the A380 in South America. I had been to this airshow a number of times since 1982 (the first time with a small training twin-jet designed by Microturbo, the Microjet 200). There was a great atmosphere. When the weather's good over the Andes (which is often the case late morning during that season), the VIP flight can go ahead, with a glass of Champagne in your hand (I'm talking about the passengers!) at some 23,000 ft, with two wide turns around the Aconcagua, the highest mountain in the Andes. Even for VIPs who've seen it all, seeing this mountain summit so close remained memorable.

* * *

Oshkosh!

Only aviation enthusiasts (or possibly kids who've worn 'OshKosh B'gosh' jeans) have probably heard of this town in Wisconsin. I never found out where the clothing was actually manufactured. In any case, the tradition goes back before Blériot and the Wrights as the words "since 1895" are proudly shown on the garments.

Oshkosh is the biggest fly-in convention in the world where you can see some 12,000 different airplanes, including 4,000 amateur constructions, many warbirds (planes from the Second World War that have been restored). Some 500,000 visitors come in over the week. Everything is on sale there whether for vintage or recent aircraft, all sorts of spare parts for fixed wing aircraft, ULMs or helicopters and, generally, anything whatever to do with aviation. The whole gathering is organised by the Experimental Aircraft Association (EAA) that brings all American amateur airplane builders together. But the fair really does go beyond amateur construction.

I had so many good memories of Oshkosh and really wanted to go back there. In 1981, when I was still an engineer at the Flight Tests Centre, I did an aerobatics show with Denis Legrand on the Cricri plane. Cricri is the smallest twin-engine airplane in the world. It has an empty weight of about 160 lb and a maximum takeoff weight of around 330 lb. The plane was designed by a brilliant Aerospatiale engineer, Michel Colomban. At that time, Cricri planes were underpowered as each engine only delivered 11 hp. Shortly afterwards, more reliable 15 hp engines were installed. Considering the available power, our manoeuvrability was limited. We couldn't go through the usual programme of our 'Les Porthos' formation, being limited to just two loops, a few rolls in formation and crossovers. Even the smallest American planes were doing acrobatics with 200 hp while we were stuck with 22 hp each! In any case, our act was a great success, probably due to the effect of surprise.

Having already given a display with the smallest aircraft at Oshkosh, I was really keen to make a mark with the biggest! It's still fun to show off, even after sixty years of age!

Going to Oshkosh during the development and certification period, in 2005 and 2006, was clearly out of the question. Then in 2007, we had discussions among ourselves and a favourable decision was in the offing. Our US subsidiary was pushing hard in favour of taking part. John Leahy too, who knew the general aviation world well, was backing the idea. But apart from them, few people at Airbus could see the interest in attending a show where no contracts were signed. Some people hadn't understood that this was a tremendous promotional vehicle for aircraft in the United States. To settle the problem, the only aircraft available during that period was undergoing works for various modifications. Then, in 2008, with the budget restrictions of Power8, there was no question of going.

In 2009, the idea was brought up again by Airbus North America (AINA) who wanted the aircraft to take part but lacked the budget. Luckily, Marc Fontaine, Corporate Secretary and Chief of Staff of Tom Enders, knew this world well and thought we should participate. So, thanks to him, everything was organised. He even convinced Tom to come along in the aircraft to be around for the first few days.

• See photo n°35 – page 246

At that time, I was no longer at flight tests management as Louis Gallois had asked me to take over the Product Safety department. However, I was still pursuing my test pilot activities. As soon as the news of the A380's participation in Oshkosh came out, Jacques Rosay offered me the chance to go and do the demonstrations. Great news!

Organisation was co-ordinated by the usual Airbus and AINA teams, but more especially by Terry Lutz, an American test pilot, hired a few years previously and also an amateur aircraft builder. He was the co-pilot during presentations. While the EAA rolled out the red carpet for us to come, we encountered the usual difficulties with the FAA. In particular, having given them a whole load of documents, including a chapter of the Airbus Flight Test Guide describing in detail what to do and what not to do in a demonstration flight, they told us that this section was too old (2003) and that limitations for transport aircraft in the USA were 200 ft minimum altitude and 30 degrees maximum bank. We replied that at all airshows around the world, minimum altitude is 500 or 600 ft and bank 60 degrees for transport aircraft. Also, everything that had been written in the Flight Test Guide was fully applicable to the A380. In response to their request, I rapidly delved into my flight logs to report that I had done that presentation 177 times (apart from rehearsals and validations) on all fly-by-wire Airbus aircraft. In addition, since I had developed the demo, it must have been done more than 500 times by all the demonstration pilots. In addition to this, I announced to the President of the EAA and Airshow Director, Tom Poberezny, that I would not be doing the demonstration with a limitation to 30 degrees bank as this was not 'calibrated' accordingly and I didn't have time to prepare another. As the AINA President was a former FAA Board Member, we were able to settle things at the highest level. We thus received a fifteen page document explaining all the rules to perform that demonstration. We settled for a maximum bank of 45 degrees, which was no problem. Luckily, other airshows around the world are not hindered by so much bureaucracy. It's strange to reflect that this should happen in the United States, where people tend to be pretty pragmatic in daily life.

We took off for Milwaukee on Monday 27 July 2009, with thirty people on board aircraft # 4. Indeed, before reaching Oshkosh, we were to make a stopover to disembark the passengers as the regulations (and common sense) require you to have the minimum crew for an in flight demonstration. On arrival, there were speeches and visits of the aircraft as usual.

We took off for Oshkosh on the afternoon of Tuesday 28 July. The weather was fine and the demonstration went well before a huge crowd that had been waiting for us.

Things started to turn sour on landing. Indeed, we had been advised to touch down if possible facing north and to make a fairly short landing to avoid having to tow the aircraft and to keep closely to the runway axis so as not to raise dry grass with the thrust reversers. The distance between the start of the runway and the taxiway was 5,200 ft. There were gusts of crosswind at a little more than 20 kt. I have to admit that I made a bit of a mess of it with too low a speed over the runway threshold. Coming down at 10.5 ft a second, this was the roughest landing I'd ever made in my life. There had to be a day for it! The limit stands at just 10 ft a second, but given the fact that the aircraft was extremely light, there were no technical consequences. But some commentators, Boeing pilots in particular, had a field day on the Web. “The A380, hard to control”, “Wings too flexible”, and so on. Some people forgave me, noting that landing in those conditions was no easy matter. I hesitated about replying, but finally I kept quiet; it's always best to let polemics die down by themselves. I could have done better, but then ‘Nobody's perfect!’ I was also itching to ask those Boeing pilots whether, with their airplanes, they could make slow passes with 15 degrees angle of attack and at 105 kt in complete safety and whether they were also ready to make tight turns at the same angle of attack and 45 degrees bank as I did during the demonstration. Clearly, none of their aircraft were capable of that. Perhaps greater modesty would have been in order, with greater consideration for the A380.

On arrival, there were the usual speeches, and the runway team supervised by Gérard Desbois and Didier Ronceray immediately took over the circuit of visits. Indeed, in addition to the traditional guided visits in small groups, the so-called ‘VIP visits’, we had for the first time organised a non-guided visit. This gave access to the main deck behind the wing, allowed you to move towards the rear, go up to the upper deck by the half-winding stairway, walk to the front of the aircraft, go back down to the main deck and leave just in front of the right wing. Those visitors didn't have access to either the cockpit or the flight engineers' station. On this circuit, the rate of flow was about 600 visitors an hour, between 10 a.m. and 4 p.m. The queue was often long, with about 600 to 1,000 people, meaning an hour to an hour and a half waiting, but everything was perfectly ordered and disciplined.

• See photo n°36 – page 247

For late afternoon on Wednesday 29 July, I'd promised to conduct VIP visits for customers of Daher-Socata (ex-Socata), who were purchasers of TBM 700 and TBM 850, fast single turboprop aircraft. I was keen to look after them well in consideration of my having worked with this company for the development of that aircraft. They were supposed to be at most 50 in number. But in the end, 150 turned up, families and friends included! That evening, I also gave a conference at the Theatre in the Wood on A380 testing with a large number of videos being projected. The evening ended under the wing of the A380, still with the TBM purchasers. Harrison Ford, an aviator but not a Daher-Socata (nor Airbus!) customer, was present. The two US Airways pilots who had put down the A320 in the River Hudson in January 2009 (thus saving the passengers' lives) also came along.

Thursday 30 July was also devoted to visits. To round off the day, a gala evening was organised, in casual dress, where an auction was held to raise money to provide flying opportunities for young people. This was an excellent EAA initiative. To give you an idea of the scale of things, a tour in the A380 simulator in Toulouse was awarded for 30,000 dollars. In total, between two and three million dollars were thus brought in for those encouragement programmes.

Visits came to an end around midday on Friday 31 July. About 10,000 people had come on board with standard and VIP visits. The aircraft was prepared for the demonstration flight. In the absence of a taxiway suited to the A380, we had to push back some 5,200 ft to take off facing north. Again, the demonstration was followed by a huge crowd. We then landed at Milwaukee to wait for the rest of the team before setting off for Toulouse late in the afternoon.

Airbus Management fully understood what a great idea it had been to come to Oshkosh and show the aircraft to US aviation connoisseurs.

SALES TOURS

Considering Airbus Management's requirement to have the A380 certified in record time, you would have thought that the absolute priority would have been granted to that objective. In fact, that was far from being the case as a pretty long sales tour was organised relatively early in the programme.

When such an aircraft is at the initial stage in development, each 'useful' day devoted to a visit or a demonstration requires about three days interruption in development work to prepare it, for the ferry flight, for the visit itself and return to standard to resume test flights.

For the sales tours on our development aircraft # 1 and # 4, we had installed a mini-cabin with thirty business class seats on the upper deck. These seats had been recovered from other programmes and were mismatched. They were intended for the entire team ensuring logistics, engineering, operations, organisation, sales, marketing, and so on. They were also used by crew members to take a rest during long flights as there was no crew rest area as are to be found on series aircraft.

* * *

The first round of visits to our customers took place in November 2005, with stopovers in Singapore (Singapore Airlines), Australia (Qantas), Kuala Lumpur (Malaysia Airlines) and finally Dubai (Emirates) for the Airshow.

The whole story started off badly. Firstly, we had a failure on an engine on aircraft # 1 (Monday 24 October) shortly after take-off, with Jacques (see the ENGINE TESTING chapter). Troubleshooting was a lengthy affair and changing the various elements of the fuel circuit failed to provide a solution. The following Saturday, during a short meeting with a Rolls-Royce team, we decided to replace what we called the gear box (a series of accessories driven by the engine). It was the week-end and if we resorted to the usual freight forwarders, we had no chance of having the replacement equipment needed before Tuesday, meaning another day lost. We then checked whether one of our test aircraft was available to go over on Monday to East Midland and Rolls-Royce to recover that part. We only had our development A340-600 available at the time. Unfortunately, it was equipped with tyres of different types in a non-certified configuration. Unable to get in touch with A340 programme management, I decided to replace the non-standard tyres and plan the excursion for Monday morning. By 1.30 p.m. we were back with the gear box we needed and a quarter of an hour later assembly was under way. There was some banter around this and a joker renamed our team 'EVHL' as a kind of amalgam between the French acronym for Flight Tests and DHL. But we were happy as we had probably managed to safeguard the departure for Singapore early the following week. But we were far from the end of our problems.

While engine # 1 was being prepared, rumours were going round that engines # 2 and # 4 had to be replaced. So, on the afternoon of Wednesday 2 November, the Chief Engineer of the Rolls-Royce Trent 900 programme came over to Toulouse to tell us that there was unacceptable risk in flying with those two engines for such a long tour. A bolt had come loose on the Rolls-Royce test bench and had been sucked into the engine. That engine had run for 300 hours and those installed on the aircraft had run for 400. His view was that failure could be imminent. We made mild protests. But what could we possibly do with so much insistence from the engine manufacturer? We simply had to accept the engines being replaced by others with less flying hours.

We were confronted by a new problem: the aircraft would not be ready for the three days of festivities planned to take place in Singapore. We considered a number of possibilities including going directly to Australia where a big celebration was planned with Qantas and where there was no chance of changing the dates. In that case, on the outward trip, Singapore and the cities of South-East Asia had to be avoided with a stopover at the Isle of the Reunion. Visits at Singapore Airlines and Malaysia Airlines could then be made at the end of the tour. Another solution was to stay just one day in Singapore and not visit Malaysia Airlines as we absolutely had to be there in time for opening of the Dubai Airshow. We couldn't make a decision without consulting the various airlines. Finally, the option retained was to spend just one day in Singapore on the outbound journey, to go to Kuala Lumpur at the end of the tour and send the other development aircraft to Dubai.

However, our technical problems were far from being over. On Tuesday 8 November, as the aircraft came out from engines replacement works, an oil leak appeared on the nose landing gear. Replacement was the only possible solution and that meant two more days delay.

* * *

John Leahy had decided to visit each customer with an aircraft at least in part under their colours. Although the underlying reasons for this are unclear, it is likely that the idea stemmed from a special request from Qantas late in September, when they wanted to celebrate their 85th anniversary in Brisbane with an A380 bearing their logo on the fuselage. As we couldn't possibly re-paint the airplane on each stopover, the only solution was to use huge stickers. For Qantas, the airline wanted these stickers above the wing. All the technicians had considered that this sort of installation involved major risks. A memo was issued by the team responsible for co-ordinating these activities, the MAP, to draw Management's attention to the issue, but the decision was upheld.

At flight tests, we were strongly against placing these stickers. Indeed, once the habit had been adopted, it seemed obvious that there would be no way out of doing the same for all other tours, with the huge loss of time that would involve. We also suspected that jealousy would be aroused between the different airlines. We weren't mistaken in this and the difficulties with Emirates at the Dubai Airshow are a case in point.

* * *

Finally, we left for Singapore on Thursday 10 November. The flight provided an opportunity to conduct a certain number of tests, especially on data transmission. After four and a half hours of flight, as we were drawing near to Bahrain, we heard an explosion. The air outlet of one of the two packs had come free. The cabin pressure started to drop slowly, but we managed to pursue the flight using the second system.

During this flight, John Leahy was present in the cockpit for the departure and arrival. This was his first flight on the A380. John is well familiar with aircraft as he started his professional life as a pilot and had been Sales Manager with Piper Aircraft. Shortly after the departure, he confided that he had been astonished by the number of failures announced on start-up, around 45 of them. But there had been nothing untoward. That was just the state of the aircraft at the time, with a host of systems giving true or false failure warnings. But none of this stopped us from flying in complete safety.

He also very much appreciated the cockpit and how data was presented. At that time, there was much discussion as to what the A350 would be like. To avoid spending too much on the project, Airbus Management wanted simply to redo the wing of the A330 in a CFRP version, with practically no other modification to the aircraft. We pleaded on a number of occasions that the A380 cockpit be used, which seemed to be a logical solution. But this idea was systematically refused by some managers. We argued, for example, that the A330 navigation computer interface was practically identical to that of the A310, whose first versions had been delivered towards 1980, and as the A350 would still be in service towards 2050, pilots would be using a seventy year old system. To the French members of Airbus Management, I joked that this would be rather like them having a Renault 4 CV as a company car. They smiled at that, but at the time, I failed to convince anyone! Once in the A380, John was immediately convinced and he understood that we needed to use the A380 cockpit as a basis for the A350. In his usual upfront manner, he announced at the press conference on arrival in Singapore, and then subsequent ones, that the A350 would have the A380 cockpit. He didn't bother asking for the views of programme management!

Toward the end of the flight, we slowed down considerably so as not to arrive in Singapore too early. After the press conferences, interviews and visits to the aircraft, the aircraft was put into a hangar late in the afternoon. In just a few hours, the Singapore Airlines stickers had to be removed before replacing them with Qantas ones, with the difficulties already mentioned.

At midnight, we took off for Brisbane with two well rested pilots at the controls. One had been sent over in advance by a commercial flight while the other had been saved having to attend all the festivities to get some sleep.

* * *

During the flight to Brisbane and despite an improvised repair of the air conditioning system, we had the same explosion and the same consequences.

On arrival, we prepared the next day's flight, from Brisbane to Sydney, with a series of photos and videos on the city of Sydney and more especially the Opera House. The 'photo aircraft' pilot was Clay (about 75 years old) who had a Learjet. He came along in an extremely relaxed manner with just a road map as documentation. He thought at first that he would have to manoeuvre around us while we flew straight. He finally understood that we were used to such photo opportunities and that we could also get into position in relation to his aircraft. This meant we could get a bit more out of the session than he had imagined.

After the briefing with Clay, Fernando told us all how this visit was going to work. He mentioned John Travolta's name. John Travolta has an airline pilot's licence and owns a Boeing 707 and a Gulfstream business jet that are parked right in front of his house next to his own private runway. He was the 'figurehead' pilot for Qantas and was qualified on B747. Obviously, he was to take part in all the festivities. We were followed by our usual team of cameramen and, as soon as Fernando pronounced the name Travolta, they immediately asked to film the sequence, until they got the shots they wanted!

Again we were off for an early start on Sunday 13 November for this photo opportunity flight. The key feature was obviously to fly over the coast to have the city and Opera House in the background. The air traffic controller was not very helpful, but we finally managed three passes. The result was a magnificent series of photos. We had gently made fun of Clay and his slouch, but he was a true professional. In Sydney, there was the press conference, and then VIP and Qantas personnel visits as usual.

• See photo n°37 – page 247

The next day, we set off for Melbourne. In addition to Sydney, Qantas also planned to run the A380 from this platform. Our stopover provided an opportunity to see what state of readiness the airport was in.

That evening, over dinner, we received a phone call from John Leahy asking us to have a special flight organised for John Travolta. This led to some soul-searching. We had no doubt as to John Travolta's qualities as a pilot and in addition the A380 is easy to fly. But that the whole world should know that he was the first pilot not qualified to conduct tests to land the aircraft bothered us. We therefore decided after a powwow only to let him take the controls in flight, that is outside take-off and landing phases with the excuse that the regulations so require.

The Melbourne to Brisbane flight the next day went ahead with a large number of personalities on board, including executives from Qantas and those from Airbus: Gustav Humbert, Charles Champion, and John Leahy. We had to wait a long time on arrival for all the other aircraft in front of us to land.

A number of truly memorable scenes were to follow that the participants will treasure for many years to come.

* * *

As soon as we arrived in Brisbane on 15 November 2005, we went straight over to John Travolta's personal Boeing 707 (registered N707JT). There were five of us: Gustav Humbert, John Leahy, Charles Champion, Fernando and myself. We hung around for a good hour. This was the first time I'd seen some of our Top Managers forced to wait on someone, except of course for a visiting President or Minister. John Travolta arrived with his wife in his superb limousine and dressed in his pilot's uniform. Not a glance for the Airbus team. Finally, after waiting another quarter of an hour, we were invited to join him in his aircraft. We gave him some technical information and John Leahy went off to talk about the sales side. John Travolta is clearly a real professional as he readily understood our explanations.

We then left to go up to the A380 in a convoy with a number of cars, Fernando and I following last in a minibus where there were two make-up artists. Hysteria broke out among the crowd watching the aircraft when John Travolta got out of his limousine in uniform. He made a first tour round the A380 with the journalists to recite what we had taught him.

He then got on board and had himself photographed in the pilot-in-command's seat with the window open. I must have looked a bit disgruntled and this didn't pass unnoticed by John Leahy and the President of Qantas. I explained to them that we were very much afraid that this photo, that was bound to go onto the Internet, might lead all our customers to believe that he had been the first pilot to have flown the aircraft. During our climb, towards 15,000 ft, I let him replace me so he could fly the aircraft. I went over to see the President of Qantas who then realised that John was at the controls. I explained to her that I was delighted to hand over to him, but that he shouldn't boast of having piloted the aircraft for the reasons already given. At the end of the descent, I got back into my seat. On arrival, the President repeated what I had said to her with all the necessary explanations. Everything worked fine in that respect.

The boarding bridge was brought up to the aircraft but we weren't allowed to leave yet. The two make-up artists came on board and the door was closed again. John Travolta shut himself off with them on the upper deck. This all lasted about half an hour. And during all that time Airbus Management stood around waiting. Finally, he came out first from the A380 to applause from the crowd. Interviews resumed around the aircraft and we wryly observed that they were only briefly interrupted by the make-up artists to put John's hair in order! That being said, he's a great aviation connoisseur and he did a fantastic job.

That evening, there was a big dinner to celebrate the 85th anniversary of Qantas with, of course, an interview with Travolta, then the nose of the A380, partially decked out in Qantas colours was edged into the hangar. Overnight the Qantas stickers were removed to be replaced by those of Malaysia Airlines. Then on Wednesday 16 November, towards midday, the aircraft left for Kuala Lumpur for a new demonstration.

* * *

The visit to Kuala Lumpur was much more conventional, with a press conference and visits to the aircraft. The Malaysian Prime Minister took part.

The return to Toulouse on Friday 18 November 2005 provided an opportunity to set a temporary record with a flight lasting fifteen hours, due to a very strong head-wind. On arrival, after a number of unsuccessful attempts to extend the landing gear with the normal systems, the emergency extension system had to be used. The root cause was the right wing landing gear being blocked in the retracted position.

But the demonstration tour was far from being over. We had to get the aircraft back into condition to pursue testing. Engines # 2 and # 4, that were installed before the departure, were mounted again because equipped with the entire measurement installation needed. A further week went by before we resumed flights.

* * *

Practically all route proving flights were used by the sales team to win the confidence of customers or prepare others for a future purchase of the aircraft. I won't go into these flights here.

To keep some people happy, we had to conduct visits that had not been initially scheduled. Thus, just one month before certification, we received a request from Thailand to go to Bangkok on 5 December 2006, on the King's birthday. Considering the aircraft's programme, we were hardly in favour of this, but Louis Gallois made it an order. This was with aircraft # 2. That flight nevertheless enabled us to remove a last minute reserve for certification. This related to condensation and the water droplets that fell from the ceiling during descent. The problem was not totally settled for route proving flights, but during that flight to Bangkok, the latest drainage modifications proved effective.

* * *

Aircraft # 7's cabin had many options that were different to what was provided in # 2 and we had to validate them before entry into service. This was the case, for example, with the Panasonic IFE system. It was therefore decided to re-conduct a small series of route proving flights with aircraft # 7, outside certification, with Lufthansa participation. The destinations were chosen according to the commercial objectives sought. We took advantage of this opportunity to go to the United States for the first time. The first stopover chosen was New York.

First contacts with the FAA were made late in January 2007. Discussions had got off to a bad start as this Administration did not want to receive an aircraft lacking an individual airworthiness certificate (as it was a development aircraft) with 519 passengers on board. They wanted a minimum number of passengers – probably zero! They also showed a certain reticence with respect to the pilots' qualifications and a demonstration flight scheduled to take place there. A few days later, early in February, everything seemed to have been sorted out after top level intervention. But on 7 February, more stupid requests started flooding in, as with a limitation of the number of passengers to 499, so as to keep below 500, and the demand that two distinct Permits to Fly be provided for the aircraft for the two flights scheduled with aircraft # 7 (as if we were going to change everything between those flights). The next day, further difficulties were encountered with a 15 kt crosswind limitation (whereas we had done demonstrations at more than 55 kt) and a ban for Lufthansa pilots, who didn't have type rating to fly over United States territory. Added to that were many objections as to runway and taxiway widths, separation due to wake vortices and the margin in relation to obstacles in final approach.

On 9 February 2007, after a further intervention with FAA Management, everything was settled, apart from the requirement not to have Lufthansa pilots at the controls in American airspace, which seemed unavoidable. But we had barely resolved these problems when others promptly arose, this time with the immigration authorities, who didn't want any Airbus passengers on a Lufthansa (LH) flight number. This seems complicated. The context was that Europeans arriving on a commercial flight didn't need a visa and this was true with an LH code number. However, an Airbus flight was considered to be private since a manufacturer is not a listed carrier and so all passengers had to have a visa. The immigration officials demand was therefore a bit out of the ordinary and out of step with the applicable regulations as aircraft # 7 was flying to New York with a flight number 'LHxxx'. Finally, common sense prevailed.

For the crews and those going out on a private flight, the C or D visa is compulsory. Obtaining it is a real trial. Firstly, you have to have perfect identity photos, with both ears clearly visible and no smiling. The background too must represent less than 50% of the overall surface and not be too dark. Then you have to go to the Consulate in Paris, with pay slips, bank statements, proof you've paid your rates, electricity bill and all sorts of other stuff. You then have to wait in a queue for almost an hour in front of the Consulate before being frisked as you go in with security keeping everything they consider superfluous, including your keys. You then pass before two more checks at successive counters before having to wait again before a two minute interview. All in all, a special trip to Paris and more than two and a half hours hanging around at the Consulate before getting a visa a few days later including a photo with both ears cut off!

On 13 February 2007, the Lufthansa pilots, irritated by the FAA's vexatious attitude, first of all decided not to take part in these flights. They changed their minds a few days later, while still reckoning that we had not been effective in defending the case. On 15 February, a further really serious question came in from the FAA. They wanted to know whether Airbus pilots were authorised to carry passengers. This was getting crazy! We responded by giving them our Flight Operations Manual, that provides a clear explanation. I have to admit that the French authorities hardly helped us as we still didn't have Type Rating more than two months after certification (see the TYPE RATING chapter). In the end, FAA authorisation arrived on 16 February in the form of a Special FAA Authorisation (SFA). The only restrictions were that Airbus pilots had to have more than 50 flying hours on the A380 (no problem there) and that Lufthansa pilots couldn't be at the controls over the United States.

That same day, while we were celebrating this success, a row blew up in Los Angeles due to the fact that it was not our first U.S. destination. Some media sources asserted that AINA President Allan McArtor promised that we would reserve our first landing in the States for Los Angeles. They were up in arms against the situation and the whole thing got back up to Louis Gallois who asked us to provide some explanations. Everything had already been organised and it was impossible to backtrack and switch our first destination from New York to Los Angeles. But then the idea emerged of sending out two airplanes, one to New York and the other to Los Angeles, for them to land simultaneously. For Los Angeles it had to be one of our test aircraft, # 1 or # 4. On 22 February, we made a proposal to Management on those lines. But Qantas woke up to what was going on. Indeed, it had been planned that they would be first to ensure flights to Los Angeles. The airline wanted to be brought in and see the arrival take place at their future disembarkation gate. It was hard to keep everybody happy! A few days later, Louis Gallois' agreement came in, but he didn't want the extra cost to exceed 100,000 Euros. A quick calculation showed that the sum proposed would not even cover the cost of fuel and that much more money would be needed. But, after discussion with Programme Director Mario Heinen, we finally decided to go ahead without going over these figures with the Management!

Qantas was not totally satisfied. They wanted to have stickers in their colours on the fuselage, whereas the aircraft that was to come in to New York under the Lufthansa flight number was under Airbus colours, without any distinctive sign. Obviously, we couldn't satisfy their demands. A few days later, we were asked to move the arrival forward by a few hours to allow a visit from John Travolta (yet again!) as he had other commitments later. This meant modifying organisation for the arrival of the two aircraft and the request was thus turned down.

But our problems with the FAA were far from over. On 28 February 2007, someone discovered that, in the airworthiness certificate issued by the FAA, it was forbidden to use the APU on the ground and in flight as a modification on fire resistance had not been applied on our aircraft. However, this restriction did not exist on the European certificate and, on special authorisation (SFA), it was stated that we should use the aircraft in accordance with its European airworthiness certificate. The person having raised this futile issue must have been chastised as everything was finally sorted out.

The Americans of AINA understood why we were reluctant to go to the United States as they saw all this red tape and so many contrived problems to be resolved. In all events, preparing this tour kept us all pretty busy right through to the end.

* * *

On 19 March 2007, aircraft # 1 took off from Toulouse for Los Angeles and # 7 left from Frankfurt to New York. The aim was to have near simultaneous landing in both these cities.

Aircraft # 1 had old engines and engine # 2 consumed a huge amount of oil. Above Hudson's Bay, the indication gave four litres remaining, knowing that down to zero, there would just be enough to make a safe go-around. We therefore considered cutting off that engine until the start of descent. But finally this didn't prove necessary.

We were keen to synchronise the two aircraft. While on board # 1, I exchanged a number of phone calls with Fernando, who was on board # 7 to adjust our respective speeds. In fact, on # 7, the Germans (both those with Lufthansa and those from Airbus) wanted to be first and they had cheated by setting out forty-five minutes ahead of time. However, we could still regulate things by playing on the speeds and possibly bringing aircraft # 7 to wait for a while before landing in New

York. On # 1, we were flying as fast as we could, at Mach 0.88, while the red line not to overshoot on the indicator stood at Mach 0.89. On landing in Los Angeles, despite extended radar guidance, we were thirty seconds in advance of schedule.

In New York, the picture was somewhat different due to the head-start gained on departure. The German Airbus pilot had yielded to patriotic sentiment and failed to obey the company instructions. Aircraft # 7 therefore touched down twenty minutes in advance, much to the pride of the Germans. A pity for Airbus though! Fernando was furious at this organisational cock-up. Meanwhile, the Airbus pilot concerned got a severe dressing down on his return!

On arrival in Los Angeles, we went to what is known as the 'Imperial' apron. As usual, the media were well represented and there were many photographers. I accompanied the City Mayor onto a magnificent red carpet as far as the stage where a few speeches were made by the Mayor, the President of AINA and a Qantas representative. Just behind the scene, there were the American, Californian and Australian flags. But no European flag. I was outraged, but hardly surprised. There was no point creating a diplomatic incident, but I did insist to Allan McArtor a little later that we had been shocked.

After the many visits, the aircraft was taken to the future Qantas gate. To do so, we had to allow sniffer dogs on board. Just like the first time we organised the flight for the media, the dog on the upper deck relieved itself at the top of the main stairway. It was getting to become a habit!

On arrival in New York, contrary to FAA instructions, the Lufthansa pilots wanted to remain in the pilots' seats down to 15,000 ft in descent, then, after landing, taxi the aircraft up to the disembarkation gate. This was a new source of tension in the cockpit. In addition, contrary to internal Airbus decisions not to show flags, three were waved out of the cockpit windows: the American, German and Lufthansa banners. If we had decided to make a showing of the colours, we would have chosen that of Airbus and the European and American flags. During interviews, the President of Lufthansa expressed his dissatisfaction with the FAA that had not allowed his pilots to land the aircraft.

Aircraft # 1 flew back to Toulouse the following evening after a full day of visits (including the dogs with no incident this time!) On departure, we had to back track the runway as spacing between the runway and taxiway was insufficient. Works under way were to improve that situation.

Aircraft # 7 did a round trip to Chicago, then returned to New York and a VIP flight was organised. It returned to Frankfurt on the afternoon of 21 March.

* * *

Aircraft # 7 was to leave for Hong Kong on 23 March 2007. Unfortunately, an Electric Backup Hydraulic Actuator (EBHA) on an aileron failed. Time was pressing to make the repair as shortly after arrival, a VIP flight had been scheduled. We therefore sent out a small twin-jet to Frankfurt with the replacement part. The A380 left three hours late and the VIP flight was finally only delayed by two hours. On return from Hong Kong on 25 March, there was a mere ninety minute stopover in Frankfurt before being outward bound to the States again. The stopover went well apart from a slight delay due to a non-technical issue.

For this airplane's second flight to the United States, the destination was Washington. The next day, there was a VIP flight, but the FAA representatives heeded the advice of their lawyers and refused to take part. At that time, it was part of the articles of association of the FAA that it contribute to promoting US industry. Having its members board an A380 flight could have been considered contrary to that principle! Authorisation had however been granted to fly over the centre of Washington with an accompanying aircraft, meaning we brought back some fine photos.

That same evening the aircraft returned to Frankfurt, and then Munich for a visit from members of the Lufthansa Board. Technically, the only serious problem reported on all these flights involved the frequent failures of the centralised cooling system for the galleys.

* * *

The pace of sales tours then intensified up to the end of the year 2007 as we had aircraft # 7 equipped with a cabin representative of how it could be fitted out available to us until the end of the year. On each tour, according to the case, there were VIP flights, demonstration flights for pilots and validations of airport installations for the airlines. There is no point going into the detail of each tour, but the list below gives an idea of how intense sales activity was at the time. Meanwhile, a large number of demonstrations were laid on for pilots in Toulouse on our test aircraft. This all took place in a more favourable context as we had support from the air traffic controllers. In addition, in 2007, route proving flights for aircraft # 9 equipped with Engine Alliance engines were also used for visits.

In April 2007, there was a visit lasting a few days to India for the Kingfisher Airline (that also owns the brand of beer going under the same name!) taking in Delhi and Mumbai with VIP flights for the company's guests.

Early in June, there was a VIP flight organised from Paris-Charles-de-Gaulle with Dominique Bussereau, Secretary of State for Transport and a host of personalities. As we prepared to leave for Paris, three pilots came up to me, including two from Air France in uniform. They were expecting to stay in the cockpit throughout the flight, which would deprive other guests of access and there were some 150 of us on board. They also wanted to be photographed with us on arrival, as we went down the boarding bridge, but they refused the photo if they couldn't remain in the cockpit for the entire flight. I don't know who had invited them. After negotiation, we found a compromise for the Air France pilots, with one up front on departure and the other on arrival. On leaving the aircraft, they were surprised to see that the boarding bridge was covered - so no photos!

The week-end was devoted to tests on the ground with Air France before leaving for Tokyo. The next stopovers were Sydney and Taipei, before returning to Toulouse. Demonstration flights were made in these three cities.

In August 2007, there were visits to Singapore Airlines and Emirates mainly intended to validate their installations and make certain they were ready to receive the A380. Over two consecutive days there were demonstration flights from Dubai with more than 460 passengers to check the 'turn around time'. Here we had to check that all the various tasks including disembarking and boarding the passengers, but also tidying and cleaning, unloading and loading the luggage, taking food on board, etc. could be done quickly enough. This was a useful exercise as we found out that the rear of the upper deck was not sufficiently cooled on the ground in hot climates (see the CABIN COMPLETION chapter).

Late August and early September 2007 saw a new tour in Asia with Bangkok, Chang Mai, Hanoi, Hong Kong and Seoul as destinations. A number of flights were organised on behalf of customer airlines and potential purchasers. When pushing back the aircraft in Bangkok, with VIPs on board, the left wing tip collided with a hangar door after a clumsy tractor manoeuvre. All subsequent flights were therefore made with the wing tips removed on both sides. Plane spotters who find photos of an A380 under Airbus colours without wing tips can put a date to them.

In October 2007, there was a further tour to China, taking in Canton, Shanghai and Beijing, with a VIP flight every time except for Beijing where the airfield was closed for several hours due to fog. Then return with a stopover in Istanbul.

In November 2007, there was a series of flights with Air France in the United States and Canada. Firstly, Air France took guests to Montreal, then we made a return trip from Montreal to Orlando.

The last tour on aircraft # 7 took place between 26 November and 11 December 2007 taking in Minneapolis, Los Angeles, Sydney, Buenos Aires and São Paolo. A week was spent in Sydney to prepare entry into service of the Qantas aircraft. It was worth noting that during the VIP flight at Los Angeles, the City Mayor was not entitled to fly due to a law dating back more than a century that forbids elected representatives from being offered free travel. That law had been enacted to prevent corruption when contracts were being signed with the railroads!

THE MEDIA

You can't talk about the A380 without bringing up the media. At flight tests, some of us were used to having the media people around, but their presence had never been so strongly felt, except no doubt at the time of Concorde (that I had only been involved in remotely).

Some journalists proved to be totally excessive (not all, luckily) writing about the greatest aircraft, the biggest, the one with the best performance, before going on the next day to blacken the picture just because there's been a slight delay in the schedule. No sense of nuance.

But as always, there are some real professionals, and not just in the specialised press, but also in the mainstream media. Often enough, they themselves have a pilot's training or sound aeronautical knowledge and, as we have the habit of saying amongst us: "They recognise the front of the aircraft from the tail". Unfortunately, there were also the 'pros' of the cheap headline. I remember one of them in particular whose name or newspaper I won't mention. In January 2008, the A380 had been in service with Singapore Airlines for nearly three months, flying between Singapore and Sydney. One day, the tractor that was moving the aircraft from the boarding gate had a brake failure before the engines had been started and the A380 ended up moving at slow speed into the grass with two landing gears off the asphalt. There was no damage apart from having to replace some tyres. The headline spun out for this was "Crash of an A380"! What would the journalist have written in the event of a real accident with victims? Some radio stations had also grossly overblown the phenomenon, announcing that the aircraft had quit the runway. Louis Gallois, EADS CEO, had been warned about this. All that fuss for just a tyre change!

Obviously, we, pilots and flight test engineers, were often up on the front line when it came to talking about the aircraft at a technical level. Our ethical code at flight tests forbade us from stating untruths, but we were prepared sometimes to be evasive and state that we weren't yet ready to confirm such and such a result, even if, in our minds things were already clear. In all events, we had to remain extremely cautious in our declarations.

In this context of constant exaggeration, one of the things we most feared was that films or photos get out onto the Internet. For example, the loss of a small element from the aircraft (a piece of a landing gear door to take the example recounted in a previous chapter) is not a common event, but it can easily happen during programme development. For the Design Office and ourselves, this was a non-event: we reinforce the door and the matter is settled. For the gutter press, the photo of a missing part provides them with a field day and the 'affair' becomes a European drama impugning the A380's reliability and qualities.

* * *

So how to master the matter of communication?

Airbus had no lack of good professionals who all did excellent work and with whom, at flight tests, we always maintained really good relations. I really should mention Barbara Kracht, who ran relations with the press with an iron grip. Barbara has had good knowledge of the aviation world since her youth as her father, Felix, was one of the Airbus founders. Barbara's team never spoke out of turn. The guidelines were clearly set and when there were delicate technical issues, we were regularly consulted to avoid any misunderstanding or untimely disclosure. Meanwhile, we never really had any major difficulty in the communication field: the aircraft was an aerodynamic success, test progress was tangible and the results were good. So there was no problem talking around that and avoiding more delicate matters.

* * *

As from before the very first flight, we had been followed by a small team that prepared documentaries for various TV channels. This was managed by Sylvain Pascaud and Mike Magidson. Of course, there were some specific interviews, but most of the time they simply lived and worked alongside us. During the first days, we were constantly checking whether they were there or not so as to take care about what we were saying, but then we forgot they were there at all, trusting them to cut out certain comments we made between us and that, taken out of context, could have seemed inappropriate.

They were there for the first flight and its preparatory work, some delicate or spectacular manoeuvres (VMU or emergency stop at maximum energy, for example), the first tours abroad, some airshows, etc. Of course, they couldn't be on board for flights other than sales tours or some flights with passengers, but they made good use of cameras installed at various locations around the aircraft, whether those already made available to the flight engineers or specially installed.

A large number of documentaries were prepared for various channels, much to the TV watchers' satisfaction if can be judged by the comments made afterwards.

* * *

Shortly after the first flight, star TV journalist Michel Drucker wanted to organise a major programme to be broadcast on the main TV channel France 2 just before the Paris-Le Bourget Airshow. This programme was recorded on Friday 3 June 2005 and broadcast the eve of the opening day.

We all tremendously appreciated working with Michel Drucker who was extremely approachable and straightforward, as also highly professional in preparation.

Michel would have loved to fly on the A380 before his programme, all the more so in so far as he is a pilot himself, and do some filming with a small team on board. However, to do so this would have to have been organised barely a month after the start of tests and towards flight # 18. We had every confidence in the aircraft, but this would have meant a special flight and half a day's tests being lost. And above all, we didn't want to expose ourselves to criticism and take passengers along too early in the programme. So, after discussions among ourselves, Jacques, Fernando and I decided to refuse.

The recording took place in one of the big hangars of the Lagardère Plant where the A380s are assembled and prepared for their first flights. That day, there was a cold wind and even some rain on the grandstands as the big door remained open to show the nose of the A380, with an A340 and an A318 in the background.

For us, there was too much of an element of showbiz, but we knew that was the rule of the game. So we all came onto the TV set in flight suits with boots that had been specially purchased so we would all have the same. We would have been happier with something more discreet.

This programme was a huge success and the A380 was lauded to the skies by all the media.

* * *

As we weren't communication professionals, we were interested to see how the media people actually worked. I failed to understand why some journalists prepared absolutely nothing before the interviews and put questions I reckoned to be pointless. I even did a live TV programme where just a few minutes would have been enough to have a prior discussion on interesting topics and where nothing of that came to be. But it's not my job and maybe I've got it wrong and fail to understand what the general viewing public really wants.

Nevertheless, the great professionals I met were always really conscientious in preparing their interviews. That obviously applies to Michel Drucker as just mentioned. I also have an amusing memory of another TV and radio anchorman Jean-Pierre Elkabbach who was to interview me live the day after the maiden flight, towards 8:00 a.m. on the Europe 1 radio channel. He phoned me for about a quarter of an hour the evening of the first flight towards 10 p.m, when all I wanted really was a bit of peace and quiet at home. He then called me again the next day, well before the time to go on air. He was clearly looking for something original to say so as to make the interview really stand out. And I ended up inadvertently providing him with what he was looking for when I admitted to not having listened to the news of the first flight on the car radio as I drove home, preferring instead to listen to some opera music. I liked the way he pushed me as interviewee to get out just that little extra to break with the routine.

* * *

We wanted to organise a flight for the media fairly early in programme development. Two camps very quickly emerged, with those wanting to get on with this as soon as possible on aircraft # 2 and the others who wanted to use aircraft # 7 equipped with a more attractive cabin and a much more discreet test installation.

In the camp of those wanting to get on with this really early, there was in particular Barbara Kracht's team as also flight tests. We thought we needed to fly before certification. We also reckoned that the drawbacks relating to the test installation were minor and that all those taking part in the flight would understand it was a test aircraft and not one destined for a customer.

The others, led by John Leahy, Head of Sales, really wanted the finest aircraft, and therefore # 7 and, considering its delay in coming out, that flight could only take place early in 2007. Finally, thanks to his position in the hierarchy, John had the greater weight and won out, meaning that flight took place after certification.

The date had to be set well in advance and 7 February 2007 was chosen. Right up to the last minute, at flight tests, we were uneasy about availability of that aircraft. It only came out of a whole month of modification works three days before the big day. But the flight after works showed us that, in the cabin, many of the problems had not been satisfactorily resolved. For that reason, we prepared aircraft # 2 as a backup.

Come the day, security came first and a sniffer dog was sent out on reconnaissance shortly before boarding. It must have liked the aircraft because it marked its territory on the carpet at the bottom of the rear stairway!

A press room had been installed in the magnificent and recently inaugurated Henri Ziegler delivery centre. There were about 250 people to board and two bridges were used, one to the main deck and the other to the upper deck so as to demonstrate this option.

So as to foster as much open discussion as possible, we had decided to have a large number of crew on board. The Anglo-German flight crew included Peter Chandler as pilot-in-command and Wolfgang Absmeier as second pilot. Admittedly, our flight tests world remained very much male dominated. We tried to redress the balance somewhat with Sylvie Loisel being appointed flight test engineer for this flight. In addition, Jacques Rosay, Fernando Alonso and a number of others, including myself, were available to field questions.

Two hundred people, journalists, cameramen, etc. were able to move freely throughout the cabin, trying out all types of seats. There were some amusing reactions: I remember in particular a journalist who discovered that there was a window in the first class toilets. Almost all of them then wanted to film these magnificent toilets! In fact, this feature was simply an option demanded by certain customers.

All the Airbus participants gave a whole series of interviews. But the main difficulty was in managing access to the cockpit. A lot of people wanted to take photos and films, conduct interviews, etc. and we had to limit the time devoted to each of them as we had decided not to make too long a flight. It finally lasted two hours and twenty minutes. Prior to that, there had been long discussions to decide on how long the flight was to last. Some like John Leahy and Barbara Kracht didn't want it to go beyond one hour. Others (ourselves at flight tests included) were more in favour of two hours as in one hour all you have time for is taxiing, take-off, climb, a few minutes cruise and then descent. Hardly time even to serve and taste the Champagne! And how could all the media people get a chance to have even a short spell in the cockpit?

Globally, all the participants were highly satisfied, and we were sorely in need of this good news to boost our image after the announcement of delays in delivery. They appreciated the comfort in the cabin and the low level of noise. Note that the entire economy class on this aircraft was in fact 'super-economy', meaning with extra legroom. This is the standard chosen by nearly all the first customers to offer improved comfort, even in economy class.

A MIXED BAG OF STORIES

This section tells the story of three rather special flights that were not really devoted to tests, as also the reasons that led to our development aircraft being registered, a topic that is always of interest to some spotters.

* * *

On 11 October 2006, over the Mediterranean, just as certification tests on fuel jettison on aircraft # 4 had been completed, the tests centre air traffic controller suddenly warned the pilots of the A380 that an aircraft was climbing up really fast towards them. She was clearly extremely worried, not knowing where the intruder who was beginning to move in dangerously was coming from. Finally, the crew saw two Turkish F16 fighter jets in patrol formation.

Who had asked them to intercept the A380? That remained a mystery! It was a fairly regular occurrence for French fighter aircraft on training missions to come up and join us as a friendly target during our flights, and especially those just after the airplane had come off the production line. But the air traffic controllers always first asked for our consent. We always answered positively, except when special manoeuvres were planned as part of the test campaign. Here, the F16s had come in with no warning.

On return from the flight, we soon realised that they had not been acting of their own volition but that they were under orders from a US flagship, the USS Ross. Indeed, there was an international military exercise under way and the maximum altitude granted to them in this especially reserved zone was 19,500 ft. The A380 was at that time flying at 25,000 ft. There was no real danger as the weather was fine and we were clearly visible, so avoidance was easy. The American military were probably intrigued by our aircraft following a rather unusual flight path to remain in the assigned zone that, horizontally, must roughly have covered theirs. Nevertheless, there was a matter of principle involved and their behaviour was quite shocking. What would the Americans have had to say if the French Charles de Gaulle aircraft carrier had sent up two Rafale fighters to patrol alongside the latest Boeing B787 then being tested without first warning Boeing, the USAF or the US Navy? This would no doubt have created a diplomatic incident.

To better understand what had happened, I asked the crew to file an 'airmiss', the international procedure to analyse near misses and last second avoidances. As foreign military aircraft were involved, the inquiry was conducted by the Taverny Operations Centre. It's clear that the French were extremely embarrassed by this blunder. This was serious: the US military had sent foreign fighter jets to fly inside a zone that was closed to them and ordered them to intercept a civil aircraft. It would have been enough to have warned us and the crew of the A380 would have accepted this mid-air rendezvous, at the end of testing. What's more, if an F16 had crossed behind us during the fuel jettison exercise from the dump valves located to the rear of the wings, it could well have exploded in flight.

The final report came out in May 2007. The Commander supervising the 'Brilliant Midas 2006' exercise had detected an "unknown echo with behaviour judged doubtful". That explains why he asked the F16 patrol under his orders to go and take a look, and all this without any coordination whatsoever with the French military authorities. The Commander on board the vessel had for codename 'Red Crown'! As the F16s joined the A380, he asked them to identify it. The answer was "big white aircraft... four engines...". Red Crown wanted a more detailed description, so one of the F16 pilots sent the registration number: F-WWDD. The response came back: "continue to escort". The Aix-en-Provence Air Traffic Controller was annoyed: "I've no idea with whom they're working, they're outside the zone too. I'm not keen on this at all". A few minutes later, Red Crown became more cautious and told the F16 pilots: "Do not get close to the aircraft, it may be a commercial civilian aircraft". Finally, after coordination, the F16s left the A380 alone.

The Taverny Ops Centre concluded that what had happened was "an interception of an aircraft in test flight by foreign fighter jets under guidance of a foreign seaborne control centre, without coordination with the French authorities". In addition, it transpired that the

USS Ross had no knowledge of the status of activity in the zones concerned and they had considered the movements of the A380 threatening (maybe they thought it was a Weapon of Mass Destruction!)

The commission of inquiry deplored the fact that the USS Ross failed to forward a dossier to explain matters to them. Surprised?

* * *

Nowadays, there are techniques you can use to produce near-perfect photo-montages. Such methods are used regularly to showcase the aircraft of the future. But, for the crews, nothing can replace the pleasure of seeing photos actually taken in flight, especially when it involves a patrol of a number of aircraft.

So it was a fantastic experience to get a group of four A380s to take to the sky together. That memorable day was the 30th August 2006. Having four planes scheduled on the same day, at the same time, is exceptional as it is a fairly frequent occurrence for at least one of them to be grounded for modifications. There was only one slot possible and, as luck would have it, the weather conditions were perfect. This was only the second flight of aircraft # 9.

After a short parade on the ground, mainly to make some films, the four aircraft took off one behind the other, with the shortest time possible between them so as to get the gathering going as soon as possible. Each plane had the load and the fuel necessary to pursue testing after the photo session. Aircraft # 9 was the leader for the formation. It was much heavier than the others as it was heading out for several flying hours to conduct performance measurements in cruise with a high aircraft weight. This can be seen on the photos as it had a slightly greater nose-up attitude.

The aircraft took off in the following order: # 1, # 4, # 2 and # 9. A Corvette business jet had taken the photographers and cameramen on board. We brought our airplanes together as we flew towards the Mediterranean. The first photos were taken in what air force pilots call 'echelon' to the right, meaning the aircraft were extremely close to each other and aligned on a line at 45 degrees from their axis of symmetry. The order was # 9, # 1, # 4 and # 2. This is rather like soldiers on parade, where the sergeant major wants everyone lined up in a column and only wants to see the head of the man standing in front. We finally lined up successfully and a series of great photos were taken.

We then changed formation to form a lozenge, with # 9 still leader, # 1 on the right wing, # 2 on the left wing and finally # 4 bringing up the rear just behind the leader and slightly lower.

• See photo n°38 – page 248

Readers might well wonder about how you keep in patrol formation with such a huge aircraft. It's certainly a little more difficult than on a fighter jet as the aircraft's gliding ratio makes decelerations a more delicate matter and you may have to use the airbrakes. Similarly, you have to avoid setting the engines fully to idle as the response in acceleration will then be too sluggish. Actual handling is extremely easy though, with the aircraft responding promptly to the controls. However, considering the big wingspan, distances are not easy to estimate and you need to seek a small but adequate spacing between aircraft.

After this series of photos, we returned to Toulouse in a different formation we affectionately referred to as the 'duck' with the two leading aircraft (# 9 followed by # 4) forming the head and neck of the bird, while # 1 was in echelon on the right of the # 4 and the last A380 symmetrically opposite to the left to form the body and wings. We flew in close by the city and then went along the axis of Blagnac Airport's runways at a height of about 1,500 ft. We then split up with each airplane proceeding to conduct its own tests: cruise performances for # 9, engine for # 1, smoke evacuation for # 2 and noise for # 4.

That flight will remain memorable for those who took part in it. Everyone else can just admire the magnificent photos.

* * *

Among the many demonstration flights organised for pilots, one of them was particularly memorable.

Catherine Maunoury is twice world aerobatics champion and, in 2010, she became Director of the Air and Space Museum. I hadn't seen her for some time before we met up again for a conference talk we were both to give at the Palais de la Découverte in Paris. It seemed to me important for an aviator like her and in such a prominent job to get to know the Airbus family better, so I invited her to come and fly the A380.

You can always get to know things about an aircraft on a simulator but, for a pilot, a simulator session just doesn't feel like the real thing! At Airbus, our considerable experience in demonstration flights had proved to us that a pilot can take over the controls of any of our aircraft without prior sessions on the simulator. We use simulators with our customers to gain time in addressing the so many details encountered with the different systems. With Catherine, this was superfluous.

We still had to conduct a one-hour cockpit briefing session so she could get used to the left seat. This was done the day before. I knew she had no experience of multi-engine aircraft. But, as we talked together, I realised that she had hardly ever used an artificial horizon (the instrument showing you the aircraft's attitude), as there was none on her Extra 300 aerobatics airplane. In addition, she had never used an ILS, which is the usual form of approach to Toulouse and, finally, she knew nothing about the Flight Director. This was an unprecedented although not yet hopeless situation (we'd seen other such problems!). The easy answer to this was to just switch off all these automated systems as soon as they kicked in on take-off after the landing gear had been retracted. That way, piloting was an easy matter of keeping a steady speed (shown by an indicator), the heading (also displayed), and keeping an eye on the altitude. Back to the basics!

The next day, just before the flight, there was the customary briefing to talk about what we were going to do, then start-up of the engines and take-off as planned. We were flying light (362 tonnes on departure). Let's be honest. There was some hesitation. This was understandable when you consider that on an Extra 300, a quick flip of the stick means you can perform part of a 'hesitation roll' (a manoeuvre in which the roll is deliberately and very briefly interrupted 4 or 8 times in the course of its 360°s rotation). But on an A380, the same action has no effect. Considering the inertia involved, you have to keep the controls deflected until you get the correct bank angle and attitude that are then maintained by the computers. The A380 remains agile as it can touch down on the same runways as other transport aircraft, but the responses to the controls are not so rapid due to its inertia. Taking over control of the aircraft was thus a gradual affair with increasingly tight turns.

Catherine then went on to try out the protections at high and then low speed. I took over the controls to show her an avoidance manoeuvre, something that is far from intuitive with such a big aircraft. But sometimes, if there is an imminent danger of collision, you need to react quite brutally. As you come in on approach speed, you have to pull the stick right back to nose up. The engines go to maximum thrust and, then, you pull the stick 'in the corner' to start a sharp right or left turn. This is what spectators see at the end of the flight demonstration of an Airbus, after the high angle of attack pass. It is even more impressive when seen from inside the aircraft, with a tight turn and steep climb at 115 kt and 45 degrees of bank, with the stick against the stop. It was Catherine's turn next, doing something she was more used to as she returned to her aerobatics training with the stick taken to its limits!

We flew back to the airfield. We were now unlucky with the weather as we came down through a small layer of clouds, only leaving them behind us at 800 ft. So 'vocal' navigation was required on ILS: "A little to the right, a little to the left, down a little more, a bit less..." Finally, we had the runway in view and everything became easier. Vocal guidance helped get well onto the runway axis due to the large wingspan, and then during the flare. The landing was a good one and I didn't need to touch the controls. Touch-and-go, a new circuit and then 'kiss landing', without me needing to say a word or even touch the stick.

As always in flight tests, we always do some soul-searching to try and understand why some things worked and others didn't. Catherine was an especially interesting case in point. Switching from an aerobatics aircraft to the A380 in such a short time was astonishing. We reckoned her success was due to three factors. Firstly, she had an innate gift for piloting, although that wasn't enough in itself. Next, she was quick to pick up on the advice we gave her on A380 handling techniques right from the start of the flight. Finally, her ability to constantly analyse her actions to improve, with spectacular progress being made between the start and the end of the flight. I believe these were the three very same qualities that led to her achieving such success in high level aerobatics. But putting on my Airbus cap, I can't help thinking that a major contributing factor was the ease with which the A380 can be controlled. The Design Office teams too deserve a round of applause for that result.

- See photo n°39 – page 248

* * *

Plane spotters have always been interested in aircraft registration matters and especially when it comes to prototypes. For those not in the know, a few explanations.

French test aircraft have registration starting with an F-, followed by four letters. Within certain limits, there is freedom to choose the two or three last letters. The first is practically imposed, especially for prototypes, and has to be a W. Registration of airplanes coming off the production line starts F-WW and the next letter is assigned to us by the administration by aircraft type, as with B, D or I for A320s, K or Y for A330s, and A or S for A380s. We then add the last letter as aircraft are released to fly. For prototypes, we choose the last two letters, on condition that the third has not already been allocated, and more unusually the last three where possible.

We therefore had to find something for the A380s used as development aircraft. Early in November 2002 a demand came in from Marketing Management suggesting that we reserve two registrations F-WWOW (WOW!) and F-WWDD (DD for Double Deck). A third was also needed. F-WXXL was suggested but many people were against it as it sounded like the aircraft was maybe overweight! John Leahy settled the issue: F-WWOW for the first aircraft, F-WWDD for # 4, and F-WXXL for # 2. As the third letter D had already been assigned to Airbus for the A320 series, we simply transferred F-WWDD from an A320 to an A380. The O was used by Eurocopter and they kindly gave us F-WWOW. Finally, the Civil Aviation Authority accepted that we assign F-WXXL. Thus, in a few hours, everything was settled. Later test aircraft # 9, the first to receive the Engine Alliance engines became registered under F-WWEA, with the E being 'lent' to us by ATR.

Some well-informed spotters probably noticed that the two or three letters on the nose landing gear doors, on which can usually be seen the registration or its last letters, were completely unrelated. What happened was that the test flight engineers in charge of an aircraft (sometimes over several years) had got used to placing their own initials or their trigram. Thus, looking at photos of aircraft # 1 on the ground, you can see GD on those doors, for Gérard Desbois. On aircraft # 4, there's VNO for Pascal Verneau and BB on # 9 for Bruno Bigand. A way of making the aircraft their own!

THE INAUGURAL COMMERCIAL FLIGHT

In the life of an aircraft, two flights have a very special symbolic aspect: its maiden flight and the first commercial flight. Technically, they are not the most important ones. There are others that make a greater mark on its existence. For example, those where we encounter a major problem that can have a serious impact on the design and lead-times. Others, conversely, where we realise that a difficulty has been overcome, and when, at the end of the flight, the entire Design Office and test flight team combined can feel satisfied.

But these two flights have something special about them as they mark a moment of truth. During the first flight, we discover the aircraft and, from then on, we really got to grips with the aerodynamic facts and can analyse how the various systems operate. All the assumptions and models are set aside (well, almost!). As from the first commercial flight, there is no more hiding anything from our customers, including both airlines and passengers. The aircraft has to fulfil all its objectives in terms of reliability, consumption and comfort.

Our first client was Singapore Airlines (SIA). After a delivery with great fanfare in Toulouse of aircraft # 3 at the new Henri Ziegler Delivery Centre, on 15 October 2007, SIA had organised an inaugural trip from Singapore to Sydney for Thursday 25 October to return the following afternoon. This involved two flights with passengers paying a high price for their seats to raise money for charity. Among the recipients were 'Médecins Sans Frontières' (Doctors Without Borders), organisations in Singapore and a children's hospital in Sydney. About 1.3 million dollars were garnered. Who then were the generous donators? Discussions in the aircraft suggested a number of profiles, including real aviation enthusiasts and others out for a memorable adventure. All seats had been auctioned with a record amount of 100,380 dollars for two suites on a single outbound trip. These tickets had been purchased by a young trader who had made his fortune in the dot.com world and who offered the second suite to a friend. The minimum price tag was 560 dollars for a single fare in economy class. The average price in this class for a single came to 1,000 dollars.

The truly sumptuous first class suites were the really big attraction. They were designed by a Frenchman, Jean-Jacques Coste, usually in charge of fitting out luxury yachts. As on a boat, every bit of space was used to the utmost. Each suite had an armchair and a real bed. When in flight, there can be complete isolation and privacy. The two central suites can communicate with two beds side by side. A fairly senior couple had paid 70,000 dollars for two of these outbound. The media pounced on this story when the beds were installed!

On that outbound journey, the oldest passenger was a 91 year old Singaporean. He had made a success of his import-export business and his son wanted to give him a treat by taking him along on the flight. The son in question had not done things by halves as he had booked three suites, two business class suites for his family and an economy seat for his father's nurse, all costing 56,000 dollars. On the return flight, the youngest passenger was a baby less than one year old (I don't know whether he paid for his 'seat'!). There were also people who are first flight enthusiasts. One of the passengers had been on the inaugural flight of the Boeing 747 and another that for Concorde.

Singapore Airlines had done me the honour and afforded me the great pleasure of inviting me into the cockpit for this inaugural flight. I thus found myself alongside the charming Robert Ting, our captain for this trip, Gerard Peacock, co-pilot, and Gerard Yeap, Manager for all Flight Operations at Singapore Airlines. The latter two were qualified as pilots-in-command on board the A380.

I met up with them at the check-in counter for a flight appropriately named SQ380. My status was a bit strange as I was in uniform but was travelling with an economy class ticket, plus a special authorisation to enter the cockpit. As always for the A380, there were plenty of cameras and microphones being brandished. As I was in uniform I attracted attention but every time I saw a journalist approach I told them I was just an Airbus guest and that they should talk to Captain Ting. It mostly worked and I was pretty much left alone!

At check-in I met four French people, including a couple. They didn't want to miss the opportunity. They hadn't come along together but had met up by accident. True supporters.

Here was much hubbub in the aircraft before passenger boarding. A representative of the authorities came into the cockpit to check the pilots' licences. He took his time. A great sign of trust indeed!

The aircraft had 471 seats, but not all were occupied. There were 425 supposedly commercial passengers. In fact, among them were representatives of the press who obviously hadn't paid. There were 23 cabin crew members, with six technicians ready to intervene if problems arose with cabin systems, and four pilots (including myself, though I was not entitled to take the controls unless I had a local validation of my licence).

The first to board was the auction bid recordman. He was filmed by all the TV channels present. All the passengers with suites were entitled to a swift visit of the cockpit on the ground as, contrary to what we did at Airbus during our own flights, access to the cockpit was strictly reserved for the crew during the flight, this all being part of SIA security rules.

The airplane was pushed back exactly at 8.00 a.m. local time, or 00.00 UTC. The aircraft weight was 467 tonnes and we had 125 tonnes of fuel, about half the tank capacity. Take-off was 15 minutes later. To boost the service lifetime of its engines, SIA used reduced thrusts take-off and climb, and passengers commented a number of times to me that they had found the take-off rather long.

• See photo n°40 – page 249

Shortly after the start of cruise at 37,000 ft, we had a smoke warning in the cabin crew rest compartment on the same level as the cargo holds. Luckily, this was a false alert that disappeared just a few hours later. Apart from that, the aircraft worked like clockwork through to arrival.

The cruise speed had been determined with a Cost Index (CI) of 80. The CI is a system used to optimise the cost of flights in relation to a number of parameters including the price of fuel, the flight altitude, its duration, the outlay to pay for crew, etc. That value of 80 led to cruise Mach values of 0.83 or 0.84 according to the weight and altitude. This was not the optimum for the A380 as it had been designed to fly at Mach 0.85. The crew were unwilling to change this in favour of the established flight plan. Later on, this led to a whole series of discussions between Airbus performance specialists and Singapore Airlines to get to a more appropriate cruise speed. The CI calculation turned out to be slightly inaccurate at the time.

Back in the cabin, the media were going about their business, especially in the first class with visits to the suites and interviews with passengers, Robert Ting and the SIA CEO who had his business seat. One of the journalists got special attention, Richard Quest, a star on CNN. Before the other passengers boarded, he had already spent his time pacing around the aircraft with his cell phone. He then took his seat, or rather collapsed into it, in one of the suites, still on his phone. I never saw what he had to say but I can report that he had holes in his socks!

There were plenty of cameras and microphones, and air of great enthusiasm and lots of autograph taking. Robert Ting and his CEO went round the aircraft to talk with the passengers. As we departed, Robert had mentioned the presence of one of the Airbus pilots having made the first flight. So I went round the cabin to chat, sign autographs on the booklets handed out on boarding, on tickets, aircraft models and even tee-shirts. In the economy class I came across the same atmosphere as during the Airbus Early Long Flights. The difference here though was that passengers couldn't change class, whereas during our flights, everyone could move about freely. The subdued atmosphere of the business class contrasted with the exuberance in economy. First class was different due to the media presence.

The two chefs who had prepared the meals, with a choice between European (Matt Moran) or Chinese (Sam Leong) cuisine, were on board and came along to be photographed with us in the cockpit. The wine waiter was also present. To accompany these excellent meals, he had selected prestigious wines for the suites, with Dom Pérignon rosé Champagne 1996, Château Cos d'Estournel 1982, Château Pichon Longueville Comtesse de Lalande 1982 (usually sold on at 750 Euros the bottle). I was in uniform, so I couldn't get a taste. A real pity!

The weather was poor on arrival. SIA had planned we fly over Sydney Harbour. We followed the planned flight path (through clouds unfortunately). Finally, we came in for landing. It was a 'kiss landing' by Robert with not much braking to avoid heating up the brakes. Local time was 17.30 and we still had 35 tonnes of fuel. We heard the applause even from the cockpit.

* * *

We spent a restful night in Sydney. This was my third trip here with the A380 but, usually, it was a whistle-stop visit. This time, the next morning afforded an opportunity to visit the interior of the Opera House. Unforgettable!

The following morning, the weather was fine. Unfortunately, as the time for departure loomed up, it got worse and at Sydney the rules are extremely strict in forbidding catering embarkation and refuelling if there a storms near to the airport. So, an hour and twenty minutes before departure, with the crew already on board, operations came to a halt. All we could do was turn on our computers and do a bit of paper work while waiting for a break in the weather. Finally, there was an improvement, but catering embarkation took a long time. Finally, we left an hour and fifty-three minutes late. But Airbus and Singapore Airlines had nothing to do with it!

Before leaving, we had the failure of an electro-hydraulic pump during towing. This was inconsequential and the regulations authorised us to depart as we were, even with two failed pumps.

The passengers in the suites and other classes were no longer the same as very few had booked a return trip. A young couple was installed in the two forward suites. A late departure and the length of the flight must have been good for them to take advantage of the comfort of the twin bed. In business class, there was also a young French couple. They weren't aviation enthusiasts, but the young man had dreamt about this flight. There was also a French lady mad about flying adventures.

So no more media people with the suites and all was calm. But in economy class there was still the same party atmosphere with plenty of talk going on. People were clearly happy just to be there.

Technically, the aircraft worked perfectly. This was hardly a surprise, but it was in any case a relief as unexpected problems can always crop up.

The control tower complained that our speed was too low and failed to respect the flight plan that specified Mach 0.85. We increased the Cost Index to 200 so as to obtain that cruise Mach.

Arrival was at midnight local time. The flight ended with an autoland as asked for by the Singapore authorities (administrative burdens again!)

The reception was held in the terminal which was festooned with garlands of flowers and was accompanied by music. Then, the entire crew came down to the foot of the aircraft for a photo session. Finally, we got back on board to party for a good fifteen minutes. A chance to taste some Dom Pérignon rosé and Pichon Longueville 1982, even in uniform!

The passengers and Singapore Airlines were thoroughly satisfied. They had taken part in the beginnings of a great adventure. Everyone was able to appreciate the aircraft's comfort and silence, realising that our insisting on these factors in the publicity surrounding the plane was justified. For Airbus, there was relief all round as, technically, the aircraft had handled marvellously and the pleasure shared by the passengers and the airline was our reward.

Saturday provided a day of rest for the aircraft. On Sunday 28 October 2007, the real regular service started with a round trip to Sydney a day. A new aircraft entered service marking a new era in aviation.

A FINAL WORD FOR YOUNG ENGINEERS

As we come to the end of this book, you will have realised the scale of work involved in developing and certifying an aircraft. This is nothing to do with size but is rather commensurate with the level of innovation.

Development brought both disappointments and satisfactions, but in the end, it was a success story. The A380 is an aircraft unanimously appreciated by travellers, which was the final goal. Everyone contributed to this. At flight tests, we had the chance to see all aspects of development, but all the engineers who did their share of the work can feel proud. Developing a flight controls law to enhance comfort, a highly accurate autoland system, a device that reduces the risks of runway overrun, an autopilot that automatically ensures avoidance of other airplanes are some of the achievements that provide such tremendous intellectual satisfaction to the engineers who contributed. And as the aeronautics industry fits into the long story of events, the results of all these people's works will remain there to see for decades to come. Another source of pleasure.

I hope I have aroused in young engineers the wish to join our industry. Not just Airbus is involved, but all those other companies and organisations working in aviation and the space sector, including Boeing, Dassault Aviation, Bombardier, Embraer, Airbus Helicopters, equipment manufacturers, and many others. And we must not forget general aviation.

For a long time now, aircraft have traditionally had a fuselage, two wings, an empennage and engines and would therefore appear to resemble each other. In reality, however, practically everything is different as improvements are made from one model to another in terms of aerodynamics, structure, control systems and navigation. Fuel consumption has to be reduced and noise too. Innovation is constant and there will always be new airplanes to be designed. For Airbus right now there are the various versions of the A350 and of the A320NEO. There will be others in the future.

So, if you are keen to take on challenges and aeronautics excites you, you have found your path.

THANKS

I would like to express my heartfelt thanks to all those who helped me gather together the various documents needed to fill out my personal records or who were willing to check out various sections for one reason or another, including Fernando Alonso, François Barre, Manfred Birnfeld, Paule Botargues, Wolfgang Brüggemann, Jean-Philippe Cottet, Emanuele Costanzo, Gérard Desbois, Bruno Dumont, Marc Fontaine, Jean-François Grandjean, Jacky Joye, Anne Kerrien, Florent Laporte, Patrick Libralesso, Robert Lignée, Patrick Macaire, Bernard Masdoumier, Jean Piatek, Didier Ronceray, Marc Teulier, Hugues Van der Stichel, Fabrice Villaumé and Michael Wilson.

Thanks are also due to Vincent Rouquière for his drawing and to the photo library team of Airbus France (Sylvain Ramadier) and Airbus Central Entity (Marie Bonzom).

I also greatly appreciated the work put in by Christine Hubert who proof read the entire manuscript and helped me in data searching. Thank you too to Séverine Kaci for her critical review.

GLOSSARY OF TERMS AND ABBREVIATIONS

• **Acceptance flights:** All flights conducted on an aircraft between the moment it comes off the production through to delivery to the Customer.

• **ADIS:** AFDX and Digital Injection System (AFDX: Avionic Full Duplex - the aircraft's 'Internet'). System specific to Airbus test aircraft used to modify certain parameters in the computers' software when in flight.

• **ADP (Aéroports De Paris):** Airport authority that owns and manages the major airports in and around Paris

• **Aérospatiale:** Name of the French company that was forerunner to Airbus.

• **Ailerons:** Movable control surfaces on the trailing edge of the wings that are deflected to manoeuvre the aircraft (roll or bank).

• **AINA:** Airbus North America.

• **Airbus Deutschland:** Airbus German subsidiary.

• **Airbus Flight Test School:** School created by Airbus in 1995 to train test pilots and flight test engineers.

• **Air Data Computer:** An avionics component that can determine the calibrated airspeed, Mach number and altitude trend data from the inputs provided by the aircraft's external and system sensors to return a series of calculated results critical to the control of the aircraft.

• **Aircraft minus one:** Simulator used at the very start of a programme for early design validation work.

• **Aircraft zero:** Simulator using the aircraft computers associated with the best aerodynamic model.

• **Angle of attack:** Angle between the chord line of the aircraft wing and the vector representing the relative motion between the aircraft and the atmosphere.

• **APTCAS (Autopilot Traffic Alert and Collision Avoidance System):** System designed by Airbus to automatically avoid collisions. This system is covered by a full chapter.

• **APU (Auxiliary Power Unit):** Small turbine in the tail providing electrical and pneumatic energy as required. Used at the gate and if necessary in flight.

• **Artificial horizon:** Instrument showing pitch and bank attitudes.

• **Autorotation (engine):** Rotation of the spools of a shut down engine, at high speed, due to the airflow.

• **Autothrust:** System that automatically adjusts the thrust to maintain flight conditions.

• **BCM (Back up Control Module):** On the A380, seventh flight control computer taking over in the extremely unlikely event of the six other computers (PRIM and SEC) failing.

• **Back Up Speed Scale:** Airbus back up system that provides information from the aircraft's incidence probes (angle of attack) to fly the aircraft for situations where speed data as calculated from pitot probe data (pressures) is lost.

• **Ballast:** System used to load the aircraft in order to obtain the appropriate weight and centre of gravity position for tests (water ballast on Airbus aircraft).

• **Belly fairing:** Fairing between the wing and the fuselage.

• **Brake fans:** Fans used to cool the brakes.

• **Brake To Vacate:** Automatic braking system that brings the aircraft to the optimum speed to vacate the runway at the selected taxiway.

• **BTV:** Brake To Vacate.

• **BUSS:** Back Up Speed Scale.

• **CAA:** Civil Aviation Agency.

• **Cabin Intercommunication Data System:** Computer controlling all cabin systems, including communications, air conditioning, lighting, IFE, etc.

• **Cabin virtual first flight:** First 'flight' performed on the ground with passengers aboard to check cabin operation.

• **Cabin zero:** Scale 1 mock-up of the cabin used for the development of some cabin systems.

• **Category 3:** Name given to an approach with very low visibility. Followed by autoland on Airbus aircraft.

• **CDL:** Configuration Deviation List.

• **Ceiling:** Maximum usable altitude for the aircraft. May be linked to aerodynamic or technical characteristics or the inability to climb due to lack of thrust.

• **CER (Circulation Essais Réception):** Air traffic services for flight tests in France.

• **CRI (Certification Review Item):** Means by which potential controversial subjects are documented by the Authority and the manufacturer during the certification process to provide a means for the identification and resolution of significant technical, regulatory, and administrative issues.

• **CEV (Centre d'Essais en Vol):** French Agency monitoring civil and military flight tests. Changed recently to become DGA Essais En Vol.

• **CIDS:** Cabin Intercommunication Data System.

• **Climb detent:** On an Airbus, when using the autothrust, 'Climb' is the detent for the thrust levers for the entire flight duration, except take-off, flare and on the ground. The thrust is automatically adjusted according to the commands of the crew.

• **Configuration:** A configuration is characterised by the position of the various movable elements: slats, flaps, landing gear, etc.

• **Configuration 1:** Configuration where slats are deflected and flaps up.

• **Configuration 1+F:** Slats and flaps are at minimum deflection.

• **Configuration 2:** Slats and flaps are at an intermediate position.

• **Configuration 3:** Slats are at their maximum deflection and flaps are in an intermediate position, but above configuration 2.

• **Configuration clean:** Configuration with flaps and slats retracted.

• **Configuration Deviation List:** List of concessions to permit flying with missing, failed or damaged systems and possible associated limitations.

• **Configuration Full:** Slats and flaps are at their maximum deflection. Used for a standard landing.

• **Contaminated runway:** Runway covered by substantial amounts of water, slush, ice, etc.

• **Cost index:** Used in the navigation computer to determine the optimum speed as a function of the costs of flight hour, fuel, crew, etc.

• **Cumulative noise:** Unit to quantify the global noise of an aircraft.

• **DASA:** Daimler Chrysler Aerospace Airbus. Name of the German company that was forerunner to Airbus.

• **dB:** Decibel.

• **Decision height:** Height at which the pilot must go-around if not in sight of the runway for landing.

• **Decrab:** Manoeuvre to straighten out and align the aircraft with the runway during a crosswind approach.

• **Derotation:** Time during landing between touch down of the main landing gear wheels and the nose wheels.

• **DGA Essais en vol:** New name given to CEV since 2010. See CEV.

• **DGPS (Differential Global Positioning System):** Enhancement to the Global Positioning System (GPS) that provides improved accuracy based on reference stations to broadcast the difference between the positions indicated by satellite systems and known fixed positions. Can be used to measure accurately the distance between 2 aircraft.

• **Dihedral:** With zero bank, angle between each wing and the horizontal plane.

• **Direct law:** Flight controls law where surface deflections are proportional to the pilot's inputs.

• **DOA (Design Organisation Approval):** Agreement granted by the Administration to an aircraft manufacturer for aircraft design, development and certification.

• **Drag:** Force due to relative airflow opposing the forward motion of the aircraft.

• **Dynamic pressure:** Pressure resulting from a medium in motion. Depends on the density and velocity of the fluid.

• **Early Long Flights:** First flights scheduled early in the programme with full complement of passengers aboard.

• **EASA (European Aviation Safety Agency):** Body responsible for aircraft certification.

• **EBHA (Electrical Backed up Hydraulic Actuator):** Hydraulic servo-control with electrical back up. General principle similar to EHA.

• **EGPWS (Enhanced Ground proximity Warning System):** Advanced version of GPWS associated with a database to warn against risks of collision with the ground or an obstacle.

• **EGT (Exhaust Gas Temperature):** The temperature of the exhaust from the jet engines.

• **EHA (Electro Hydraulic Actuator):** Servo-control using a small electrical motor to generate the hydraulic power needed to run an actuator.

• **Electrical transient:** Short interruption in electrical power that may result from switching from one electrical system to another.

• **ELF:** Early Long Flights

• **Emergency electrical configuration:** Situation where all the electrical generators from engines and APU are lost. On the A380, electrical power will then be provided by the Ram Air Turbine.

• **Encounter:** Manoeuvre during which an aircraft penetrates the wake vortex of the preceding airplane.

• **EPNdB (Effective Perceived Noise decibel):** Unit for noise measurement on the ground taking into account discomfort, frequencies and duration.

• **EPNER (Ecole du Personnel Navigant Essais et Réception):** French school where test pilots and flight test engineers receive training.

• **Epopée:** Airbus simulator used to validate new concepts.

• **Eurocontrol:** European organisation coordinating air traffic.

- **Experimental test pilot:** Test pilot authorised to carry out all types of tests.
- **FAA (Federal Aviation Administration):** National aviation authority of the United States.
- **FADEC (Full Authority Digital Engine Control):** System consisting of a digital computer that controls all aspects of aircraft engine performance.
- **Fan:** First compressor of the engine. Has very large blades.
- **FAR 25:** US certification regulations for transport aircraft.
- **Fatigue:** Repeated application of stress that may eventually lead to structural failure.
- **FD (Flight Director):** Flight instrument overlaid on the attitude indicator that shows the pilot the attitude required to follow a certain trajectory.
- **Feed tank:** Tank providing fuel directly to the engine.
- **Flaps:** Surfaces on the rear of the wings that are deflected down on take-off and landing to reduce flying speeds.
- **Flare**: Modification of the approach trajectory when close to the ground to allow for a smooth touchdown.
- **Flight controls laws:** The A380 flight controls are all electronically controlled and hydraulically activated. Software determines the deflection of the control surfaces as a function of the pilot's actions and various parameters, including speed, altitude, centre of gravity position, etc.
- **Flight envelope:** Plot of altitude vs speed defining the performance limits within which an aircraft can operate. The demonstrated flight enveloped is explored during flight tests.
- **Flight level:** Flight altitude in hundreds of feet using standard pressure as a reference (1013 hPa).
- **Flight plan:** Route followed by the aircraft with waypoints.
- **Flight Test Guide:** Document including information on test techniques.
- **Flight test order:** Flight programme prepared by the flight test engineer describing all the tests scheduled during the flight.
- **Flight tests A and B:** Classification of flight tests. A flight A requires a higher qualification for the test pilot and the flight test engineer than a flight B.
- **Flight Warning Computer:** Computer providing all warnings and procedures for the crew.
- **Flutter:** Sustained oscillation, usually on wing, fin or tail caused by interaction of aerodynamic forces (further details in the 'Flutter' chapter).
- **Fly-by-wire:** Flight controls system using electric/electronic signalling.
- **Glide:** Vertical guidance system of the ILS.
- **Go-around:** Manoeuvre where the crew aborts a landing on final approach and sets maximum thrust before a new approach or continued flight to an alternate airport.
- **GPWS (Ground Proximity Warning System):** System designed to warn pilots if their aircraft is in immediate danger of flying into the ground or an obstacle and giving avoidance instructions to the crew.
- **Ground effect:** Increased lift that an aircraft's wings generate when close to the ground ('air cushion' effect).
- **Heavy:** Aircraft category in the classification for wake vortex separations (above 136 tons).
- **High speed taxiway:** Taxiway orientated at 45 degrees from the runway in order to vacate at higher speed.

• **High-pressure spool (engine):** Spool operating at high pressure in a turbofan engine. The airflow enters its compressor as it emerges from the low-pressure compressor, and then its turbine after the combustion chamber.

• **Horizontal stabiliser:** Horizontal surface at the aft of the fuselage. Its deflection keeps the aircraft balanced according to the flight conditions

• **ICAO (International Civil Aviation Organization):** United Nations Agency that codifies the principles and techniques of international air navigation and issues recommendations to the aeronautical community.

• **Ice shapes:** Shaped pieces of polystyrene glued to the leading edge of wings and empennages to simulate the accumulation of blocks of ice.

• **IFTT (Integrated Flight Test Team):** Team in charge of the development of a system in flight (includes crew members and Design Office engineers).

• **ILS (Instrument Landing System):** Ground system that precisely guides the aircraft laterally and vertically for manual or automatic landing.

• **Iron bird:** Aircraft zero coupled with a bench at scale 1 including all the aircraft electrical, mechanical and hydraulic systems.

• **JAA (Joint Aviation Authorities):** Was a body representing the civil aviation regulatory authorities of a number of European States who had agreed to cooperate in developing and implementing common safety regulatory standards and procedures. Since replaced by EASA.

• **JAR 25:** European certification regulations for transport aircraft.

• **Jump seat:** Seat in the cockpit between and slightly behind the two pilots' seats.

• **Knot (kt):** Speed of 1 nautical mile per hour.

• **Lapse Rate Take Off:** Test performed to check the engine parameters at maximum thrust and with a reduced climb rate to simulate an engine failure.

• **LCO (Limit Cycle Oscillation):** Natural oscillation of a surface with limited amplitude and that does not diverge.

• **LiDAR:** Laser system used to establish wake vortex characteristics.

• **Lift:** Component of the aerodynamic forces supporting an aircraft in flight.

• **Limit loads:** Maximum load anticipated from a particular condition of flight used as a basis to design the aircraft structure. Beyond this limit, deformation is possible. No failure must be observed at less than 150 % of limit loads. Computations are made under the most severe assumptions.

• **Load factor:** Ratio of a given external load to the weight of the aircraft. In cruise, its value is 1. In a turn or during a pull-up, it is greater than 1.

• **LOC (Localizer):** Lateral guidance on the ILS.

• **Low-pressure spool (engine):** Spool operating at low pressure in a turbofan engine. Includes the first compressor in the airflow and the last turbine.

• **Mach:** Unit measuring speed as a percentage of the speed of sound.

• **MAP:** 'Mise au Point': team assisting flight tests in preparing the aircraft for mid or short term operations.

• **MD:** Dive Mach. Maximum Mach demonstrated during flight tests.

• **Mechanical flight controls:** Unlike with fly-by-wire, mechanical flight controls involve pilot inputs being transmitted to the surfaces using cables and rods.

• **Medium:** Aircraft category in the classification for wake vortex separations (7 to 136 tons).

• **Megajoule:** Energy measurement unit.

• **Minima:** Decision height or minimum altitude in approach to be in sight of the runway for landing.

- **Minimum descent altitude:** Altitude at which the runway must be in sight to land in a non-precision approach (ILS excluded to keep it simple).
- **MMO:** Maximum operating Mach. Red line on the pilot's Mach display.
- **N3:** RPM of the high pressure rotating part on Rolls-Royce engines.
- **Negative load factor:** When the load factor is negative, all objects will tend to fly up to the ceiling. An aerobatic aircraft flying upside down has a load factor of -1. Severe turbulence may lead to a negative load factor on transport aircraft.
- **NM:** Nautical Mile.
- **Normal law:** Standard flight controls law. All stabilisation and protection systems are operative.
- **OPS:** Operations.
- **Pack:** Air conditioning supply system.
- **Pedals:** Action on the pedals, left and right, commands rudder deflection and, when on the ground, nose wheel steering. Braking on each landing gear (left and right) involves pushing down on the top of the pedal.
- **PIO (Pilot Induced Oscillations):** Sustained oscillations obtained when the pilot reacts by instinct to an unexpected movement of the aircraft leading to a series of corrections in opposite directions. Risks of PIO need to be eliminated during aircraft development.
- **Pitch:** Motion of the aircraft around an axis along the wings.
- **Pitch up:** Situation where the aircraft pitches up by itself without input from the pilot due to aerodynamic characteristics.
- **Pitot (sensor):** Sensor used to measure the dynamic pressure of the airflow.
- **Power8:** Airbus savings programme launched in 2006.
- **Precommand:** Immediate deflection of a surface as a function of stick displacement.
- **PRIM:** Main (primary) flight controls computer.
- **Product Safety Board:** Airbus board including high-level technical engineers and Heads of programmes that makes the main safety-related decisions for Airbus aircraft.
- **Pull up:** Manoeuvre where the pilot pulls on the stick to increase the load factor and pitch the aircraft.
- **RA (Radio Altimeter):** Radar system that accurately measures the height above the ground.
- **RAT (Ram Air Turbine):** Small propeller deployed under the wing that, in an emergency, is driven by the airflow to work an electrical generator (installed on the A380).
- **RECAT (Recategorisation):** Programme launched by ICAO to establish a new classification of aircraft for wake vortex separations.
- **Resolution Advisory:** TCAS mode giving instructions to avoid collision with another close flying aircraft.
- **Reverse:** System installed on the engine to reverse the jet flow on the ground in order to slow down the aircraft without using the brakes.
- **ROPS (Runway Overrun Prevention System):** A response to runway overrun events during the landing phase.
- **Rotation:** Phase between the pull up action of the pilot on the stick at take-off and lift-off.
- **Rotor:** Rotating part of the engine.
- **Rotor disk:** Disk attached to a spool shaft to which the blades are attached.
- **Route proving:** Operational flights performed for certification to demonstrate aircraft reliability.

• **RTL (Rudder Travel Limiter):** System limiting the deflection of the rudder in relation to speed.

• **Rudder:** Vertical surface behind the fin controlling yaw and sideslip.

• **SEC:** Secondary computer for flight controls.

• **Shock wave:** Surface of discontinuity in which the airflow changes abruptly from supersonic to subsonic.

• **Sideslip:** Angle between the plane of symmetry of the aircraft and the airflow.

• **Slats:** Movable surfaces on the leading edge of the wings deflected for take-off and landing.

• **Slide:** For emergency evacuation of the passengers.

• **Spoilers:** Device on the upper side of the wing to change the airflow and reduce or destroy the lift. They are used asymmetrically to roll the aircraft. In flight, if deflected symmetrically, the aircraft decelerates or increases its rate of descent (speed brakes). On the ground their deflection increase the loads on the landing gears and improves braking.

• **Spool (engine):** Rotating part of the engine.

• **Stall (aircraft):** When the aircraft speed is too slow, it cannot be maintained in flight due to insufficient lift. Stall is reached when the lift coefficient reaches its maximum value.

• **Stall (engine):** Airflow blockage in the engine that may be due to a number of causes. Can disappear by itself or lead to an engine shut down and possible damage.

• **Stator:** Fixed, radially disposed aerofoil blades that straighten the airflow in the turbine.

• **Strake:** Small surface on the engine nacelle creating a vortex at low speed to give more energy to the flow on the upper part of the wing and avoid flow separation.

• **Supplemental cooling:** Centralised cooling system for the galleys replacing individual refrigerators.

• **Test pilot B:** Under French regulations, test pilots in category B are not authorised to perform the more delicate flight tests.

• **Total pressure:** See dynamic pressure.

• **Touch and go:** Landing followed immediately by a take-off without stopping on the runway.

• **TCAS (Traffic Collision Avoidance System):** System installed on aircraft giving information on all other air traffic and that calculates collision avoidance manoeuvres when needed.

• **Trim:** Adjustment of an aircraft's controls to achieve stability in flight.

• **Trim tank:** Fuel tank in the horizontal stabiliser.

• **Turbine:** Rotating part of the engine driven by hot gases.

• **Turbofan:** Jet engine in which part of the compressed air does not go through the combustion chamber. All jet engines are turbofans nowadays.

• **Turn coordination:** System minimising sideslip during a turn, thanks to rudder deflection.

• **Type Board Meeting:** High-level meeting to check the status of the certification documents and major items before issue of the Certificate of Airworthiness.

• **Type rating:** Pilot qualification to fly a certified aircraft.

• **V1:** Decision speed at take-off. In case of engine failure before V1 the aircraft must be stopped. Above V1, take-off continues.

• **V2:** Climb speed with one engine out after take-off.

• **VD:** Dive speed. Maximum speed demonstrated in flight.

- **VMC:** Minimum control speed with a failed engine.
- **VMCA:** VMC after take-off.
- **VMCG:** VMC on the ground.
- **VMCL:** VMC in approach.
- **VMCL-2:** VMC with 2 engines out on the same side.
- **VMO:** Maximum operating speed. Red line on the speed display of the pilot.
- **VMU:** Minimum take-off speed with the rear of the fuselage touching the runway (tail skid for the tests).
- **VR:** Rotation speed. Speed at which the pilot pulls on the stick during take-off.
- **Wake vortex:** Vortex created by the lift of the wing and propagating far behind the aircraft (see detailed explanation in the relevant chapter).
- **Yaw:** Angular rotation of the aircraft about the vertical axis.

CONTENTS

From the same author

* * *

From flight school to the A380 (Ed. JPO)

* * *

Functional check flights (Ed. JPO)

Cet ouvrage a été achevé d`imprimer
par Pulsio Sarl, Paris en mai 2015
Imprimé en UE